INDOOR NAVIGATION STRATEGIES FOR AERIAL AUTONOMOUS SYSTEMS

INDOOR NAVIGATION STRATEGIES FOR AERIAL AUTONOMOUS SYSTEMS

PEDRO CASTILLO GARCÍA
Sorbonne Universités, Université de Technologie de Compiègne, CNRS, UMR 7253 Heudiasyc, Compiègne, France

LAURA ELENA MUNOZ HERNANDEZ
Rennes, France

PEDRO GARCÍA GIL
Universidad Politécnica de Valencia, ISA, Valencia, Spain

ELSEVIER

AMSTERDAM • BOSTON • HEIDELBERG • LONDON
NEW YORK • OXFORD • PARIS • SAN DIEGO
SAN FRANCISCO • SINGAPORE • SYDNEY • TOKYO

Butterworth-Heinemann is an imprint of Elsevier

BH

Butterworth-Heinemann is an imprint of Elsevier
The Boulevard, Langford Lane, Kidlington, Oxford OX5 1GB, United Kingdom
50 Hampshire Street, 5th Floor, Cambridge, MA 02139, United States

Library of Congress Cataloging-in-Publication Data
A catalog record for this book is available from the Library of Congress

British Library Cataloguing-in-Publication Data
A catalogue record for this book is available from the British Library

ISBN: 978-0-12-805189-4

For information on all Butterworth-Heinemann publications
visit our website at https://www.elsevier.com

Publisher: Joe Hayton
Acquisition Editor: Sonnini Yura
Editorial Project Manager: Ana Claudia Garcia
Production Project Manager: Mohana Natarajan
Designer: Greg Harris

Typeset by VTeX

CONTENTS

ABOUT THE AUTHORS

Pedro Castillo García received his BS degree in Electromechanical Engineering from the Instituto Tecnológico de Zacatepec, Morelos, Mexico, in 1997, the MSc degree in Electrical Engineering from the Centro de Investigación y de Estudios Avanzados (CINVESTAV), Mexico, in 2000, and the PhD degree in Automatic Control from the Université de Technologie de Compiègne, France, in 2004. He has held visiting positions at the University of Sydney, Australia (2004), at the Massachusetts Institute of Technology (MIT) in 2005, at the Universidad Politécnica de Valencia, Spain (2005). He received the best PhD thesis of Automatic Control award from club EEA, France, in 2005. P. Castillo received his HDR (Habilitation à Diriger des Recherches) degree from the Université de Technologie de Compiègne, France in January 16th, 2014. He has held a visiting position at the LAFMIA UMI CNRS 3175 CINVESTAV-IPN, Mexico, from December 2012 to November 2014. At the moment, he is a researcher at the French National Research Foundation (CNRS), in the Laboratory Heudiasyc, at the Université de Technologie de Compiègne, France. He has co-authored one book for Springer-Verlag and co-authored more than 25 papers in international journals. His research topics cover real-time control applications, nonlinear dynamics and control, aerial vehicles, vision and underactuated mechanical systems.

Laura Elena Munoz Hernandez was born in Hidalgo, Mexico. She obtained her BS degree in Electronics and Telecommunications Engineering in 2005 and her MSc degree in Automation and Control in 2007 from the Universidad Autónoma del Estado de Hidalgo, Mexico. In 2012 she obtained her PhD degree in Automatic Control from the Université de Technologie de Compiègne, France. During her PhD studies she had scientific internships at the Universidad Politécnica de Valencia, Spain, in 2012. She currently works as a research and development engineer for a start-up in France. Her research interests cover real-time control applications, embedded control systems, robust nonlinear control, optimal control, and vision and control of autonomous vehicles.

Pedro García Gil was born in Valencia, Spain. In 2007 he obtained his PhD in Control Systems and Industrial Computing from the Universidad Politécnica de Valencia, Spain. He is currently an Assistant Professor of

Automatic Control at the Universidad Politécnica de Valencia. He has been a visiting researcher at the Lund Institute of Technology, Lund, Sweden, in 2006, at Université de Technologie de Compiègne, Compiègne, France, in 2007, at University of Florianopolis, Brazil, in 2010, at the University of Sheffield, UK, in 2014, and at the University of Hangzhou, China, in 2016. He has co-authored more than 20 papers in the top impact journals. His research interests are within the broad area of time delay systems, embedded control systems, and control of autonomous vehicles.

PREFACE

The purpose of this book is to give necessary and sufficient theoretical basis for those interested in working with Unmanned Aerial Vehicles (UAVs). Likewise, it provides, for those working in this area, a substantial theoretical and practical complement to their work.

When we work in autonomous navigation for aerial vehicles, it is common to consider ideal cases, i.e., full knowledge of the states, ideal measurements of the sensors, and non-external (or known) perturbations. Nevertheless, in flight tests this is not the case. The book's benefits to the audience are several: first of all, we propose three different approaches to mathematically represent the dynamics of an aerial vehicle. In one of these methodologies the quaternion technique is used to solve the singularity problem in UAVs. Secondly, detailed information is provided about how to fuse inertial data for attitude estimation with the results comparable to those of an expensive and commercial IMU (Inertial Measurement Unit). In addition, the book proposes substantial theoretical and practical validation to improve dropped or noisy signals. This part is crucial when using commercial sensors in handmade aerial prototypes. The UAV localization problem in this book is tackled by proposing an observer-control scheme using only the basic sensors in a drone.

For those dedicated to control systems, different control strategies, from classical to modern algorithms, haven been proposed using various approaches. One of the goals settled while writing this book was to give to the reader a wide spectrum of techniques so he/she could choose the most appropriate for his/her needs. Algorithm design considers a possible outdoor application, i.e., robustness with respect to unknown perturbations such as wind. Last but not least, three tools are given to improve autonomous navigation or to assist the manual pilot. The first one considers specific tasks in defined conditions (time, velocity, etc.) to generate a trajectory and follows it using an UAV. The second tool is used to avoid crashes when obstacles are present along the trajectory. The last approach considers a situation when the UAV is performing a semi-autonomous mission and its pilot suddenly loses sight of the vehicle. Here the vehicle states are sent to the pilot to provide him/her with more information about the vehicle's flight conditions, by means of a haptic joystick, and to improve its performance (teleoperation mode).

A quadrotor vehicle is chosen as an aerial configuration in this book due to its popularity among researchers. However, all the algorithms proposed may be adapted to work with different aerial configurations. Our results are validated in simulations, real time or flight tests on different platforms (commercial or handmade aerial prototypes) and with different sensors, and this makes the book valuable.

ACKNOWLEDGMENTS

Authors acknowledge support from the CONACyT under basic research grants for PhD Scholarships. This work has been sponsored by the French government research *programme Investissements d'avenir* through the Robotex Equipment of Excellence (ANR-10-EQPX-44). The authors would like to thank G. Sanahuja from Heudiasyc Lab in France for his help in realizing the experiments.

We thank Ana Claudia A. Garcia, Sonnini R. Yura, Mohanambal Natarajan and the edition team at Elsevier for guidance and patience throughout the publishing process, as well as their team for carefully reviewing this work during this process.

P. Castillo, L.E. Munoz & P. García
July 2016

PART 1

Background

Civil and military applications of Unmanned Aerial Vehicles (UAVs) have increased considerably for many years. The hovering ability of Vertical Take-Off and Landing (VTOL) vehicles makes them a suitable choice for many indoor and outdoor applications. The design of controllers and estimators that allow for attitude stabilization, autonomous flight, path following, avoiding obstacles, etc. have been the focus of several groups not only in the research but also in the hobbyist community, which has resulted in significant and interesting breakthroughs in the UAV field. The first part of the book is dedicated to an overview of mathematical definitions and UAV models used throughout the book. The objective is to introduce general concepts so that a person with no experience in this field could quickly learn the basics whereas an experienced person could either use it as a reference or to extend his/her knowledge.

In addition, a state-of-the-art of the topics covered in the chapters is presented in this first part. We focus our study on the quadrotor UAV configuration, yet the literature review is not limited to items used in quadrotors and gives an overall picture of the different topics covered in the book.

CHAPTER 1

State-of-the-Art

Since the first automatically controlled flight of an aircraft in 1916, civil and military applications of Unmanned Aerial Vehicles (UAVs[1]) have increased considerably. Although the notion of using UAVs has been around the last 20 years, civil applications have emerged with the development of new technologies [1]. Fixed wing UAVs are the most popular vehicles and have been widely used for years in surveillance missions. The hovering ability of Vertical Take-Off and Landing (VTOL) vehicles makes them a suitable choice for applications in which the UAV must often be able to hold at quasi-stationary flight [2].

In practical applications the attitude in a UAV is automatically stabilized via an on-board controller, while an operator through a remote control system generally controls its position. The design of position controllers that allow autonomous flights has been the focus of several groups in the research community, which has resulted in significant and interesting breakthroughs in this field.

UAVs that can autonomously operate in outdoor/indoor environments are envisioned to be useful for a variety of applications including surveillance, search, and rescue. For all these applications an imperative need for UAV autonomy is the ability of self-localization in the environment. Indeed, precise localization is crucial in order to achieve high performance flight and to interact with the environment.

The quadrotor[2] has become the most popular VTOL vehicle and is a useful prototype for learning about aerodynamic phenomena and control of aerial vehicles. Unlike conventional helicopters, this vehicle has constant pitch blades and is controlled varying only the angular speed of each rotor. The same principle applies to other rotorcraft configurations. The popularity of this configuration has grown so much that today it is the prototype most used in research laboratories and/or events of aerial vehicles.

In this book the quadrotor configuration was adopted for studying and solving current or emergent problems when flying aerial vehicles

[1] Also named drone, aerial drone.

[2] Also named quadcopter, four rotor rotorcraft.

Indoor Navigation Strategies for Aerial Autonomous Systems.
DOI: http://dx.doi.org/10.1016/B978-0-12-805189-4.00002-0

in (semi-)autonomous mode. In this way, some challenges, present when working with UAVs, have been addressed.

Thus the main subjects covered in this book include: (a) mathematical representation of the dynamics of an aerial vehicle, (b) measurement and control of the attitude of a VTOL vehicle using low cost sensors, (c) classical problems, e.g., of noise and delays, when measuring the states of a drone, (d) alternative solution to the localization problem of a drone in denied GPS environments or improvement of the position measure using standard sensors in an aerial vehicle, (e) design and validation of simple control laws using an integral component, (f) development of a new nonlinear control algorithm using fashionable techniques, (g) ability to integrate robustness in control laws, (h) trajectory generation for special aerial missions, (i) facility for avoiding obstacles when they are present along the trajectory, and finally, (j) helping the pilot by giving extra-information about the drone when it flies outside his/her visual range.

The main works of these subjects are described in the following. We possibly skip some other interesting references as sometimes it is not possible to include all of them.

1.1 MATHEMATICAL REPRESENTATION OF THE VEHICLE DYNAMICS

The mathematical representation of an aerial vehicle is one of the most important tasks when designing nonlinear controllers; nevertheless, in some cases it is omitted, and simple models gain popularity. Every researcher working with UAVs proposes or modifies the dynamic equations according his/her interest or the facility for developing the control scheme. In this way, several models can be found in the literature, some of them include aerodynamic effects or vehicle's characteristics, making these more interesting for the UAV control community.

Linear models are also used for control design or observer schemes, the advantage of these representations is that algorithms can be easily obtained and sometimes implemented. Another characteristic of linear models (or simplified models) is that in some cases all vehicle dynamics can be separated into subsystems. The most used linear representation for these subsystems is a chain of integrators in a cascade.

The classical approaches to mathematically represent an aerial vehicle are the Euler–Lagrange and Newton–Euler techniques. In both the aerial vehicle is generally studied as a rigid body moving in a 3D space. In the

Euler–Lagrange approach, the idea is to obtain the Lagrangian using the potential and kinetic energies. Here sometimes it is not evident how to include aerodynamic effects for the beginners in the subject. The Newton–Euler technique is maybe more intuitive; therefore, more aerial vehicle representations are derived from this approach. Aerodynamic effects such as blade rotor flapping or drag phenomena can be easily included and analyzed with this approach. The drawback of both approaches is that they can include a singularity in aggressive maneuvers.

Works related to the previous approaches are, e.g., [1–12].

A new trend in modeling a UAV is the quaternion approach which gives no singularity and an easy design and implementation of the control algorithms; nevertheless, it is less intuitive to understand and sometimes the reader can get lost. Therefore, not many authors have adopted the quaternion-based representation for modeling aerial vehicles probably because, in contrast to the Euler angle representation, they lack a direct intuitive visualization of the rotation.

Nevertheless, some researchers have noticed that the advantages of quaternions exceed the disadvantages. For example, [13] presents a full attitude control based on quaternions using a state feedback from the axis–angle representation of the orientation, [14] introduces a mathematical model for a hexarotor vehicle using the Euler–Lagrange approach, [15] proposes a technique for stabilizing the position–yaw tracking of quadrotors and, although the authors use a traditional Euler angles representation for the vehicle's attitude, their control for the z-axis is presented using a quaternion approach. Other control techniques such as optimal, LQR, and feedback linearization are presented in [16], [17], and [18].

Other authors have noticed that the dual quaternion approach can provide advantages when stabilizing both orientation and position of a rigid body, [19] presents an introduction to dual quaternions, and a modeling of a quadrotor UAV using the Lagrangian formulation, [20] proposes a feedback control without linear and angular velocity feedback using dual quaternions, and finally, [21] proposes a feedback regulator that globally stabilizes a fully actuated system using dual quaternions.

In Chap. 1, three methodologies to represent the dynamic of an aerial vehicle are presented: the Euler–Lagrange, the Newton–Euler, and the quaternion approaches. We mainly focus our study on the quadcopter configuration; however, these approaches can also be used for other configurations of UAVs. The Euler–Lagrange methodology is essentially used in the ideal case, i.e., without perturbations and uncertainties in the model.

In the Newton–Euler approach, the aerodynamic and flapping effects are taken into account. The third methodology, the quaternions, is introduced as a new solution to the singularity problem that could appear in other formalisms.

1.2 ATTITUDE ESTIMATION USING INERTIAL SENSORS

Achieving autonomous flight for a drone could be divided in two steps. The first one includes attitude stabilization and it is essentially used for semi-autonomous flights, also named remotely operated flights. The second one deals with the complete system, i.e., the attitude and position, and the goal is to realize autonomous missions. The second step is accomplished if the first is performed.

Thus, disregarding the control strategy, a high-performance attitude tracking subsystem is a prerequisite for developing any other high-level controlling task. A good example of this statement can be found in [22] where a full control (vision, collision avoidance, landing/taking-off) is developed relying on the attitude control. Therefore, the key state variables to be estimated are the angular position and rate, as they are primary variables used in attitude control of the vehicle [23].

Inertial Measurement Units (IMUs), which are the core of lightweight robotic applications, have experienced proliferation, resulting in cheaper and more accurate devices [24]. The emergence of cheaper IMUs makes UAVs available for civil purposes like ground traffic inspection [25], forest fire monitoring [26], or real-time irrigation control [27]. Low-cost IMUs (prices below 100 USD) have poor performance in terms of bias stability, nonlinearities, and signal-to-noise ratio than those on the market for industrial applications. A comparison of a wide range of IMUs can be found in [28].

A challenge and interesting task is to obtain a reliable attitude estimation using low-cost sensors. The difficulty comes from the low performance of the sensors which restricts the quality of the resulting estimation. And it is definitely interesting because the problem of obtaining an accurate attitude estimation is crucial and it usually represents a large portion of the cost of a UAV [24].

In Chap. 2, the goal is to demonstrate that by using low-cost sensors and an appropriate observer it is possible to achieve good attitude estimation comparable with a commercial and expensive IMU. The sensor fusion problem consists in obtaining an optimal estimate of the required vehicle state variables with the direct measurements from multiple sensors. There are many possible solutions to this problem, i.e., Kalman Filters (KF) [29,30] or complementary filters [31–33]. Despite the lack of convergence and optimality guarantees, the Extended Kalman Filter (EKF) has been the workhorse of real-time spacecraft attitude estimation for quite some time [34]. The earliest application [35] was published in 1970 by Farrell, and several others have followed since.

Lefferts, Shuster, and Markley published a thorough review of the topic [36] in 1982, and since then Markley, Crassidis, and several others have kept Kalman filtering an active topic of research in the space industry [37–39]. Several different attitude representations have been used in Kalman filtering with varying degrees of success.

Regarding to the rotation representation, quaternion approach is widely used in space applications because it provides a singularity-free representation. However, four parameters are necessary. A survey of nonlinear methods with quaternions is presented in [34]. Euler angles only require three parameters, at the expense of introducing a singularity at $\theta = 90°$. If there is no intention of performing acrobatic maneuvers, Euler angles are a more compact representation that reduces the computational burden of the estimation algorithm.

1.3 DELAY SYSTEMS & PREDICTORS

Delay systems is a subject studied for several years. The delay community is characterized essentially by the development of predictors schemes based on the Smith Predictor [40], and/or the Finite Spectrum Assignment (FSA) [41]. The interest of the UAV community in studying the delays in aerial vehicles has been growing for the last years.

Delays are immersed in UAVs due to different sampling periods coming from sensor measurements or even due to computation of the controllers. For example, IMU sensors include delays when estimating the angular position from the inertial sensors. As for the position measurement, the Global Positioning System (GPS) measures the position with a considerably different sampling period compared to the sample period used to compute the navigation algorithm. When using cameras for pose or translational velocity estimation (optical flow), considerable delays are also included in the closed-loop system.

Predictors schemes have only been seen as theoretical contributions without possible validation in real time for several years, as a result of its difficult implementation in real systems. The technological advances, due to faster micro-processors or miniaturization of the sensors, have driven the practical validation of these algorithms in tests-bed or prototypes, and new challenges related to the delay systems begin to be studied by the UAV control community.

In Chap. 3, the delay problem in a quadcopter vehicle is addressed, and two schemes based on predictors are proposed and validated in real time. The first one improves the attitude of a VTOL aircraft, while the second compensates delays in the control input. Both schemes have been validated in simulations and real time.

References related to this topic can be found in [42–45,41,46–51].

1.4 DATA FUSION FOR UAV LOCALIZATION

One of the main current fields of research in UAVs is the positioning problem. Some works on indoor position controlling approaches for flying objects based on localization systems (VICON, Optitrack) or Simultaneous Localization and Mapping (SLAM) have already been reported in the literature. However, most of them either depend on a highly accurate and expensive localization system that needs to be deployed manually [52,53] or they need a high amount of computing processing, mostly off-board [54].

Once more, fusing data from different sensors seems to be a very good solution to this open and interesting problem. Fusing data from different sensors improves the performance of the overall sensing system. For example, for aerial navigation outdoors, fusion of GPS measurements with INS (Inertial Navigation Systems) by means of filtering techniques corrects the inherent error accumulation of dead-reckoning and increases the localization precision required for UAV missions. However, GPS-based localization remains impractical when the signal is noisy, perturbed, or indoors. Hence for these cases GPS becomes totally impractical. Alternative methods for localization in indoor environments are thus necessary for autonomous navigation.

Some works fuse information from two or more sensors, for example, in [55] the authors introduce a computer vision algorithm to hold a quadrotor aircraft in hover position using a low-cost video system. This system is based on a Kalman Filter for data-fusion algorithm that includes measures from inertial and visual sensors. Similarly, in [56] a real-time vision based algorithm for 5 degree-of-freedom pose estimation and set-point control for a Micro Aerial Vehicle (MAV) is proposed. The vision system uses only a camera mounted on-board and two concentric circles as landmarks. In addition, the authors have shown that by using a calibrated camera, conic sections, and the assumption that yaw is controlled independently, it is possible to determine the 6 degree-of-freedom pose of the MAV.

Likewise, in [57] the authors have proposed two visual algorithms for UAV control. The first one is based on the detection and tracking of planar structures from an on-board camera, while the second one is based on the detection and 3D reconstruction of the position of the UAV based on an external camera system. Both strategies are tested experimentally with a VTOL UAV, and the results show good behavior of the visual systems. In the test, the vehicle position is computed using the vision algorithm with landmarks, and the velocity is estimated using a Kalman filter. In a similar manner, in [58] a vision system for pose estimation and stabilizing control of a model helicopter has been proposed. The method consists in using both ground and on-board cameras. A vision algorithm is used to obtain the relative position and orientation of the helicopter with respect to the ground camera.

In [59] a model-based vision system for the position control of a quadrotor is introduced. In this work the vehicle is equipped with a camera pointing downward and with an IMU. The vision system estimates the position of the vehicle, and the translational velocity is estimated using a Kalman filter.

In [60] a method for trajectory tracking and waypoint navigation of a quadrotor is proposed. The methodology includes a vision-based camera system and an exact method for outdoor positioning using the Real-Time Kinematic Global Positioning System (RTK-GPS). The main result of the authors is the design of a low-cost DGPS (Differential Global Positioning System) for outdoor localization of the quadrotor and the enhancement of a trajectory tracking control algorithm based on a vision based camera system with landmarks.

In [61] a new localization methodology based on the PLICP (Point-to-Line Iterative Closest/Corresponding Point) algorithm is proposed to provide accurate position estimation of the vehicle. The autonomous hovering and trajectory tracking control are achieved via on-board sensor and processors. Real-time flight tests are implemented under different scenarios including indoor hovering and trajectory tracking between buildings. These experiments show that safe autonomous fight control is achieved for the quadrotor when it is flying in an unknown environment where GPS signals are very weak or absent.

The authors of [62] propose an algorithm that computes vehicle pose from the observation of three point features in a single camera image, once the angles are estimated from IMU measurements. The proposed algorithm is based on the minimization of a cost function and is evaluated on both synthetic and real data. The very low computational cost of the proposed approach makes it suitable for pose control in tasks such as hovering and autonomous take-off and landing. In [63] the author studies and compares nonlinear Kalman Filtering and Particle Filtering methods for estimating the state vector of UAVs through the fusion of sensor measurements. Here, the authors propose the use of the estimated state vector in a control loop for autonomous navigation and trajectory tracking by the UAVs. The proposed nonlinear controller is derived according to the flatness-based control theory. The performance of the nonlinear control loop is evaluated through simulation tests.

The authors of [64] present a control, navigation, localization and mapping solution for an indoor quadrotor UAV system. Three main sensors are used on-board the quadrotor platform, namely an inertial measurement unit, a downward-looking camera, and a scanning laser range finder. Complete flight tests were carried out to verify fidelity and performance of the navigation solution. In [65] an evaluation system for verifying algorithm of UAV control and self-localization in GPS-denied environments, instead of the VICON motion capture system that is being widely used as ground truth of the quadcopter position control, is developed and its feasibility is evaluated. The feasibility of the algorithm is validated in experiments employing two kinds of quadcopters.

In [66] the author proposes a control algorithm based on range-only measurement such that a UAV can circumnavigate an unknown target at the desired distance under a GPS-denied environment. The design of the control algorithm is divided into two steps: First, a control algorithm based on range and range rate measurements was proposed to solve the circumnavigation problem. Second, an estimated range rate, obtained from a sliding mode range rate estimator, was used to replace the actual range rate measurement needed in the control algorithm proposed in the first step. The proposed algorithms were tested only in simulations.

As it can be appreciated, there exist several systems to locate a UAV, most of them are mainly based on the fusion of the position (obtained by GPS, laser or vision) and inertial sensors. The fusion techniques are essentially employed to estimate the missing states which are, for all reported works, the translational velocities. The effectiveness of this estimation will depend largely on the quality of the sensors. On the other hand, among the estimation algorithms reported in literature, KF is the most popular approach [36,67]. Nevertheless, this problem it is still of great interest to a large part of the control research community.

In Chap. 5, the challenge is to develop a 3D indoor localization system that can be a low-cost and accurate solution for UAV applications. The main idea is to employ the minimal number and the most classical sensors for a UAV, i.e., an IMU, camera, and ultrasonic sensor, to perform autonomous trajectory tracking missions. The originality of the approach lies in estimating 3D localization on-board and with good precision. The proposed data fusion estimation with the EKF was compared, in flight tests, with the measurements coming from the OptiTrack system. This indoor localization system provides, with good precision, the 6 DOF (Degree-Of-Freedom) position and orientation of the quadrotor in real time.

1.5 CONTROL & NAVIGATION ALGORITHMS

Manual flight of aerial prototypes is difficult and not so intuitive. The goal of the UAV control community is to facilitate manual flight tests when adding an embedded system containing mainly sensors and algorithms to stabilize the vehicle. New sensors are smaller and cheaper; nevertheless, the precision in the measurement sometimes is affected by noise or bad estimation. This problem is addressed in Chaps. 3–5.

Constructing algorithms to stabilize a vehicle is a hard task for beginners working with aerial vehicles. The UAV control community is not attractive for the amateur or industrial UAV community. The main reason is the proposed complex and difficult control algorithms that are sensitive to uncertainties in the system or noise in the measurements. In addition, the tuning gains and implementation of these controllers are sometimes discouraging.

Linear controllers are most used by the amateur or industrial UAV community and sometimes beginning researchers. The disadvantage of these controllers is the tendency to diverge in the presence of nonlinear perturbations or different conditions far from the equilibrium point. The necessity to improve these algorithms appears, and simple nonlinear controllers emerge as possible solutions.

Numerous papers dedicated to flight controller design for the quadrotor vehicle have been addressed using several methods, varying from conventional PD (Proportional Derivative) control to more advanced techniques, such as feedback linearization, backstepping, etc. [68–72].

Bounded controllers were designed to limit the control inputs and to preserve the system. Nonlinear controllers based on saturation functions have become popular, and several algorithms have been developed recently. Stability analysis and stabilization of systems presenting saturations have been studied since the late 1960s due to their frequent occurrence in engineering and scientific problems [73,74]. In 1992 Teel gave explicit expressions of globally asymptotically stabilizing feedbacks for chains of integrators [75], an open problem at the time. This important result on linear systems, later extended in [76,77], has inspired many researchers, see, e.g., [78,73,79,71,80–83]. Controllers using bounded input approaches owe their success to the fact that they solve theoretical [73,76] as well as applied problems [77]. They are also the origin of the control design methodology called forwarding [84,85] and of many useful results in the literature, see, for example, [78,86–88].

In Chap. 6, two controllers based on the saturation functions and the backstepping techniques are addressed. These controllers have the characteristic (and the main difference with the previously described) to include an integral component that improves, in several cases, the closed-loop performance in the presence of constant perturbations or uncertainties in the model. Simulations and practical validation demonstrate good performance of the algorithms.

Robustness issues can be critical for the rotorcraft due to the complex aerodynamic effects (which makes it difficult to obtain an accurate dynamic model), such as the errors from sensors and the external disturbances (e.g., wind). The first two issues have been studied in the context of linear control and more recently using nonlinear techniques. Nevertheless, only few studies have been done in the case where the quadrotor vehicle is affected by wind.

Robust controllers can be obtained for design or also when adding observers, predictors, or nonlinear components in the control scheme. Sliding Mode (SM) algorithms become "fashionable" due to their robustness properties and quick convergence to the desired values in finite time.

The main idea of SM control is based on the introduction of a "custom-design" function, named the sliding variable. As soon as the suitably designed sliding variable becomes zero, it defines a sliding manifold or surface; this is usually realized by means of a high-frequency control switching. The properly designed sliding variable yields good closed-loop system performance, while the system's trajectories belong to the sliding manifold. The idea of SMC is to steer the system trajectory to the properly chosen sliding manifold and then to maintain motion on the manifold thereafter by means of the control [89]. Within the sliding techniques, it is well known that High Order Sliding Mode (HOSM) is a powerful technique that can be used for systems with the relative degree greater than 1, see [90–92]. A system motion in a sliding mode is robust to matched disturbances and uncertainties.

A variety of successful control applications based on the traditional (relative degree 1) and high order sliding mode algorithms have been reported in the literature. Such applications range from missile guidance, underwater vehicles, and robotic manipulators to overhead crane, see, for example, [93,94]. As we can see, in these works the sliding mode approach provides different techniques for the design of output based control [95,96], differentiators [97], observers, disturbance rejection and estimation, etc.

Among other related references is, for example, Raffo et al. [98] that introduced an integral predictive and robust nonlinear H_{∞} control strategy for a quadcopter. This control strategy was designed under the consideration of external disturbances, including the wind, acting on all degrees of freedom. Alexis et al. [99] developed a Constrained Finite Time Optimal Controller (CFTOC) for attitude set-point maneuvers of a quadrotor operating under severe wind conditions. A two-loop approach using input–output linearization to design a nonlinear controller for a quadrotor dynamics was proposed by [100]; however, only simulations were performed to show the effectiveness of the control law in the presence of input disturbances. Likewise, Besnard et al. [101] presented simulation results to show the performance of a sliding mode control (SMC) driven by a sliding mode disturbance observer (SMDO) applying to a quadrotor vehicle subject to external disturbances and model uncertainties.

Despite the large number of references using the sliding mode techniques that we could find in the literature, only a few have been proved in aerial prototypes. In Chap. 7, a new sliding mode control is proposed to stabilize the attitude of a quadrotor vehicle. The design of the controller is realized using the Lyapunov analysis [102–105]. In addition, a theoretical procedure is included in the chapter to better tune the gain parameters. Experimental results have demonstrated good performance of the proposed algorithm.

More references of the SM methodology can be found in [89,106–109]. Similar references related to robust controllers are, for example, the following:

In [110] the design and implementation of a robust flight control system for an unmanned helicopter is presented. The robust flight control system consists of a three-layer control architecture which includes (i) an inner-loop controller designed using the H_∞ control technique to internally stabilize the aircraft and at the same time yield good robustness properties with respect to external disturbances, (ii) a nonlinear outer-loop controller to effectively control the helicopter position and yaw angle in the overall flight envelope, and lastly, (iii) a flight-scheduling layer for coordinating flight missions.

Likewise in [111], the authors present the design of a controller based on the block control technique combined with the super twisting control algorithm for trajectory tracking of a quadrotor. The wind parameter resulting from the aerodynamic forces is estimated in order to ensure robustness against these unmatched perturbations. The performance and effectiveness of the proposed controller are tested in a simulation study taking into account external disturbances.

In [112] the authors addressed a robust control based on finite-time Lyapunov stability controller and proved it with the backstepping method for the position and attitude of a small rotorcraft unmanned aerial vehicle subjected to bounded uncertainties and disturbances. The proposed controller combines the advantage of the backstepping approach with finite-time convergence techniques to generate a control law which guarantees faster convergence of the state variables to their desired values and compensates for the bounded disturbances. Simulation results are presented to corroborate the effectiveness of the proposed control method.

Naldi and Marconi [113] presented a control strategy to robustly accomplish the hover to level flight transition maneuvers for a class of V/STOL aerial vehicles. By means of a path-following formulation of the control problem, the proposed flight control strategy is shown to be able to overcome the presence of possible wind disturbances affecting the flight dynamics of the vehicle. On the other hand, [114] deals with an altitude controller of a quadrotor with an unknown total mass of the structure. An Adaptive Robust Control (ARC) is utilized to compensate the UAV for the parametric uncertainty. Then Lyapunov based stability analysis shows that the proposed control design guarantees asymptotic tracking error for the UAVs altitude control. Time-based numerical simulation results are presented to illustrate good tracking performance of the designed control law.

Even though there are a lot of modern control techniques that can be adopted for autopilot applications, classical designs remain a very interesting and effective approach for autopilot designs. In Chap. 8 we address, theoretically and experimentally, the stabilization problem of a four-rotor rotorcraft in the presence of lateral wind by designing bounded control laws that are robust with respect to unknown and bounded disturbances. The proposed nonlinear control strategy is simple and uses the ideas introduced in [78] and [75]. Also in this chapter a predictive feedback algorithm in combination with the Uncertainty Disturbance Estimator (UDE) is developed. This controller increases the control performance in closed loop even in the presence of unknown and external perturbations. It will be proved in Chap. 4 that the UDE algorithm also improves the closed-loop performance even in the presence of delay in the measurements.

The main advantage of both control strategies, in Chap. 8, is that they can be considered as linear controllers with a few tuning parameters, and therefore their implementation in quadrotor vehicles is easy. This fact makes them an interesting choice for controlling quadrotors in real applications, dealing with model uncertainties and persistent disturbances with outstanding performance [115].

1.6 TRAJECTORY GENERATION & TRACKING

Once the hovering aerial vehicle is stabilized, it is possible to give a set of points simulating a desired trajectory or mission. If there are no restrictions on the vehicle's parameters, then the problem is reduced to a regulation control. Some missions, as inspection or surveillance, require that the aerial vehicle passes through some specific points with specific conditions (velocity, time, etc.) and in this case trajectory generation taking into account the vehicle's parameters is necessary.

The challenge in Chap. 9 is to develop a trajectory generation algorithm for a quadcopter vehicle to minimize its arrival time to a desired point taking into account its physical constraints. These restrictions can be regarded as limitations on flight path angle, velocity, and their rate of change, respectively, as well as the rate of change of heading angle.

The problem of minimal time trajectory is equivalent to the minimal path length problem proposed by Dubins in [116]. He proved the existence of the shortest path between two specified points for a vehicle moving in a 2D plane with constant velocity. In addition, he showed that the optimal paths are a combination of at most three segments. These portions take the form of circular arcs and straight lines. Based on Dubins' results, the authors in [117–119] have proposed algorithms to generate feasible paths taking into account kinematic and tactical constraints imposed on the UAV. A dynamic trajectory smoothing algorithm in the horizontal plane was presented in [120]. The idea is to smooth path segments in order to obtain an extremal trajectory with explicit consideration of kinematic constraints. A generalization of the previous method is shown in [121], where a 3D trajectory planner based on initial trimmed elementary trajectories was suggested. Similarly, other solutions based on splines were proposed to compute feasible trajectories in [122–125].

In the second part of this chapter, we deal with the trajectory tracking problem. In fact, many prior works in the area of quadrotor control and tracking can be found in the literature. The methods developed in those works range from the standard linear controller to advanced nonlinear control strategies. For the trajectory tracking problem a simple nonlinear controller developed in Chap. 8 is used.

1.7 OBSTACLE AVOIDANCE

Recently, UAVs have been deployed for various search and surveillance missions. One of the main objectives of a mission is to generate non-conflicting paths such that the UAV can follow. A path is usually described in terms of waypoints that the UAV has to visit sequentially. The path planning problem is to generate non-conflicting "paths" in terms of waypoints taking various environmental and physical constraints of the UAV into account. For UAV applications where the UAV has to fly in obstacle-rich environments, the computational complexity of the path planing algorithm is an interesting issue that needs to be addressed since the UAV is in motion at high velocity.

Path planning is an important branch in mobile robot research. It refers to how a robot searches a path without collisions from the original to the target place based on some certain evaluation standards. The Artificial Potential Field (APF) method is widely used for autonomous mobile robot path planning due to its efficient mathematical analysis and simplicity. The application of this method, however, is often associated with the local minima problem which occurs when the total force acting on a robot is summed up to zero although the robot has not reached its goal position, yet. In order to solve the problem, some previous researches have been done constructing potential fields free from the local minima problem.

One novel technique to solve the minima problem uses the limit cycle strategy to avoid static obstacles. The main advantage of this method is the low computational cost which makes this technique specially useful for small robots with limited computational power. In short, the method relies on creating a circular limit cycle that a robot should follow in the proximity of an obstacle to avoid collision.

In many service applications, mobile robots need to share their work areas with obstacles. Avoiding moving obstacles with unpredictable direction changes, such as humans, is more challenging than avoiding moving obstacles whose motion can be predicted. Precise information on the future moving directions of humans is unobtainable for use in navigation algorithms. Furthermore, humans should be able to pursue their activities unhindered and without worrying about the robots around them. An Enhanced Virtual Force Field-based mobile robot navigation algorithm (EVFF) is presented in [126] for avoiding moving obstacles with unpredictable direction changes. This algorithm may be used with both holonomic and nonholonomic robots. It incorporates improved virtual force functions and an improved method for selecting the sense of the detour force to better avoid moving obstacles.

In [127] the authors introduce path planning algorithms using Rapidly-exploring Random Trees (RRTs) to generate paths for multiple UAVs in real time, from given starting locations to goal locations in the presence of static, pop-up, and dynamic obstacles. Generating nonconflicting paths in obstacle-rich environments for a group of UAVs within a given short time window was their challenging task. The difficulty was increased when the turn radius constraints of the vehicles had to be compatible with the corridors where those intended to fly. The authors solved this problem by quickly generating a path using RRT and taking into account the kinematic constraints of the UAVs. Simulation studies were carried out to show the performance of their algorithm.

By analyzing the shortcoming of the artificial potential field methods for robot path planning, in [128] the authors proposed an obstacle avoidance method based on gravity chain. They supposed that there is a rubber band which connects the beginning and the end in the obstacle potential field space. The role of the rubber band was played by the potential field power, and the authors built a model to simulate the shape of the rubber band. Then the algorithm generated a steer angle tangent to the rubber band instead of the angle of artificial potential field. By putting effective obstacle avoidance information into potential field through gravity chain, the authors solved the problem of the artificial potential field method which often converged to local minima, hardly reached the end, and had oscillatory movement. The simulation results showed that the proposed method was correct and effective.

In [129] a method for robot motion planning in dynamic environments was introduced. The method consists in selecting avoidance maneuvers to avoid static and moving obstacles in the velocity space, based on the current positions and velocities of the robot and obstacles. The algorithm can be considered a first order method since it does not integrate velocities to yield positions as functions of time. Avoidance maneuvers are generated by selecting robot velocities outside of the velocity obstacles, which represent the set of robot velocities that would result in a collision with a given obstacle that moves at a given velocity, at some future time. To ensure that the avoidance maneuver is dynamically feasible, the set of avoidance velocities is intersected into the set of admissible velocities defined by the robots' acceleration constraints.

The problem of determining a collision-free path for a mobile robot moving in a dynamically changing environment was addressed in [130]. By explicitly considering a kinematic model of the robot, a family of feasible trajectories and their corresponding steering controls were derived in closed form and expressed in terms of one adjustable parameter for the purpose of collision avoidance. Then a new collision-avoidance condition was developed for the dynamically changing environment which consisted of a time criterion and a geometrical criterion, and it had explicit physical meaning in both the transformed space and the original working space. By imposing the avoidance condition, one can determine one (or a class of) collision-free path(s) in closed form. Such a path meets all boundary conditions, is twice differentiable, and can be updated in real time once a change in the environment is detected. The simulation results have demonstrated that the algorithm is effective!

In [131] a novel navigation method combining the limit-cycle and vector field methods for navigating a mobile robot while avoiding unexpected obstacles in the dynamic environment was developed. The limit-cycle method was applied to avoid obstacles placed in front of the robot while the vector field method was applied to avoid obstacles placed on the side of the robot. The proposed method was tested using Pioneer 2-DX mobile robot. Through both simulation and experiment, the effectiveness of proposed methods for navigating a mobile robot in the dynamic environment was demonstrated.

In [132] a new algorithm incorporating ellipsoidal limit cycles was presented for avoiding static 3D obstacles by using 3D limit cycles. The ellipsoidal limit cycles represent a safety zone around an obstacle. By using ellipsoidal limit cycles, the shape of the obstacles can be better represented. Ellipsoidal limit cycles are also useful when there is a preferred direction in which the obstacle should be avoided, e.g., to prevent passing in front of the obstacle. Further, the limit cycle method was combined with the velocity obstacle approach in order to avoid moving obstacles in 2D and 3D. To ensure that the robot chooses the optimal rotation direction with multiple and/or moving obstacles, a tree search is performed, yielding the globally (resolution) optimal rotation direction around each obstacle.

Similarly, in [133] the problem of generating flight paths for uniform surveillance of a region, using the limit cycle concept was addressed. The generated flight path covers whole area of a circle with desirable radius and needed density of the trajectories. Then by tracking such trajectories, UAVs are able to get the precise aerial photos or movies (or another task which needs such trajectories). The proposed real-time flight path was divided into three main phases; in the first phase, the UAV takes off from the launch station and moves near of the center of the considered circular sightly area for getting aerial photos or movies. In the second phase, the flying object tracks a uniform helicoidal trajectory to reach the circular limit cycle and to completely scan a circular area. In the last phase, when the scanning of the considered area has been accomplished, the UAV comes back to the launch station following a trajectory produced similarly that in the first phase. Simulation results have illustrated good performance of the proposed approach.

Related works can be found in [134–137].

In Chap. 10 the obstacle avoidance for aerial vehicles is addressed. The algorithm is based on the APF method. Background material about this method is also included in the chapter. The minima problem is solved using virtual obstacles and the limit cycle technique. Simulation and experimental results have demonstrated good performance of the controller even in the presence of minima problems.

1.8 TELEOPERATION

The introduction of autopilots capable of automatically regulating the drone's attitude to its stable position, or any other desired reference, was the first big step towards remotely operated UAVs. Using an attitude stabilization controller embedded in the drone allows the pilot to only provide angular references to displace the UAV, instead of directly controlling the motor's speeds, which results in a much complicated task.

To this day, the most common kind of teleoperation mode, and the base for many others, consists in controlling the pitch and roll angle references, the yaw's rate of change, and the desired altitude speed. This mode is usually referred to as stabilization mode. A simplified version of this stabilization mode consists in controlling the movements of the drone with respect to the pilot's point of view regardless of the UAV's orientation. This is easier and more intuitive than controlling the drone with respect to its own reference frame, especially for new users.

Another kind of operation mode allows the pilot to control angular rates of change instead of angles. This is useful in performing aggressive or acrobatic flight, such as doing looping. Note that such an operation mode requires more expertise and awareness from the human user, and special treatment must be taken for the system singularities.

Others commonly used flight modes attempt to improve the level of autonomy of the system by adding more information from sensors, or application based behaviors. For example, it is normal to help the pilot to control the altitude by including an altitude sensor such as an ultrasound range finder or a pressure sensor. The inclusion of a position estimation, for example, from a GPS (Global Positioning System), may improve the level of autonomy of the system, and allow for other operation modes where the mission of the human user is reduced to monitoring the results and the status of the UAV. Some examples are the so-called loiter mode, Return-to-Launch (RTL), way-points following, and autonomous trajectory tracking.

In loiter mode the UAV attempts to maintain the current position, heading, and altitude, but the pilot my take manual control at any time to change the current states of the system. In RTL mode the drone will return to its home position automatically. Automatic missions can be preprogramed for path following using way-points, or trajectory tracking of a time varying trajectory, like a circle or a lemniscate. If a relative position with respect to an object of interest is available instead of a global one, a follow-me mode may be used to track the target.

Works related with this topic are the following:

In [138] a novel concept of haptic queuing for an airborne obstacle avoidance task is proposed. The novel queuing algorithm was designed in order to appear 'natural' to the operator, and to improve the human–machine interface without directly acting on the actual aircraft commands. An experimental evaluation of two different Haptic aiding concepts for obstacle avoidance is presented in the paper. The authors used the approach called the Direct Haptic Aid (DHA) class and the Indirect Haptic Aid (IHA) class for their algorithm.

The authors of [139] described the design and theoretical evaluation of a novel AFF (Artificial Force Field), i.e., the parametric risk field, for teleoperation of an UAV. The field allows adjustments of the size, shape, and force gradient by means of parameter settings which determine the sensitivity of the field. Computer simulations were conducted to evaluate the effectiveness of the field for collision avoidance for various parameter settings, and the obtained results indicated that the novel AFF performs the collision avoidance function more effectively than potential fields known from the literature.

Similarly, in [140] the authors investigated the application of force–stiffness feedback, a combination of force offset and extra spring load, to a haptic interface for UAV teleoperation with time delay. Force feedback was based on applying force offsets inducing operator stick deflections to guide the UAV away from obstacles. Likewise, in [141] a bilateral teleoperation of underactuated aerial vehicles was addressed using force feedback. The state stability of the teleoperation loop with respect to bounded or dissipative virtual environment forces was established based on Lyapunov analysis.

In [142] the authors proposed an approach that modulates the set point for the vehicle's controller based on the user input energy, estimated potential energy and vehicle's kinetic energy. Incorporating the novel approach with the Dynamic Kinesthetic Boundary, the human operator can better perceive the environment where the robot is deployed through the rich spatial haptic cues rather than an onset gradual single force vector.

The authors of [143] introduced an intuitive teleoperation scheme for an untrained user to safely operate a wide range of VTOL UAVs in a cluttered environment. This scheme includes a force-feedback algorithm that enables the user to feel the texture of the environment. In addition, the authors introduced a novel mapping function to teleoperate the UAV in an unlimited workspace in the position control mode with a joystick which has limited workspace. An obstacle avoidance strategy was designed to autonomously modify the position set point of the UAV independently of the pilot's commands. The algorithms were validated in experiments using an haptic joystick and a hexacopter UAV equipped with a 2D laser range scanner.

In [144] a novel, simple, and effective approach for teleoperation of aerial robotic vehicles with haptic feedback was developed. The approach is based on energetic considerations and uses the concepts of network theory and port-Hamiltonian systems. In addition, the authors provided a general framework for addressing problems such as mapping the limited stroke of a "master" joystick to the infinite stroke of a "slave" vehicle while preserving passivity of the closed-loop system in the face of potential time delays in communication links and limited sensor data. For more references see [145–147].

Chapter 11 is devoted to introducing a collision-free haptic teleoperation scheme to facilitate the control of a quadcopter vehicle. The user controls the UAV by giving angle references from a haptic joystick. A control algorithm is implemented in the vehicle in order to assist the user in his/her task; the obtained scheme is more intuitive and furnishes collision-free flight for the quadcopter. A vision algorithm is used to estimate the position of the aerial vehicle and also possible and defined obstacles. Experimental results demonstrated good performance of the teleoperation scheme.

REFERENCES

1. G.M. Hoffmann, H. Huang, S.L. Waslander, C.J. Tomlin, Precision flight control for a multi-vehicle quadrotor helicopter testbed, Control Engineering Practice 19 (9) (2011) 1023–1036.
2. R. Lozano, Unmanned Aerial Vehicles Embedded Control, John Wiley–ISTE Ltd, 2010.
3. R.W. Prouty, Helicopter Performance, Stability, and Control, Krieger Publishing Company, 2001.
4. P. Pounds, R. Mahony, P. Corke, Modelling and control of a quad-rotor robot, in: Proceedings of the Australasian Conference on Robotics and Automation, Auckland, New Zealand, 2006.
5. P. Pounds, J. Gresham, R. Mahony, J. Robert, P. Corke, Towards dynamically favourable quadrotor aerial robots, in: Proceedings of the Australasian Conference on Robotics and Automation, Canberra, Australia, 2004.
6. G. Hoffmann, H. Huang, S. Waslander, C. Tomlin, Quadrotor helicopter flight dynamics and control: theory and experiment, in: Proceedings of the AIAA Guidance, Navigation, and Control Conference, USA, 2007.

7. H. Goldstein, Classical Mechanics, Addison Wesley Series in Physics, Addison–Wesley, USA, 1980.
8. P. Castillo, R. Lozano, A.E. Dzul, Modelling and Control of Mini-Flying Machines, Advances in Industrial Control, Springer-Verlag, London, UK, 2005.
9. B. Etkin, L. Duff Reid, Dynamics of Flight, John Wiley and Sons, New York, USA, 1959.
10. R. Lozano, B. Brogliato, O. Egeland, B. Maschke, Passivity-Based Control System Analysis and Design, 2nd edition, Communications and Control Engineering Series, Springer-Verlag, 2006.
11. K. Nonami, F. Kendoul, S. Suzuki, W. Wang, D. Nakazawa, Autonomous Flying Robots: Unmanned Aerial Vehicles and Micro Aerial Vehicles, Springer, 2010.
12. D.G. Hull, Fundamentals of Airplane Flight Mechanics, Springer, Berlin, 2007.
13. E. Fresk, G. Nikolakopoulos, Full quaternion based attitude control for a quadrotor, in: European Control Conference (ECC), Zürich, 2013, pp. 3864–3869.
14. A. Alaimo, V. Artale, C. Milazzo, A. Ricciardello, L. Trefiletti, Mathematical modeling and control of a hexacopter, in: International Conference on Unmanned Aircraft Systems (ICUAS), Atlanta, USA, IEEE, 2013, pp. 1043–1050.
15. A. Sanchez-Orta, V. Parra-Vega, C. Izaguirre-Espinosa, O. Garcia-Salazar, Position–yaw tracking of quadrotors, Journal of Dynamic Systems, Measurement, and Control 137 (2015) 061011.
16. W. Jian, A. Honglei, L. Jie, W. Jianwen, M. Hongxu, Backstepping-based inverse optimal attitude control of quadrotor, International Journal of Advanced Robotic Systems 10 (2013) 1–9.
17. E. Reyes-Valeria, R. Enriquez-Caldera, S. Camacho-Lara, J. Guichard, LQR control for a quadrotor using unit quaternions: modeling and simulation, in: International Electronics Conference on Communications and Computing (CONIELECOMP), Cholula, Mexico, 2013, pp. 172–178.
18. Y. Long, S. Lyttle, N. Pagano, D.J. Cappelleri, Design and quaternion-based attitude control of the omnicopter MAV using feedback linearization, in: Proceedings of the ASME 2012 International Design Engineering Technical Conferences and Computers and Information in Engineering Conference, Chicago, IL, USA, August 12–15, 2012.
19. Y. Jing, C. ZhiHao, W. Yingxun, Modeling of the quadrotor UAV based on screw theory via dual quaternion, in: AIAA Modeling and Simulation Technologies (MST) Conference, Boston, MA, USA, 2013.
20. F. Nuno, P. Tsiotras, Simultaneous position and attitude control without linear and angular velocity feedback using dual quaternions, in: American Control Conference, Washington, DC, USA, 2013.
21. X. Wang, C. Yu, Feedback linearization regulator with coupled attitude and translation dynamics based on unit dual quaternion, in: IEEE International Symposium on Intelligent Control (ISIC), Yokohama, Japan, IEEE, 2010, pp. 2380–2384.
22. S. Bouabdallah, R. Siegwart, Full control of a quadrotor, in: IEEE/RSJ International Conference on Intelligent Robots and Systems (IROS), San Diego, CA, USA, IEEE, 2007, pp. 153–158.
23. R. Mahony, V. Kumar, P. Corke, Multirotor aerial vehicles: modeling, estimation, and control of quadrotor, IEEE Robotics & Automation Magazine 19 (2012) 20–32.
24. H. Chao, Y. Cao, Y. Chen, Autopilots for small unmanned aerial vehicles: a survey, International Journal of Control, Automation, and Systems 8 (2010) 36–44.

25. L. Mejias, J.F. Correa, I. Mondragón, P. Campoy, COLIBRI: a vision-guided UAV for surveillance and visual inspection, in: International Conference on Robotics and Automation, Roma, Italy, IEEE, 2007, pp. 2760–2761.
26. D.W. Casbeer, R. Beard, T. McLain, S.-M. Li, R.K. Mehra, Forest fire monitoring with multiple small UAVs, in: Proceedings of the American Control Conference, Portland, Oregon, USA, 2005, pp. 3530–3535.
27. H. Chao, M. Baumann, A. Jensen, Y. Chen, Y. Cao, W. Ren, M. McKee, Band-reconfigurable multi-UAV-based cooperative remote sensing for real-time water management and distributed irrigation control, in: IFAC World Congress, Seoul, Korea, 2008.
28. H. Chao, C. Coopmans, L. Di, Y. Chen, A comparative evaluation of low-cost IMUs for unmanned autonomous systems, in: Conference on Multisensor Fusion and Integration for Intelligent Systems (MFI), Salt Lake, UT, USA, IEEE, 2010, pp. 211–216.
29. D. Simon, Optimal State Estimation: Kalman, H_∞, and Nonlinear Approaches, Wiley, 2006.
30. S.-G. Kim, J.L. Crassidis, Y. Cheng, A.M. Fosbury, J.L. Junkins, Kalman filtering for relative spacecraft attitude and position estimation, Journal of Guidance, Control, and Dynamics 30 (2007) 133–143.
31. R. Mahony, T. Hamel, J.-M. Pflimlin, Nonlinear complementary filters on the special orthogonal group, IEEE Transactions on Automatic Control 53 (2008) 1203–1218.
32. M. Euston, P. Coote, R. Mahony, J. Kim, T. Hamel, A complementary filter for attitude estimation of a fixed-wing UAV, in: IEEE/RSJ International Conference on Intelligent Robots and Systems (IROS), Nice, France, IEEE, 2008, pp. 340–345.
33. L. Benziane, A.E. Hadri, A. Seba, A. Benallegue, Y. Chitour, Attitude estimation and control using linearlike complementary filters: theory and experiment, IEEE Transactions on Control Systems Technology PP (99) (2016) 1–8.
34. J.L. Crassidis, F.L. Markley, Y. Cheng, Survey of nonlinear attitude estimation methods, Journal of Guidance, Control, and Dynamics 30 (2007) 12–28.
35. J.L. Farrell, Attitude determination by Kalman filtering, Automatica 6 (1970) 419–430.
36. E.J. Lefferts, F.L. Markley, M.D. Shuster, Kalman filtering for spacecraft attitude estimation, Journal of Guidance, Control, and Dynamics 5 (1982) 417–429.
37. J.L. Crassidis, F.L. Markley, Predictive filtering for nonlinear systems, Journal of Guidance, Control, and Dynamics 20 (1997) 566–572.
38. J.L. Crassidis, F.L. Markley, Unscented filtering for spacecraft attitude estimation, Journal of Guidance, Control, and Dynamics 26 (2003) 536–542.
39. F.L. Markley, J.L. Crassidis, Y. Cheng, Nonlinear attitude filtering methods, in: AIAA Guidance, Navigation, and Control Conference, San Francisco, CA, USA, 2005.
40. O.J. Smith, Closer control of loops with dead time, Chemical Engineering Progress 53 (1957) 217–219.
41. A. Manitius, A.W. Olbrot, Finite spectrum assignment problem for systems with delays, IEEE Transactions on Automatic Control 24 (1979) 541–552.
42. V.K.D. Mellinger, N. Michael, Trajectory generation and control for precise aggressive maneuvers with quad-rotors, The International Journal of Robotics Research 31 (5) (2010) 664–674.
43. A.L. Chan, S.L. Tan, C.L. Kwek, Sensor data fusion for attitude stabilization in a low cost quadrotor system, in: International Symposium on Consumer Electronics, Singapore, Singapore, 2011.

44. J. Normey-Rico, E.F. Camacho, Dead-time compensators: a survey, Control Engineering Practice 16 (2008) 407–428.
45. Z. Artstein, Linear systems with delayed controls: a reduction, IEEE Transactions on Automatic Control 27 (1982) 869–879.
46. J. Richard, Time-delay systems: an overview of some recent advances and open problems, Automatica 39 (2003) 1667–1694.
47. P. Garcia, P. Castillo, R. Lozano, P. Albertos, Robustness with respect to delay uncertainties of a predictor–observer based discrete-time controller, in: Proceedings of the 45th IEEE Conference on Decision and Control (CDC), San Diego, CA, USA, IEEE, 2006, pp. 199–204.
48. A. Gonzalez, A. Sala, P. Garcia, P. Albertos, Robustness analysis of discrete predictor-based controllers for input-delay systems, International Journal of Systems Science 44 (2) (2013) 232–239.
49. P. Albertos, P. García, Predictor–observer-based control of systems with multiple input/output delays, Journal of Process Control 22 (2012) 1350–1357.
50. A. Gonzalez, P. Garcia, P. Albertos, P. Castillo, R. Lozano, Robustness of a discrete-time predictor-based controller for time-varying measurement delay, Control Engineering Practice 20 (2012) 102–110.
51. J.E. Normey-Rico, E. Camacho, Control of Dead-Time Processes, Springer, 2007.
52. G. Ducard, R. D'Andrea, Autonomous quadrotor flight using a vision system and accommodating frames misalignment, in: Proceedings of IEEE International Symposium on Industrial Embedded Systems (SIES), Switzerland, IEEE, 2009, pp. 261–264.
53. R. Zhang, X. Wang, K.Y. Cai, Quadrotor aircraft control without velocity measurements, in: Proceedings of the 48th IEEE Conference on Decision and Control, 2009 held jointly with the 2009 28th Chinese Control Conference, CDC/CCC, Shanghai, China, 2009.
54. J. Eckert, R. German, F. Dressler, On autonomous indoor flights: high-quality real-time localization using low-cost sensors, in: Proceedings of IEEE International Conference on Communications (ICC), Ottawa, Canada, 2012.
55. D.M.M. Bošnak, S. Blazic, Quadrocopter control using an on-board video system with off-board processing, Robotics and Autonomous Systems 60 (2012) 657–667.
56. D. Eberli, D. Scaramuzza, S. Weiss, R. Siegwart, Vision based position control for MAVs using one single circular landmark, Journal of Intelligent & Robotic Systems 61 (2011) 495–512.
57. C. Martínez, I.F. Mondragón, M.A. Olivares-Méndez, P. Campoy, On-board and ground visual pose estimation techniques for UAV control, Journal of Intelligent & Robotic Systems 61 (2011) 301–320.
58. E. Altug, C. Taylor, Vision-based pose estimation and control of a model helicopter, in: Proceedings of the International Conference on Mechatronics, Istanbul, Turkey, IEEE, 2004, pp. 316–321.
59. E.M.C. Teuliere, L. Eck, N. Guenard, 3D model-based tracking for UAV position control, in: International Conference on Intelligent Robots and Systems (IROS), Taipei, Taiwan, IEEE/RSJ, 2010, pp. 1084–1089.
60. U. Pilz, W. Gropengießer, F. Walder, J. Witt, H. Wernner, Quadrocopter localization using RTK-GPS and vision-based trajectory tracking, in: International Conference on Intelligent Robotics and Applications (ICRA), Aachen, Germany, 2011.

61. Y. Song, B. Xian, Y. Zhang, X. Jiang, X. Zhang, Towards autonomous control of quadrotor unmanned aerial vehicles in a GPS-denied urban area via laser ranger finder, Optik 126 (2015) 3877–3882.
62. C. Troiani, A. Martinelli, C. Laugier, D. Scaramuzza, Low computational-complexity algorithms for vision-aided inertial navigation of micro aerial vehicles, Robotics and Autonomous Systems 69 (2015) 80–97.
63. G.G. Rigatos, Nonlinear Kalman Filters and Particle Filters for integrated navigation of unmanned aerial vehicles, Robotics and Autonomous Systems 60 (2012) 978–995.
64. F. Wang, J. Cui, B. Chen, T. Lee, A comprehensive UAV indoor navigation system based on vision optical flow and laser FastSLAM, Acta Automatica Sinica 39 (2013) 1889–1899.
65. J.-Y. Baek, S.-H. Park, B.-S. Cho, M.-C. Lee, Position tracking system using single RGB-D camera for evaluation of multi-rotor UAV control and self-localization, in: International Conference on Advanced Intelligent Mechatronics, Busan, Korea, IEEE, 2015, pp. 1283–1288.
66. C. Yongcan, UAV circumnavigating an unknown target under a GPS-denied environment with range-only measurements, Automatica 55 (2015) 150–158.
67. S.-G. Kim, J.L. Crassidis, Y. Cheng, A.M. Fosbury, J.L. Junkins, Kalman filtering for relative spacecraft attitude and position estimation, Journal of Guidance, Control, and Dynamics 30 (1) (2007) 133–143.
68. T. Dierks, S. Jagannathan, Output feedback control of a quadrotor UAV using neural networks, IEEE Transactions on Neural Networks 21 (2010) 50–66.
69. H. Voss, Nonlinear control of a quadrotor micro-UAV using feedback-linearization, in: Proceedings of IEEE International Conference on Mechatronics, Changchun, China, 2009, pp. 1–6.
70. T. Madani, A. Benallegue, Control of a quadrotor mini-helicopter via full state backstepping technique, in: Proceedings of the 45th IEEE Conference on Decision and Control, San Diego, CA, USA, 2006.
71. P. Castillo, R. Lozano, A. Dzul, Stabilization of a mini rotorcraft with four rotors, IEEE Control Systems Magazine 25 (2005) 45–55.
72. A. Tayebi, S. McGilvray, Attitude stabilization of a VTOL quadrotor aircraft, IEEE Transactions on Control Systems Technology 14 (2006) 562–571.
73. H.J. Sussmann, E.D. Sontag, Y. Yang, A general result on the stabilization of linear systems using bounded controls, IEEE Transactions on Automatic Control 39 (1994) 2411–2425.
74. A.T. Fuller, In the large stability of relay and saturation control systems with linear controllers, International Journal of Control 10 (1969) 457–480.
75. A.R. Teel, Global stabilization and restricted tracking for multiple integrators with bounded controls, Systems & Control Letters 18 (3) (1992) 165–171.
76. A.R. Teel, Semi-global stabilization of minimum phase nonlinear systems in special normal forms, Systems & Control Letters 19 (3) (1992) 187–192.
77. A.J. Teel, R. Andrew, Semi-global stabilization of the "ball-and-beam" using "output" feedback, in: American Control Conference, San Francisco, CA, USA, 1993.
78. P.C.G. Sanahuja, A. Sanchez, Stabilization of *n* integrators in cascade with bounded input with experimental application to a VTOL laboratory system, International Journal of Robust and Nonlinear Control 20 (2010) 1129–1139.

79. P. Castillo, A. Dzul, R. Lozano, Real-time stabilization and tracking of a four-rotor mini rotorcraft, IEEE Transactions on Control Systems Technology 12 (4) (2004) 510–516.
80. P. Castillo, P. Albertos, P. Garcia, R. Lozano, Simple real-time attitude stabilization of a quad-rotor aircraft with bounded signals, in: 45th IEEE Conference on Decision and Control, San Diego, CA, USA, IEEE, 2006, pp. 1533–1538.
81. E.N. Johnson, S.K. Kannan, Nested saturation with guaranteed real pole, in: Proc. American Control Conference, Denver, Colorado, USA, 2003.
82. Y. Yang, H.J. Sontag, E.D. Sussmann, Global stabilization of linear discrete-time systems with bounded feedback, Systems & Control Letters 30 (1997) 273–281.
83. N. Marchand, A. Hably, Global stabilization of multiple integrators with bounded controls, Automatica 41 (2005) 2147–2152.
84. M. Jankovic, R. Sepulchre, P.V. Kokotovic, Constructive Lyapunov stabilization of nonlinear cascade systems, IEEE Transactions on Automatic Control 41 (1996) 1723–1735.
85. F. Mazenc, L. Praly, Adding an integration and global asymptotic stabilization of feedforward systems, IEEE Transactions on Automatic Control 41 (1996) 1559–1578.
86. G. Kaliora, A. Astolfi, A simple design for the stabilization of a class of cascaded nonlinear systems with bounded control, in: IEEE Conference on Decision and Control, Orlando, Florida, USA, 2001.
87. G. Kaliora, A. Astolfi, On the stabilization of feedforward systems with bounded control, Systems & Control Letters 54 (2001) 263–270.
88. L. Marconi, A. Isidori, Robust global stabilization of a class of uncertain feedforward nonlinear systems, Systems & Control Letters 41 (4) (2000) 281–290.
89. V. Utkin, Sliding Modes in Control and Optimization, Springer-Verlag, 1992.
90. A. Levant, Sliding order and sliding accuracy in sliding mode control, International Journal of Control 58 (1993) 1247–1263.
91. A. Levant, Quasi-continuous high-order sliding-mode controllers, IEEE Transactions on Automatic Control 50 (2006) 1812–1816.
92. A. Levant, Robust exact differentiation via sliding modes technique, Automatica 34 (3) (1998) 379–384.
93. A. Pisano, S. Scodina, E. Usai, Load swing suppression in the 3-dimensional overhead crane via second-order sliding-modes, in: The 11th International Workshop on Variable Structure Systems, Mexico, 2010, pp. 452–457.
94. G. Bartolini, A. Pisano, Global output-feedback tracking and load disturbance rejection for electrically-driven robotic manipulators with uncertain dynamics, International Journal of Control 76 (2003) 1201–1213.
95. V.A.L. Freguela, L. Fridman, Output integral sliding mode control to stabilize position of a Stewart platform, Journal of the Franklin Institute 349 (2012) 1526–1542.
96. M.T. Angulo, L. Fridman, A. Levant, Output-feedback finite-time stabilization of disturbed LTI systems, Automatica 48 (2012) 606–611.
97. A. Levant, High-order sliding modes: differentiation and output feedback control, International Journal of Control 76 (2003) 1924–1941.
98. G.V. Raffo, M.G. Ortega, F.R. Rubio, An integral predictive/nonlinear H_∞ control structure for a quadrotor helicopter, Automatica 46 (2010) 29–39.
99. K. Alexis, G. Nikolakopoulos, A. Tzes, Constrained optimal attitude control of a quadrotor helicopter subject to wind-gusts: experimental studies, in: American Control Conference (ACC), Baltimore, MD, USA, 2010, pp. 4451–4455.

100. A. Das, K. Subbarao, F. Lewis, Dynamic inversion with zero-dynamics stabilisation for quadrotor control, IET Control Theory & Applications 3 (2009) 303–314.
101. L. Besnard, Y.B. Shtesselb, B. Landruma, Quadrotor vehicle control via sliding mode controller driven by sliding mode disturbance observer, Journal of the Franklin Institute 349 (2012) 658–684.
102. P. Khalil, Nonlinear Systems, Prentice Hall, 1996.
103. A. Polyakov, A. Poznyak, Method of Lyapunov functions for systems with higher-order sliding modes, Automation and Remote Control 72 (2011) 944–963.
104. A. Polyakov, A. Poznyak, Lyapunov function design for finite-time convergence analysis: twisting controller for second-order sliding mode realization, Automatica 49 (2009) 444–448.
105. J. Moreno, M. Osorio, Strict Lyapunov functions for the super-twisting algorithm, IEEE Transactions on Automatic Control 57 (4) (2012) 1035–1040.
106. T. Ledgerwood, E. Misawa, Controllability and nonlinear control of rotational inverted pendulum, in: Advances in Robust and Nonlinear Control Systems, ASME Journal on Dynamic Systems and Control 43 (1992) 81–88.
107. A. Lukyanov, Optimal nonlinear block-control method, in: Proceedings of the 2nd European Control Conference, Groningen, Netherlands, 1993, pp. 1853–1855.
108. A. Lukyanov, S.J. Dodds, Sliding mode block control of uncertain nonlinear plants, in: Proceedings of the IFAC World Congress, San Francisco, CA, USA, 1996, pp. 241–246.
109. V. Utkin, J. Guldner, J. Shi, Sliding Mode Control in Electromechanical Systems, Taylor and Francis, 1999.
110. G. Cai, B.M. Chen, X. Dong, T.H. Lee, Design and implementation of a robust and nonlinear flight control system for an unmanned helicopter, Mechatronics 21 (2011) 803–820.
111. L. Luque-Vegan, B. Castillo-Toledo, A.G. Loukianov, Robust block second order sliding mode control for a quadrotor, Journal of the Franklin Institute 349 (2012) 719–739.
112. M.R. Mokhtari, B. Cherki, A new robust control for mini rotorcraft unmanned aerial vehicles, ISA Transactions 56 (2015) 86–101.
113. R. Naldi, L. Marconi, Robust control of transition maneuvers for a class of V/STOL aircraft, Automatica 49 (2013) 1693–1704.
114. B.-C. Min, J.-H. Hong, E.T. Matson, Adaptive robust control (ARC) for an altitude control of a quadrotor type UAV carrying an unknown payloads, in: 11th International Conference on Control, Automation and Systems, Gyeonggi-do, Korea (South), IEEE, 2011, pp. 1147–1151.
115. R. Sanz, P. Garcia, Q. Zhong, P. Albertos, Robust control of quadrotors based on an uncertainty and disturbance estimator, Journal of Dynamic Systems, Measurement, and Control 138 (2016) 071006–071013.
116. L.E. Dubins, On curves of minimal length with a constraint on average curvature and with prescribed initial and terminal positions and tangents, American Journal of Mathematics 79 (1957) 497–517.
117. G. Ambrosino, M. Ariola, U. Ciniglio, F. Corraro, E. De Lellis, A. Pironti, Path generation and tracking in 3-D for UAVs, IEEE Transactions on Control Systems Technology 17 (2009) 980–988.
118. H. Wong, V. Kapila, R. Vaidyanathan, UAV optimal path planning using C-C-C class paths for target touring, in: The 43rd IEEE Conference on Decision and Control, Nassau, Bahamas, 2004, pp. 1105–1110.

119. G. Yang, V. Kapila, Optimal path planning for unmanned air vehicles with kinematic and tactical constraints, in: The 41st IEEE Conference on Decision and Control, Las Vegas, Nevada, USA, 2002, pp. 1301–1306.
120. E.P. Anderson, R.W. Beard, T.W. McLain, Real-time dynamic trajectory smoothing for unmanned air vehicles, IEEE Transactions on Control Systems Technology 13 (2005) 471–477.
121. D. Boukraa, Y. Bestaoui, N. Azouz, Three dimensional trajectory generation for an autonomous plane, International Review of Aerospace Engineering 1 (2008) 355–365.
122. T. Berglund, H. Jonsson, I. Söderkvist, An obstacle avoiding minimum variation B-spline problem, in: Proceedings of International Conference on Geometric Modeling and Graphics, London, UK, 2003, pp. 156–161.
123. E. Dyllong, A. Visioli, Planning and real-time modifications of a trajectory using spline techniques, Robotica 21 (2003) 475–482.
124. K.B. Judd, T.W. McLain, Spline based path planning for unmanned air vehicles, in: AIAA Guidance, Navigation and Control Conference and Exhibit, Montreal, Canada, 2001.
125. B. Vazquez G., H. Sossa A., J.L. Díaz-de León S., Auto guided vehicle control using expanded time B-splines, in: IEEE International Conference on Systems, Man, and Cybernetics, San Antonio, TX, 1994, pp. 2786–2791.
126. L. Zeng, G. Bone, Mobile robot navigation for moving obstacles with unpredictable direction changes, including humans, Advanced Robotics 26 (2012) 1841–1862.
127. M. Kothari, I. Postlethwaite, D.-W. Gu, Multi-UAV path planning in obstacle rich environments using Rapidly-exploring Random Trees, in: Proceedings of the 48th IEEE Conference on Decision and Control, 2009 held jointly with the 2009 28th Chinese Control Conference, CDC/CCC, Shanghai, China, 2009.
128. L. Tang, S. Dian, G. Gu, K. Zhou, S. Wang, X. Feng, A novel potential field method for obstacle avoidance and path planning of mobile robot, in: 3rd IEEE International Conference on Computer Science and Information Technology (ICCSIT), vol. 9, Chengdu, China, 2010, pp. 633–637.
129. P. Fiorini, Z. Shiller, Motion planning in dynamic environments using velocity obstacles, The International Journal of Robotics Research 17 (2012) 760–772.
130. Z. Qu, J. Wang, C.E. Plaisted, A new analytical solution to mobile robot trajectory generation in the presence of moving obstacles, IEEE Transactions on Robotics 20 (2004) 978–993.
131. M.S. Jie, J.H. Baek, Y.S. Hong, K.W. Lee, Real time obstacle avoidance for mobile robot using limit-cycle and vector field method, in: Knowledge-Based Intelligent Information and Engineering Systems, Springer Berlin Heidelberg, Berlin, Heidelberg, 2006, pp. 866–873.
132. A. Aalbers, Obstacle avoidance using limit cycles, Master thesis, Faculty of Mechanical, Maritime and Materials Engineering, Department Delft Center for Systems and Control, 2013.
133. A. Hakimi, T. Binazadeh, Application of circular limit cycles for generation of uniform flight paths to surveillance of a region by UAV, Open Science Journal of Electrical and Electronic Engineering 2 (2015) 36–42.
134. S.S. Ge, Y.J. Cui, New potential functions for mobile robot path planning, IEEE Transactions on Robotics and Automation 16 (2000) 615–620.

135. Y. Chen, G. Luo, Y. Mei, J. Yu, X. Su, UAV path planning using artificial potential field method updated by optimal control theory, International Journal of Systems Science 47 (6) (2016) 1407–1420.
136. J.O. Kim, P.K. Khosla, Real-time obstacle avoidance using harmonic potential functions, IEEE Transactions on Robotics and Automation 8 (1992) 338–349.
137. O. Khatib, Real-time obstacle avoidance for manipulators and mobile robots, in: Proceedings of the IEEE International Conference on Robotics and Automation, vol. 2, St Louis, Missouri, USA, 1985, pp. 500–505.
138. S.M.C. Alaimo, L. Pollini, J.P. Bresciani, H.H. Bülthoff, Evaluation of direct and indirect haptic aiding in an obstacle avoidance task for tele-operated systems, in: Proceedings of the 18th World Congress of The International Federation of Automatic Control, Milano, Italy, 2011.
139. T. Lam, M. Mulder, M. van Paasen, Haptic feedback for UAV tele-operation – force offset and spring load modification, in: IEEE International Conference on Systems, Man, and Cybernetics, Taipei, Taiwan, 2006.
140. T. Lam, M. Mulder, M.M. van Paassen, J.A. Mulder, F.C.T. van der Helm, Force-stiffness feedback in UAV tele-operation with time delay, in: AIAA Guidance, Navigation, and Control Conference, Chicago, Illinois, USA, 2009.
141. H. Rifa, M. Hua, T. Hamel, P. Morin, Haptic-based bilateral teleoperation of underactuated Unmanned Aerial Vehicles, in: Proceedings of the 18th IFAC World Congress, Milano, Italy, 2011.
142. X. Hou, C. Yu, F. Linag, Z. Lin, Energy based set point modulation for obstacle avoidance in haptic teleoperation of aerial robots, in: Proceedings of the 19th IFAC World Congress, 47(3), Cape Town, South Africa, 2014, pp. 11030–11035.
143. S. Omari, M.-D. Hua, G. Ducard, T. Hamel, Bilateral haptic teleoperation of VTOL UAVs, in: Proc. IEEE International Conference on Robotics and Automation (ICRA), Karlsruhe, Germany, 2013, pp. 2393–2399.
144. S. Stramigioli, R. Mahony, P. Corke, A novel approach to haptic teleoperation of aerial robot vehicles, in: Proc. IEEE Int. Conf. Robot. and Autom., Singapore, Singapore, 2010.
145. H.W. Boschloo, T.M. Lam, M. Mulder, M.M. van Paassen, Collision avoidance for a remotely-operated helicopter using haptic feedback, in: Proc. 2004 IEEE Int. Conf. Syst. Man Cybern., The Hague, Netherlands, 2004, pp. 229–235.
146. A. Brandt, M. Colton, Haptic collision avoidance for a remotely operated quadrotor UAV in indoor environments, in: IEEE International Conference on Systems, Man, and Cybernetics, Istanbul, Turkey, 2010.
147. A.Y. Mersha, A. Rüesch, S. Stramigioli, R. Carloni, A contribution to haptic teleoperation of aerial vehicles, in: Proc. IEEE/RSJ Int. Conf. Intell. Robots Syst. (IROS), Vilamora, Portugal, 2012, pp. 3041–3042.

CHAPTER 2

Modeling Approaches*

The quadrotor is a useful prototype for learning about aerodynamic phenomena in hovering aerial vehicles. Conventional helicopters modify the lift force by varying the collective pitch. They use a mechanical device known as swashplate to change the rotor blades' pitch angle in a cyclic manner so as to obtain the pitch and roll control torques of the vehicle. The swashplate interconnects the servomechanisms and the blades' pitch links. In contrast, the quadrotor does not have a swashplate and has constant pitch blades. In a quadrotor we can only vary the angular speed of each of the four rotors. The quadrotor vehicle is an underactuated mechanical system since it has four inputs and six degrees of freedom. It has four motors which give special characteristics to it. In the classical configuration the front and rear motors rotate counter-clockwise, while the other two motors rotate clockwise, whereby the gyroscopic effects and aerodynamic torques tend to cancel in hovering. This chapter introduces three approaches to obtain the dynamic model of a quadrotor aerial vehicle. These approaches can also be used for other configurations of UAVs.

2.1 FORCE AND MOMENT IN A ROTOR

Blade element theory is particularly useful for studying airfoil and rotor performance. The forces and moments developed on a uniform wing are modeled by the lift force, drag force, and pitching moment [1]. For a rotor with angular velocity ω, the linear velocity at each point along the rotor is proportional to the radial distance from the rotor shaft. Thus the following equations [2] can be stated for the entire rotor i:

$$f_{M_{i_z}} = C_{T_i} \rho A_p\, r^2 \omega_i^2, \tag{2.1}$$

$$\tau_{M_i} = C_{Q_i} \rho A_p r^3 \omega_i^2, \tag{2.2}$$

where $f_{M_{i_z}}$ is the total thrust produced by rotor i acting perpendicularly to the rotor plane along the z_i axis and without blade flapping effect, τ_{M_i} de-

* The results in this chapter were developed in collaboration with H. Abaunza and J. Cariño from the UMI LAFMIA 3175, Mexico.

Indoor Navigation Strategies for Aerial Autonomous Systems.
DOI: http://dx.doi.org/10.1016/B978-0-12-805189-4.00003-2

scribes the rotor torque, r represents the rotor radius, ρ denotes the density of air, and A_p defines the propeller disk area. C_{T_i} and C_{Q_i} are dimensionless thrust and rotor torque coefficients. These coefficients can be obtained using the blade element theory [2–4].

Nevertheless, it is common to represent previous equations as proportional to the square of the angular speed ω, that is, $f_{M_{iz}} \approx k_{f_i}\omega_i^2$ and $\tau_{M_i} \approx k_{\tau_i}\omega_i^2$ where k_{f_i} and k_{τ_i} represent aerodynamics coefficients of the propeller. Note that, since a rotor can only turn in a fixed direction, the produced force $f_{M_{iz}}$ is always positive.

Three approaches are developed in the following; they are the most popular used in the UAV control community. The Euler–Lagrange formalism is presented for the ideal case, i.e., without perturbations and/or uncertainties in the model. In the Newton–Euler methodology, uncertainties in the model and external perturbations such as aerodynamics effects are explained and introduced. Finally, the quaternion approach is developed. This technique is becoming popular in the UAV and control community because it introduces a smart solution for avoiding singular positions in the aerial vehicle and brings simpler mathematical expressions without losing generality.

2.2 EULER–LAGRANGE APPROACH

The dynamical model of the quadrotor is described in this section using the Euler–Lagrange methodology. This model is obtained by representing the aerial vehicle as a solid body evolving in a three-dimensional space and subject to the main thrust and three torques.

The generalized coordinates of the vehicle for the Euler–Lagrange formalism are $\mathbf{x}_{quad} = (\xi, \eta) \in \mathbb{R}^6$, where $\xi = (x, y, z) \in \mathbb{R}^3$ denotes the position vector of the center of mass relative to a fixed inertial frame $\mathcal{E}$, and $\eta = (\psi, \theta, \phi) \in \mathbb{R}^3$ defines the Euler angles (yaw, pitch, and roll angles, respectively) which represent the quadrotor attitude. The Lagrangian is defined as

$$L(q, \dot{q}) = T_{trans} + T_{rot} - U,$$

where $T_{trans} = \frac{m}{2}\dot{\xi}^T\dot{\xi}$ describes the translational kinetic energy, $T_{rot} = \frac{1}{2}\Omega^T \mathrm{I}\, \Omega$ denotes the rotational kinetic energy, $U = mgz$ represents the potential energy, z is the altitude, m denotes the mass, Ω introduces the angular velocity vector, I defines the inertia matrix, and g means the acceleration vector due to gravity. The angular velocity vector Ω in the body frame $\mathcal{B}$ is related to the generalized velocities $\dot{\eta}$ (in the region where the Euler angles are valid) by means of the standard kinematic relationship $\Omega = W_\eta \dot{\eta}$ [5]. Therefore, $T_{rot} = \frac{1}{2}\dot{\eta}^T \mathbb{J} \dot{\eta}$ with $\mathbb{J} = \mathbb{J}(\eta) = W_\eta^T \mathrm{I}\, W_\eta$ and

$$W_\eta = \begin{bmatrix} -\sin\theta & 0 & 1 \\ \cos\theta\sin\phi & \cos\phi & 0 \\ \cos\theta\cos\phi & -\sin\phi & 0 \end{bmatrix}, \qquad \mathrm{I} = \begin{bmatrix} I_{xx} & 0 & 0 \\ 0 & I_{yy} & 0 \\ 0 & 0 & I_{zz} \end{bmatrix},$$

where I_{ii} denotes the moment of inertia with respect to the ith axis.

The matrix $\mathbb{J}(\eta)$ acts as the inertia matrix for the full rotational kinetic energy of the vehicle expressed in terms of η. Then the mathematical equations that represent the dynamics of the aerial vehicle are obtained using the following equation:

$$\frac{d}{dt}\frac{\partial L}{\partial \dot{\mathbf{x}}_{quad}} - \frac{\partial L}{\partial \mathbf{x}_{quad}} = \begin{bmatrix} F_\xi \\ \tau \end{bmatrix}, \tag{2.3}$$

where F_ξ denotes the translational external forces applied to the aerial vehicle, $\tau \in \mathbb{R}^3$ represents the external torques, and R is the rotation matrix representing the orientation of the vehicle relative to a fixed inertial frame. R can vary depending on the rotations made.

2.2.1 Quadrotor Dynamic Model

From Fig. 2.1 and considering that the rotor force is only in the z-axis, it can be written as $\hat{F} = u_z E_z$, where $E_z = [0\ 0\ 1]^T$ and u_z is the main thrust directed out of the top of the vehicle expressed as $u_z = \sum_{i=1}^{4} f_{M_{iz}}$. If this vector force is represented in the inertial frame using the rotation matrix then $F_\xi = R\hat{F}$.

In the quadrotor vehicle, forward pitch motion is obtained by increasing the speed of the rear motor M_2 while reducing the speed of the front motor M_4. Likewise, roll motion is obtained using the lateral motors. Yaw motion is obtained by increasing the torque of the front and rear motors while decreasing the torque of the lateral motors. These motions can be accomplished while keeping the total thrust constant. Therefore, the generalized torques are

$$\tau = \begin{bmatrix} \tau_\psi \\ \tau_\theta \\ \tau_\phi \end{bmatrix} \triangleq \begin{bmatrix} \sum_{i=1}^{4} \tau_{M_i} \\ (f_{M_{3z}} - f_{M_{1z}})\ell \\ (f_{M_{2z}} - f_{M_{4z}})\ell \end{bmatrix},$$

where ℓ is the distance between the motors and the center of gravity.

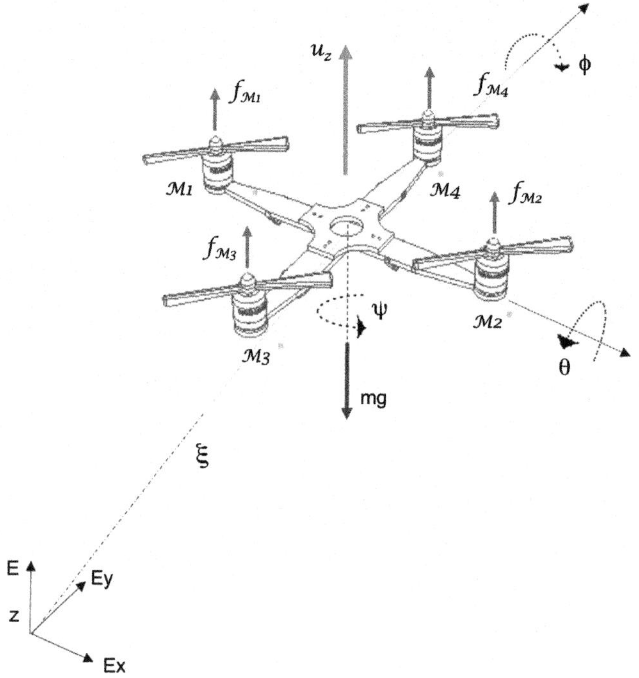

Figure 2.1 Quadcopter scheme in an inertial frame.

Developing (2.3), the Euler–Lagrange equation for the translation motion can be written as

$$m\ddot{\xi} + mgE_z = F_\xi,$$

and for the η coordinates we have

$$\mathbb{J}\ddot{\eta} + \left(\dot{\mathbb{J}} - \frac{1}{2}\frac{\partial}{\partial \eta}\left(\dot{\eta}^T\mathbb{J}\right)\right)\dot{\eta} = \tau.$$

Therefore, rewriting the two previous equations, it follows that

$$m\ddot{\xi} = F_\xi - mgE_z, \quad (2.4)$$

$$\mathbb{J}\ddot{\eta} = \tau - C(\eta, \dot{\eta})\dot{\eta}, \quad (2.5)$$

where $C(\eta, \dot{\eta}) = \dot{\mathbb{J}} - \frac{1}{2}\frac{\partial}{\partial \eta}\left(\dot{\eta}^T\mathbb{J}\right)$ is referred to as the Coriolis term which contains the gyroscopic and centrifugal terms. Expanding Eq. (2.5) is not an easy task, and in several works the full inertia matrix $\mathbb{J}$ is considered diagonal and the Coriolis matrix is, in general, neglected.

The Coriolis and inertial matrices can be obtained from (2.3) for the η dynamics. Therefore, rewriting the attitude dynamics, it yields

$$\frac{d}{dt}\left[\Omega^T \mathrm{I} \frac{\partial \Omega}{\partial \dot{\eta}}\right] - \Omega^T \mathrm{I} \frac{\partial \Omega}{\partial \eta} = \tau.$$

Then $\frac{\partial \Omega}{\partial \dot{\eta}} = W_\eta$, and thus, $\Omega^T \mathrm{I} \frac{\partial \Omega}{\partial \dot{\eta}} = \begin{bmatrix} b_1 & b_2 & b_3 \end{bmatrix}$ with

$$\begin{aligned}
b_1 &= -I_{xx}(\dot{\phi} s_\theta - \dot{\psi} s_\theta^2) + I_{yy}(\dot{\theta} c_\theta s_\phi c_\phi + \dot{\psi} c_\theta^2 s_\phi^2) + I_{zz}(\dot{\psi} c_\theta^2 c_\phi^2 - \dot{\theta} c_\theta s_\phi c_\phi), \\
b_2 &= I_{yy}(\dot{\theta} c_\phi^2 + \dot{\psi} c_\theta s_\phi c_\phi) - I_{zz}(\dot{\psi} c_\theta s_\phi c_\phi - \dot{\theta} s_\phi^2), \\
b_3 &= I_{xx}(\dot{\phi} - \dot{\psi} s_\theta),
\end{aligned}$$

with s_θ and c_θ standing for $\sin(\theta)$ and $\cos(\theta)$, respectively. Differentiating $\left(\Omega^T \mathrm{I} \frac{\partial \Omega}{\partial \dot{\eta}}\right)$ with respect to time, we get

$$\begin{aligned}
\dot{b}_1 &= -I_{xx}(\ddot{\phi} s_\theta + \dot{\phi}\dot{\theta} c_\theta - \ddot{\psi} s_\theta^2 - 2\dot{\psi}\dot{\theta} s_\theta c_\theta) + I_{yy}(\ddot{\theta} c_\theta s_\phi c_\phi - \dot{\theta}^2 s_\theta s_\phi c_\phi - \dot{\theta}\dot{\phi} c_\theta s_\phi^2 \\
&\quad + \dot{\theta}\dot{\phi} c_\theta c_\phi^2 + \ddot{\psi} c_\theta^2 s_\phi^2 - 2\dot{\psi}\dot{\theta} s_\theta c_\theta s_\phi^2 + 2\dot{\psi}\dot{\phi} c_\theta^2 s_\phi c_\phi) \\
&\quad + I_{zz}(\ddot{\psi} c_\theta^2 c_\phi^2 - 2\dot{\psi}\dot{\theta} s_\theta c_\theta c_\phi^2 - 2\dot{\psi}\dot{\phi} c_\theta^2 s_\phi c_\phi \\
&\quad - \ddot{\theta} c_\theta s\phi c_\phi + \dot{\theta}^2 s_\theta s_\phi c_\phi + \dot{\theta}\dot{\phi} c_\theta c_\phi^2 - \dot{\theta}\dot{\phi} c_\theta c_\phi^2), \\
\dot{b}_2 &= I_{yy}(\ddot{\theta} c_\phi^2 - 2\dot{\theta}\dot{\phi} s_\phi c_\phi + \ddot{\psi} c_\theta s_\phi c_\phi - \dot{\psi}\dot{\theta} s_\theta s_\phi c_\phi + \dot{\psi}\dot{\phi} c_\theta c_\phi^2 - \dot{\psi}\dot{\phi} c_\theta s_\phi^2) \\
&\quad - I_{zz}(\ddot{\psi} c_\theta s_\phi c_\phi - \dot{\psi}\dot{\theta} s_\theta s_\phi c_\phi - \dot{\psi}\dot{\phi} c_\theta s_\phi^2 + \dot{\psi}\dot{\phi} c_\theta c_\phi^2 - \ddot{\theta} s_\phi^2 - 2\dot{\theta}\dot{\phi} s_\phi c_\phi), \\
\dot{b}_3 &= I_{xx}(\ddot{\phi} - \ddot{\psi} s_\theta - \dot{\psi}\dot{\theta} c_\theta).
\end{aligned}$$

Similarly, we have

$$\frac{\partial \Omega}{\partial \eta} = \begin{bmatrix} 0 & -\dot{\psi} c_\theta & 0 \\ 0 & -\dot{\psi} s_\theta s_\phi & -\dot{\theta} s_\phi + \dot{\psi} c_\theta c_\phi \\ 0 & -\dot{\psi} s_\theta c_\phi & -\dot{\psi} c_\theta s_\phi - \dot{\theta} c_\phi \end{bmatrix}$$

and then $\Omega^T \mathrm{I} \frac{\partial \Omega}{\partial \eta} = \begin{bmatrix} h_1 & h_2 & h_3 \end{bmatrix}$, where

$$\begin{aligned}
h_1 &= 0, \\
h_2 &= -I_{xx}(\dot{\psi}\dot{\phi} c_\theta - \dot{\psi}^2 s_\theta c_\theta) - I_{yy}(\dot{\psi}\dot{\theta} s_\theta s_\phi c_\phi + \dot{\psi}^2 s_\theta c_\theta s_\phi^2) \\
&\quad - I_{zz}(\dot{\psi}^2 s_\theta c_\theta c_\phi^2 - \dot{\psi}\dot{\theta} s_\theta s_\phi c_\phi), \\
h_3 &= I_{yy}(-\dot{\theta}^2 s_\phi c_\phi - \dot{\psi}\dot{\theta} c_\theta s_\phi^2 + \dot{\psi}\dot{\theta} c_\theta c_\phi^2 + \dot{\psi}^2 c_\theta^2 s_\phi c_\phi) \\
&\quad + I_{zz}(-\dot{\psi}^2 c_\theta^2 s_\phi c_\phi + \dot{\psi}\dot{\theta} c_\theta s_\phi^2 - \dot{\psi}\dot{\theta} c_\theta c_\phi^2 + \dot{\theta}^2 s_\phi c_\phi).
\end{aligned}$$

Notice that

$$\tau \;=\; \begin{bmatrix} \tau_\psi \\ \tau_\theta \\ \tau_\phi \end{bmatrix} = \begin{bmatrix} \dot{b}_1 - h_1 \\ \dot{b}_2 - h_2 \\ \dot{b}_3 - h_3 \end{bmatrix}.$$

Thus, grouping terms and using (2.5), it follows that

$$\mathbb{J}(\eta) \;=\; \begin{bmatrix} I_{xx}s^2\theta + I_{yy}c^2\theta s^2\phi + I_{zz}c^2\theta c^2\phi & c\theta c\phi s\phi(I_{yy} - I_{zz}) & -I_{xx}s\theta \\ c\theta c\phi s\phi(I_{yy} - I_{zz}) & I_{yy}c^2\phi + I_{zz}s^2\phi & 0 \\ -I_{xx}s\theta & 0 & I_{xx} \end{bmatrix} \tag{2.6}$$

and

$$C(\eta, \dot{\eta}) \;=\; \begin{bmatrix} c_{11} & c_{12} & c_{13} \\ c_{21} & c_{22} & c_{23} \\ c_{31} & c_{32} & c_{33} \end{bmatrix},$$

where

$$\begin{aligned}
c_{11} &= I_{xx}\dot{\theta}s_\theta c_\theta + I_{yy}(-\dot{\theta}s_\theta c_\theta s_\phi^2 + \dot{\phi}c_\theta^2 s_\phi c_\phi) - I_{zz}(\dot{\theta}s_\theta c_\theta c_\phi^2 + \dot{\phi}c_\theta^2 s_\phi c_\phi),\\
c_{12} &= I_{xx}\dot{\psi}s_\theta c_\theta - I_{yy}(\dot{\theta}s_\theta s_\phi c_\phi + \dot{\phi}c_\theta s_\phi^2 - \dot{\phi}c_\theta c_\phi^2 + \dot{\psi}s_\theta c_\theta s_\phi^2)\\
&\quad + I_{zz}(\dot{\phi}c_\theta s_\phi^2 - \dot{\phi}c_\theta c_\phi^2 - \dot{\psi}s_\theta c_\theta c_\phi^2 + \dot{\theta}s_\theta s_\phi c_\phi),\\
c_{13} &= -I_{xx}\dot{\theta}c_\theta + I_{yy}\dot{\psi}c_\theta^2 s_\phi c_\phi - I_{zz}\dot{\psi}c_\theta^2 s_\phi c_\phi,\\
c_{21} &= -I_{xx}\dot{\psi}s_\theta c_\theta + I_{yy}\dot{\psi}s_\theta c_\theta s_\phi^2 + I_{zz}\dot{\psi}s_\theta c_\theta c_\phi^2,\\
c_{22} &= -I_{yy}\dot{\phi}s_\phi c_\phi + I_{zz}\dot{\phi}s_\phi c_\phi,\\
c_{23} &= I_{xx}\dot{\psi}c_\theta + I_{yy}(-\dot{\theta}s_\phi c_\phi + \dot{\psi}c_\theta c_\phi^2 - \dot{\psi}c_\theta s_\phi^2) + I_{zz}(\dot{\psi}c_\theta s_\phi^2 - \dot{\psi}c_\theta c_\phi^2 + \dot{\theta}s_\phi c_\phi),\\
c_{31} &= -I_{yy}\dot{\psi}c_\theta^2 s_\phi c_\phi + I_{zz}\dot{\psi}c_\theta^2 s_\phi c_\phi,\\
c_{32} &= -I_{xx}\dot{\psi}c_\theta + I_{yy}(\dot{\theta}s_\phi c_\phi + \dot{\psi}c_\theta s_\phi^2 - \dot{\psi}c_\theta c_\phi^2) - I_{zz}(\dot{\psi}c_\theta s_\phi^2 - \dot{\psi}c_\theta c_\phi^2 + \dot{\theta}s_\phi c_\phi),\\
c_{33} &= 0.
\end{aligned}$$

Notice that Eqs. (2.4) and (2.5) represent the mathematical model of an aerial vehicle having four rotors. These equations are also valid for other aerial configurations as long as the external forces and torques are rewritten. Observe that in this approach no aerodynamic effects are taken into account. In the following subsection, these effects will be included.

2.3 NEWTON–EULER APPROACH

The general mathematical model describing the dynamics of an aircraft evolving in a three-dimensional space is obtained by representing the aircraft as a solid body which is subject to non-conservative forces $F_\xi \in \mathbb{R}^3$ expressed in an inertial frame $\mathcal{E}$ and torques $\tau \in \mathbb{R}^3$ applied to its center of mass, specified with respect to the body frame $\mathcal{B}$, and by using the Newton–Euler approach [6–8]:

$$\dot{\xi} = V, \qquad m\dot{V} = F_\xi, \tag{2.7}$$

$$\dot{R} = R\hat{\Omega}, \qquad \mathrm{I}\dot{\Omega} = -\Omega \times \mathrm{I}\Omega + \tau, \tag{2.8}$$

where V is the linear velocity expressed in $\mathcal{E}$, $\hat{\Omega}$ describes the antisymmetric matrix of Ω, and I signifies the constant inertia matrix around the center of mass.

In this approach, external perturbations (e.g., wind) and uncertainties in the model (e.g., blade flapping) are considered. This model contains aerodynamic effects and could be used for research purposes.

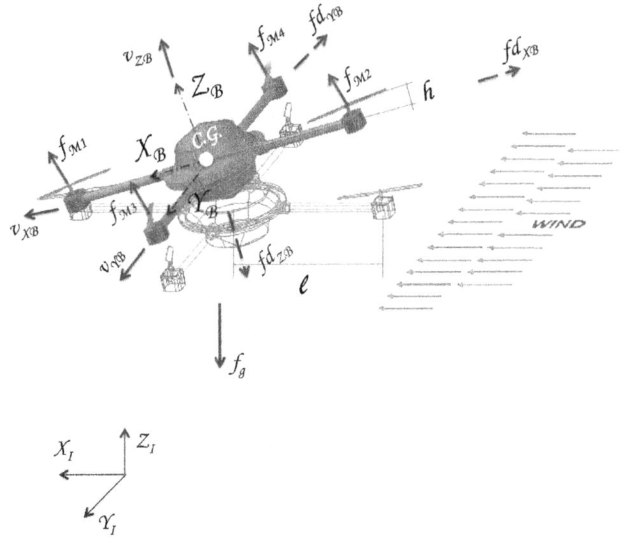

Figure 2.2 The quadrotor in the presence of crosswind.

2.3.1 Quadrotor Dynamic Model in the Presence of Wind

Consider the aerial vehicle in the presence of lateral wind. Then from Fig. 2.2 it can be concluded that

$$F_\xi = R\left(\hat{F} + f_d\right) + f_g, \tag{2.9}$$

where $f_g = -mgE_z$ defines the gravitational force applied to the vehicle, g denotes the acceleration due to gravity, and f_d describes the vector vehicle drag force in $\mathcal{B}$. The main vector force produced by the rotors is defined as $\hat{F} = [u_x \; u_y \; u_z]^T$ and will be explained in the following subsection. Notice that for the ideal case $u_x, u_y = 0$ and $u_z = \sum f_{M_i}$.

Uncertainties in the Model – Blade Flapping

A rotor in translational flight undergoes an effect known as blade flapping. This aerodynamic effect causes an imbalance in lift, inducing an up and down oscillation of the rotor blades. Then the rotor plane is sometimes not aligned with the X_B–Y_B plane of the translation, see Fig. 2.3. Hence, (2.1) becomes

$$f_{M_i} = k_{f_i}\omega_i^2 \begin{pmatrix} -\sin(a_{s_i}) \\ \cos(a_{s_i})\sin(b_{s_i}) \\ \cos(b_{s_i})\cos(a_{s_i}) \end{pmatrix}, \tag{2.10}$$

where a_{s_i} and b_{s_i} describe the longitudinal and lateral harmonic flapping angles of rotor i. This implies two external forces, in the axis x and y, respectively, thus the vector external force yields $\hat{F} = [u_x \; u_y \; u_z]^T$; for more details see [2] and [4].

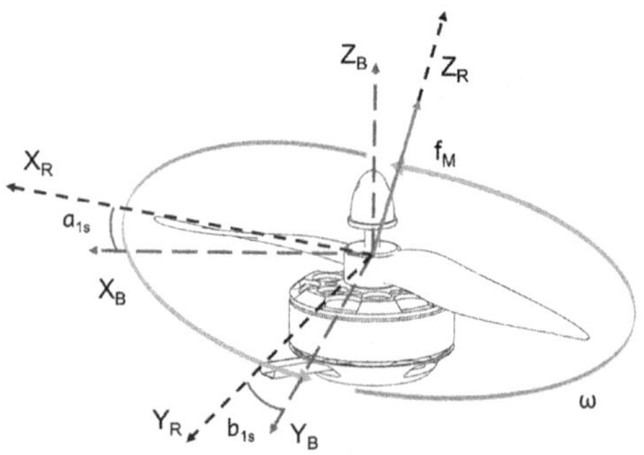

Figure 2.3 Blade rotor flapping.

Aerodynamic Effects

The drag force acting on an aerial vehicle passing through air always points in the opposite direction to the vehicle's instantaneous direction of motion. Thus, without loss of generality, it can be approximated as

$$f_{d_k} \approx C_{d_k} \rho A_k v_k (v_{w_k} - v_k), \qquad k: X_B, Y_B, Z_B, \tag{2.11}$$

where C_{d_k} is the quadrotor drag coefficient, A_k means the contact frontal area in the k-axis, v_{w_k} defines the components of the wind velocity in $\mathcal{B}$, and v_k represents the velocity of the quadrotor center of mass along the k-axis. For given air conditions, shape, and inclination of the vehicle, we must determine a value for C_{d_k} to determine drag. Determining the value of the drag coefficient is more difficult than determining the lift coefficient because of the multiple sources of drag. In this book, we will consider the drag forces as unknown disturbances.

Torques

The moments produced by the rotor flapping comprise two components: the rotor hub stiffness and the thrust vector acting around a displacement from the vehicle center of gravity (CG)

$$\tau_{r_i} = k_\beta a_{s_i} + \mathbf{r}_i \times f_i, \tag{2.12}$$

where k_β is the stiffness of the rotor blade and $\mathbf{r}_i$ denotes the vector from the CG to each rotor. It is defined as

$$r_1 = \begin{pmatrix} 0 & \ell & h \end{pmatrix}, \qquad r_3 = \begin{pmatrix} \ell & 0 & h \end{pmatrix}, \tag{2.13}$$

$$r_2 = \begin{pmatrix} 0 & -\ell & h \end{pmatrix}, \quad r_4 = \begin{pmatrix} -\ell & 0 & h \end{pmatrix}, \tag{2.14}$$

where h is the distance between the CG and the origin where the force rotor is applied.

Due to the quadrotor's bilateral symmetries, moments generated by lateral deflections of the rotor plane cancel. Therefore, the only flapping moment is created by the backward tilt of the rotor plane through a deflection angle a_{s_i} with the longitudinal thrust. In addition, the physical stiffness of a rotor is typically ignored in flyer analysis and the rotor stiffness is modeled purely as a centrifugal term. Moreover, current configurations of quadrotors do not incorporate the hub-springs; as a result, (2.12) can be substantially simplified.

Therefore, the total torque τ is written as

$$\tau = \sum_{i=1}^{4}(\tau_{M_i} + \tau_{r_i}) + \tau_d, \tag{2.15}$$

where τ_d represents the drag moment vector. In order to preserve the sign of rotation for counter-rotating rotors, τ_{M_i} is taken as

$$\tau_{M_i} = C_{Q_i}\rho A_p r^3 \omega_i|\omega_i|\hat{\mathbf{k}}. \tag{2.16}$$

Introducing (2.9) into (2.7) and (2.15) into (2.8), it follows that

$$m\ddot{\xi} = R\left(\sum_{i=1}^{4} f_i + f_d\right) + f_g, \tag{2.17}$$

$$\mathbf{I}\dot{\Omega} = -\Omega \times \mathbf{I}\Omega + \sum_{i=1}^{4}(\tau_{M_i} + \tau_{r_i}) + \tau_d. \tag{2.18}$$

Eq. (2.17) represents the translational dynamics of the aerial vehicle whilst Eq. (2.18) introduces its rotational dynamics.

Designing a controller based on this model requires parameters of the physical system to be specified. Most of these values are dictated by the flight performance of the system; some, most importantly h, can be chosen freely. The error associated with each parameter (flexibility of the propellers, h, etc.) defines the envelope of the plant model dynamic response [3,9]. However, some robust control laws obtained from linear or simple nonlinear models are able to "absorb" these errors and control or stabilize the system. Indeed, if the model is simplified, these parameters are usually considered as nonlinear uncertainties in the model.

2.4 QUATERNION APPROACH

Quaternions are a very useful mathematical tool for representing the rotation of a rigid body, they have great advantages over the more commonly used Euler angles' representation due to, e.g., lack of singularities/discontinuities and mathematical simplicity. Simultaneous rotation and translation (also called transformation) can be described using unitary quaternions if the position is rotated to the inertial frame. Dual quaternions are another mathematical tool used to describe the transformation of a rigid body, they also offer mathematical simplicity. The mathematical model of the aerial vehicle using unitary and dual quaternion approaches is presented in this section. Both models can be used for UAV tasks such as aggressive maneuvers, looping, etc. This approach is not intuitive and sometimes could be difficult to understand. For these reasons and in order to better clarify it, the main background and mathematical operations with quaternions are also presented in the following.

2.4.1 Modeling Using Quaternions

Quaternions are "hypercomplex" numbers, which means that they have three imaginary units $\hat{\imath}$, $\hat{\jmath}$, and $\hat{k}$ instead of one compared to the complex numbers. They can be used to describe rotations in three-dimensional space in a very simple mathematical and computational way. While many methods use trigonometric functions which are nonlinear and suffer from numerical inaccuracy, quaternion rotations are simple since they only need multiplications, divisions, and sums to be implemented.

Notation & Quaternion Operations

In this section, overlined letters represent vectors in 3D space $\bar{(\cdot)} \in \mathbb{R}^3$. A quaternion, $\mathbf{q}$, is a four-tuple that belongs to the quaternion space $\mathbb{H}$. It can be seen as a number that contains one real part and three imaginary parts multiplied by their corresponding imaginary units $\hat{\imath}, \hat{\jmath}$, and $\hat{k} \in \mathbb{I}$, i.e., $\mathbf{q} := q_0 + q_1\hat{\imath} + q_2\hat{\jmath} + q_3\hat{k}$, where $q_0, q_1, q_2, q_3 \in \mathbb{R}$. As there are three different imaginary parts, these are often viewed as a vector in $\mathbb{R}^3$ space. Thus, $\mathbb{R}^3$ can be seen as a subspace of $\mathbb{H}$, and an $\mathbb{R}^3$ vector can be considered a pure imaginary quaternion:

$$\mathbf{q} = q_0 + \begin{bmatrix} q_1 \\ q_2 \\ q_3 \end{bmatrix} = q_0 + \bar{q}, \qquad \bar{q} \in \mathbb{R}^3. \tag{2.19}$$

Product. The quaternion product between quaternions $\mathbf{q}, \mathbf{p} \in \mathbb{H}$, expressed as a sum between a scalar real part and an imaginary vector, $\mathbf{q} = q_0 + \bar{q}, \mathbf{p} = p_0 + \bar{p}$, is defined in the following manner $\mathbf{q} \otimes \mathbf{p} := (q_0 + p_0 - \bar{q} \cdot \bar{p}) + (q_0\bar{p} + p_0\bar{q} + \bar{q} \times \bar{p})$. Some properties can be seen from this definition. One of the most important is that quaternion product is not commutative which means that $\mathbf{q} \otimes \mathbf{p} \neq \mathbf{p} \otimes \mathbf{q}$. This is because of the same non-commutativity property of the cross-product used in the definition.

Sum. The sum of quaternions $\mathbf{q}$ and $\mathbf{p}$ is simply the sum of each of its elements, thus $\mathbf{q} + \mathbf{p} := q_0 + p_0 + \bar{q} + \bar{p}$. The set of all quaternions with addition and multiplication operations defines a non-commutative division ring; see [10].

Conjugate. The conjugate of quaternion $\mathbf{q}$ is denoted as $\mathbf{q}^* := q_0 - \bar{q}$. The conjugate of a product of quaternions is $(\mathbf{q} \otimes \mathbf{r})^* = \mathbf{r}^* \otimes \mathbf{q}^*$. This can be proved by expanding the corresponding products.

Norm. The norm of a quaternion is defined by $||\mathbf{q}||^2 := \mathbf{q} \otimes \mathbf{q}^* = q_0^2 + q_1^2 + q_2^2 + q_3^2$.

Inverse. The quaternion product forms a closed-loop group, that is, the product of two non-null quaternions is another quaternion. This means that for any non-null quaternion there exists an inverse quaternion $\mathbf{q}^{-1} := \frac{\mathbf{q}^*}{||\mathbf{q}||}$ such that $\mathbf{q} \otimes \mathbf{q}^{-1} = \mathbf{q}^{-1} \otimes \mathbf{q} = 1$.

Derivative. Let r be any given vector (quaternion with zero scalar part) fixed in an initial reference frame. Let r' be the same vector but rotated to another reference frame such that

$$r' = \mathbf{q}^{-1} \otimes r \otimes \mathbf{q}. \tag{2.20}$$

If we differentiate (2.20), then $\dot{r}' = \dot{\mathbf{q}}^{-1} \otimes r \otimes \mathbf{q} + \mathbf{q}^{-1} \otimes r \otimes \dot{\mathbf{q}}$. Then using the previous equations, we can rewrite $\dot{r}' = \dot{\mathbf{q}}^{-1} \otimes \mathbf{q} \otimes r' + r' \otimes \mathbf{q}^{-1} \otimes \dot{\mathbf{q}}$. Since $\mathbf{q}$ is a unitary quaternion, $\mathbf{q}^{-1} \otimes \mathbf{q} = 1$ and $\dot{\mathbf{q}}^{-1} \otimes \mathbf{q} + \mathbf{q}^{-1} \otimes \dot{\mathbf{q}} - 0$. Then it follows that $\dot{r}' = r' \otimes \mathbf{q}^{-1} \otimes \dot{\mathbf{q}} - \mathbf{q}^{-1} \otimes \dot{\mathbf{q}} \otimes r'$. Now the scalar part (real part) of $\mathbf{q}^{-1} \otimes \dot{\mathbf{q}}$ is $Re(\mathbf{q}^{-1} \otimes \dot{\mathbf{q}}) = \dot{\mathbf{q}}_0\mathbf{q}_0 + \dot{\mathbf{q}}_1\mathbf{q}_1 + \dot{\mathbf{q}}_2\mathbf{q}_2 + \dot{\mathbf{q}}_3\mathbf{q}_3 = 0$. Therefore the product $\mathbf{q}^{-1} \otimes \dot{\mathbf{q}}$ is a vector (a quaternion with the real part equal to zero). Then, it follows that

$$\dot{r}' = r' \otimes \mathbf{q}^{-1} \otimes \dot{\mathbf{q}} - \mathbf{q}^{-1} \otimes \dot{\mathbf{q}} \otimes r' = 2(\mathbf{q}^{-1} \otimes \dot{\mathbf{q}}) \times r'. \tag{2.21}$$

Notice that $\dot{r}'$ is the translational velocity of the vector; thus, by definition, $\dot{r}' = \Omega \times r'$, where Ω is the rotational velocity of r', and so $\Omega \times r' = 2(\mathbf{q}^{-1} \otimes \dot{\mathbf{q}}) \times r'$. Since r' can be any vector, the expression is reduced to

$$\Omega = 2(\mathbf{q}^{-1} \otimes \dot{\mathbf{q}}) \Rightarrow \dot{\mathbf{q}} = \frac{1}{2}\mathbf{q} \otimes \Omega. \tag{2.22}$$

Unit quaternions. If $\mathbf{q}$ has unit norm $||\mathbf{q}|| = 1$, then it can be called unitary quaternion. Unitary quaternions are often used to represent rotations in 3D space because they offer some advantages over other methods of representation, for example, they lack singularities. They do not have gimbal-lock effect, and they offer mathematical and computational simplicity because all operations need only multiplications and sums.

Euler–Rodríguez equation of rotation. In his theorem for rotation of rigid bodies, Euler stated that any rotation of a rigid body can be expressed as a rotation with respect to a fixed axis and a certain amount or angle. This rotation in a 3D space can be represented using unitary quaternions as

$$\bar{p}' = \mathbf{q}^{-1} \otimes \bar{p} \otimes \mathbf{q} = \mathbf{q}^* \otimes \bar{p} \otimes \mathbf{q}, \tag{2.23}$$

$$\mathbf{q} = \cos\left(\frac{\alpha}{2}\right) + \bar{u}\sin\left(\frac{\alpha}{2}\right), \tag{2.24}$$

where $\bar{p} \in \mathbb{R}^3$ is a 3D vector in the original reference frame, $\bar{p}'$ represents the same vector as $\bar{p}$ but now with respect to a new reference frame. $\bar{u} \in \mathbb{R}^3$ describes the direction of the axis of rotation and α defines the angle of rotation around the axis of rotation. In Eq. (2.24) it can be seen that a double quaternion product can be used to rotate any vector from one reference frame into another, and this rotation does not affect the vector's magnitude.

It can be seen that a quaternion $\mathbf{q}$ gives the same rotation as the quaternion $-\mathbf{q}$. If we imagine the fixed axis $\bar{u}$, it can be easily seen that two rotation magnitudes with respect to this axis can translate to the same orientation, those are α and $-2\pi + \alpha$ since $\mathbf{q} = \cos\left(\frac{\alpha}{2}\right) + \bar{u}\sin\left(\frac{\alpha}{2}\right)$ and $-\mathbf{q} = \cos\left(\frac{-2\pi+\alpha}{2}\right) + \bar{u}\sin\left(\frac{-2\pi+\alpha}{2}\right)$. This duality can be used to assure that the rotation is applied with the smallest magnitude possible.

Axis–Angle representation. From the Euler–Rodríguez formula (2.24) it is possible to obtain a direct relationship between the so-called "Axis–Angle" representation and unit quaternions for a certain rotation in order to express the rotation as a single vector $\bar{\alpha} \in \mathbb{R}^3$, with $||\bar{\alpha}|| = \alpha$ representing the rotational magnitude and $\bar{u} = \frac{\bar{\alpha}}{||\bar{\alpha}||}$. This representation gives a more intuitive way to understand the rotation of the rigid body, see Fig. 2.4. This relationship can be represented by the quaternion logarithmic mapping which is given by

$$\ln \mathbf{q} := \begin{cases} \ln ||\mathbf{q}||, & \text{if } ||\bar{q}|| = 0, \\ \ln ||\mathbf{q}|| + \frac{\bar{q}}{||\bar{q}||} \arccos \frac{q_0}{||\mathbf{q}||}, & \text{if } ||\bar{q}|| \neq 0. \end{cases}$$

It can be seen that if we are working only with unitary quaternions ($||\mathbf{q}|| = 1$), then the logarithmic mapping is reduced to

$$\ln \mathbf{q} := \begin{cases} 0, & \text{if } ||\bar{q}|| = 0, \\ \frac{\bar{q}}{||\bar{q}||} \arccos q_0, & \text{if } ||\bar{q}|| \neq 0. \end{cases}$$

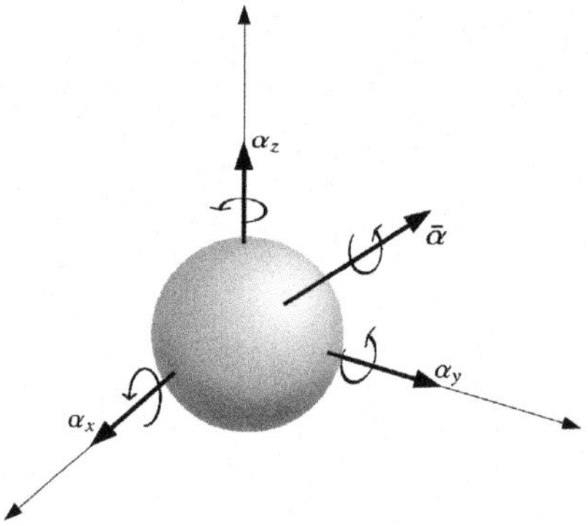

Figure 2.4 Axis–angle representation of a rigid body rotation.

This yields a relationship between a quaternion and its axis–angle representation written as

$$\bar{\alpha} = 2 \ln \mathbf{q}. \tag{2.25}$$

From this logarithmic equivalence, it follows

$$\tfrac{d}{dt}\bar{\alpha} = 2\tfrac{d}{dt}\ln \mathbf{q} = 2\mathbf{q}^{-1}\otimes\dot{\mathbf{q}} \ \rightarrow \ \dot{\mathbf{q}} = \tfrac{1}{2}\mathbf{q}\otimes\dot{\bar{\alpha}}, \tag{2.26}$$

then, (2.22) and (2.27) yield a relationship between the axis–angle representation and the angular velocity without loss of generality, namely

$$\dot{\bar{\alpha}} = \Omega = 2\tfrac{d}{dt}\ln \mathbf{q}. \tag{2.27}$$

The inverse relationship that can return a quaternion from its axis–angle representation is called quaternion exponential mapping; this is the inverse of the quaternion logarithmic mapping defined as

$$\mathbf{e}^{\mathbf{q}} := \begin{cases} \mathbf{e}^{q_0} & \text{if } ||\bar{q}|| = 0, \\ \mathbf{e}^{||\mathbf{q}||}\left(\cos\frac{||\mathbf{q}||}{2} + \frac{\bar{q}}{||\bar{q}||}\sin\frac{||\mathbf{q}||}{2}\right) & \text{if } ||\bar{q}|| \neq 0. \end{cases} \tag{2.28}$$

2.4.1.1 Quadcopter Dynamic Model Using Quaternions

A quadcopter or quadrotor can be considered as a rigid body that has 6 DOF, but only 4 are stable, this is because the platform cannot move in the orientation states without affecting its position. The force that the motors apply to the vehicle with respect to the body reference frame is considered acting only in the z-axis (no blade flapping effect is considered), but its orientation in the inertial frame can be changed by controlling the orientation of the platform.

As for any rigid body, the state vector of the quadrotor can be expressed as $\mathbf{x}_{quad} := \begin{bmatrix} \xi & \dot{\xi} & \mathbf{q} & \Omega \end{bmatrix}^T$ where $\mathbf{q} = q_0 + \begin{bmatrix} q_1 & q_2 & q_3 \end{bmatrix}^T$ defines the vehicle's orientation with respect to the inertial frame, represented as a unit quaternion, and $\Omega = \begin{bmatrix} \omega_x & \omega_y & \omega_z \end{bmatrix}^T$ represents the rotational velocity in the body frame. As stated before, the dynamic model can be separated into two subsystems, the first corresponds to the rotational and the second to the translational dynamics.

Quaternion Rotational Model

The state vector for the rotational dynamics is defined as

$$\mathbf{x}_{rot} := \begin{bmatrix} \mathbf{q}^T & \Omega^T \end{bmatrix}^T. \tag{2.29}$$

Differentiating the above equation and using (2.22) and (2.8), we can write

$$\dot{\mathbf{x}}_{rot} = \begin{bmatrix} \dot{\mathbf{q}} \\ \dot{\Omega} \end{bmatrix} = \begin{bmatrix} \frac{1}{2}\mathbf{q} \otimes \Omega \\ \mathrm{I}^{-1}\left(\tau - \Omega \times \mathrm{I}\,\Omega\right) \end{bmatrix}. \tag{2.30}$$

Equilibrium points. Equilibrium points are defined as constant solutions to a differential equation. In order to find an equilibrium point, it is necessary to find the states and control inputs $\mathbf{x}^*, \tau^*$ where $\dot{\mathbf{x}} = 0$. Taking this into account, the equilibrium points of the attitude system (2.30) can be calculated when $\dot{\mathbf{x}}_{rot} = \bar{0}$, thus it follows that

$$\mathbf{x}^*_{rot} = \begin{bmatrix} \mathbf{q}^* \\ \bar{0} \end{bmatrix}, \qquad \tau^* = \bar{0}. \tag{2.31}$$

This corroborates the intuitive approach for finding the equilibrium points: when the vehicle's orientation is constant, the torques and the angular velocities must be equal to zero.

Quadcopter Translational Dynamic

The state variable for the translational model is given by

$$\mathbf{x}_{pos} = \begin{bmatrix} \xi^T & \dot{\xi}^T \end{bmatrix}^T. \tag{2.32}$$

According to Newton's equations, the total force that acts on the body in the inertial frame can be obtained by multiplying the acceleration and mass. Observe that F_t is expressed in the body frame; thus, rewriting it in the inertial frame, we have

$$\mathbf{q} \otimes F_t \otimes \mathbf{q}^* = m\ddot{\xi}.$$

Then, the total force that acts on the vehicle could consist of the sum of the control force $\hat{F}$ and the external forces F_{ext} expressed by $F_t = \hat{F} + F_{ext}$. Since the quadrotor is an under-actuated system, the only force considered in the body reference frame that can be used to control the platform is the trust force vector u_z in the z-axis. In this study F_{ext} corresponds to the gravity force which acts with respect to the inertial frame (external disturbances could be added if necessary for further studies), thus

$$\dot{\mathbf{x}}_{pos} = \begin{bmatrix} \dot{\xi} \\ \ddot{\xi} \end{bmatrix} = \begin{bmatrix} \dot{\xi} \\ \mathbf{q} \otimes \frac{F_u}{m} \otimes \mathbf{q}^* + \bar{g} \end{bmatrix}, \tag{2.33}$$

where $\bar{g} = \begin{bmatrix} 0 & 0 & g \end{bmatrix}^T$ corresponds to the gravity's vector.

Finally, the complete model for the quadrotor is

$$\dot{\mathbf{x}}_{quad} = \frac{d}{dt} \begin{bmatrix} \xi \\ \dot{\xi} \\ \mathbf{q} \\ \Omega \end{bmatrix} = \begin{bmatrix} \dot{\xi} \\ \mathbf{q} \otimes \frac{F_u}{m} \otimes \mathbf{q}^* + \bar{g} \\ \frac{1}{2} \mathbf{q} \otimes \Omega \\ \mathrm{I}^{-1} \left(\tau - \Omega \times \mathrm{I} \Omega \right) \end{bmatrix}. \tag{2.34}$$

Since the thrust vector is fixed in the body frame and the attitude is variable according to Eq. (2.30), the position can be controlled using the attitude subsystem, and thus the platform can be globally stabilized.

2.4.2 Dual Quaternions Model

Dual numbers. A dual number is defined as $\hat{a} = a + \epsilon b$, with $a, b \in \mathbb{R}$, $\epsilon \neq 0$, $\epsilon^2 = 0$, where ^ denotes that $\hat{a}$ is a dual number, consisting of a real part a and a dual part b.

Dual vectors. Dual vectors are a generalization of dual numbers where both real and dual parts are n-dimensional vectors. In this work, we will use three-dimensional vectors when referring to dual vectors.

Let $\hat{\bar{v}} = \bar{v}_r + \bar{v}_d \epsilon$ and $\hat{\bar{k}} = \bar{k}_r + \bar{k}_d \epsilon$ be dual vectors with $\bar{v}_r, \bar{v}_d, \bar{k}_r, \bar{k}_d \in \mathbb{R}^3$. Then their **dot product** is given by $\hat{\bar{k}} \cdot \hat{\bar{v}} = K_r \bar{v}_r + K_d \bar{v}_d \epsilon$, where K_r and K_d are diagonal 3×3 matrices with diagonal entries k_{r1}, k_{r2}, k_{r3} and k_{d1}, k_{d2}, k_{d3}, respectively.

Dual quaternions. Dual quaternions are dual numbers with both real and dual parts given by quaternions, i.e., $\hat{q} = \mathbf{q}_r + \mathbf{q}_d \epsilon$, with $\mathbf{q}_r, \mathbf{q}_d \in \mathbb{H}$.

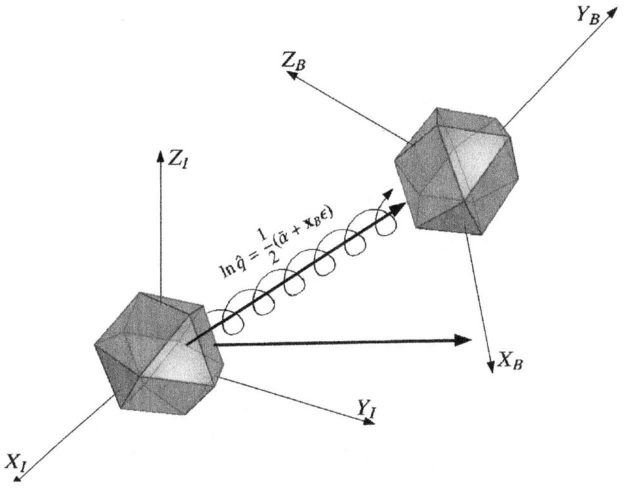

Figure 2.5 Logarithmic representation of a rigid body transformation.

Operations for Dual Quaternions

Sum. Let $\hat{q}_1$ and $\hat{q}_2$ be dual quaternions, then $\hat{q}_1 + \hat{q}_2 = \boldsymbol{q}_{1_r} + \boldsymbol{q}_{2_r} + [\boldsymbol{q}_{1d} + \boldsymbol{q}_{2d}]\epsilon$.

Product. The multiplication between dual quaternions is defined as $\hat{q}_1 \otimes \hat{q}_2 = \boldsymbol{q}_{1_r} \otimes \boldsymbol{q}_{2_r} + [\boldsymbol{q}_{1_r} \otimes \boldsymbol{q}_{2d} + \boldsymbol{q}_{1d} \otimes \boldsymbol{q}_{2_r}]\epsilon$.

Norm. The norm of a dual quaternion is defined as $||\hat{q}||^2 = \hat{q} \otimes \hat{q}^*$. Note that if $||\hat{q}||^2 = 1 + 0\epsilon$, then $\hat{q}$ is called a *unitary dual quaternion*.

Conjugation. The conjugate of a dual quaternion is defined as $\hat{q}^* = \boldsymbol{q}_r^* + \boldsymbol{q}_d^*\epsilon$. Since in this work we are dealing only with unitary dual quaternions, we can say that $\hat{q}^* = \hat{q}^{-1}$.

Logarithmic mapping. A dual quaternion can be transformed by $\ln\hat{q} = \frac{1}{2}(\bar{\alpha} + \mathbf{x}_B\epsilon)$, where $\bar{\alpha} = 2\ln\boldsymbol{q}$ represents the body's rotation given by a unit quaternion logarithmic mapping and $\mathbf{x}_B$ denotes the position vector in the body frame. This yields a relationship between a dual quaternion and the screw representation of simultaneous rotation and translation as can be seen in Fig. 2.5:

$$\bar{\alpha} + \mathbf{x}_B = 2\ \ln\hat{q}. \tag{2.35}$$

Dual quaternion derivative. When a dual quaternion is representing a simultaneous rotation and translation, it is defined as $\hat{q} \triangleq \boldsymbol{q} + \frac{\boldsymbol{q}\otimes\mathbf{x}_B}{2}\epsilon$, where

q represents the orientation of the body. The derivative of this equation is

$$\begin{aligned}
\dot{\hat{q}} &= \dot{q} + \frac{1}{2}\left[\dot{q} \otimes \mathbf{x}_B + q \otimes \dot{\mathbf{x}}_B\right]\epsilon \\
&= \frac{1}{2} q \otimes \Omega + \left[\frac{1}{4} q \otimes \Omega \otimes \mathbf{x}_B + \frac{1}{2} q \otimes \dot{\mathbf{x}}_B\right]\epsilon \\
&= \frac{1}{2} q \otimes \Omega + \left[\frac{1}{2} q \otimes (\Omega \times \mathbf{x}_B) + \frac{1}{4} q \otimes \mathbf{x}_B \otimes \Omega + \frac{1}{2} q \otimes \dot{\mathbf{x}}_B\right]\epsilon \\
&= \frac{1}{2}\left(q + \frac{q \otimes \mathbf{x}_B}{2}\epsilon\right) \otimes \left(\Omega + [\Omega \times \mathbf{x}_B + \dot{\mathbf{x}}_B]\epsilon\right).
\end{aligned}$$

Define

$$\hat{\zeta} \triangleq \Omega + [\Omega \times \mathbf{x}_B + \dot{\mathbf{x}}_B]\epsilon, \tag{2.36}$$

where $\hat{\zeta}$ is the *twist* dual vector (combination of angular and translational velocities). Finally, we obtain the expression for the derivative of a dual quaternion

$$\dot{\hat{q}} = \frac{1}{2}\hat{q} \otimes \hat{\zeta}. \tag{2.37}$$

Unitary dual quaternions. Let $\hat{q}$ be a dual quaternion given by $\hat{q} = q + \frac{q \otimes \mathbf{x}_B}{2}\epsilon$, where q is the unitary quaternion that defines the attitude of the body and $\epsilon^2 = 0, \epsilon \neq 0$. If $\hat{q}$ is a unitary dual quaternion, then it can be used to simultaneously describe the rotation and translation of the body with respect to the inertial frame.

An advantage of using this representation, contrary to the separated rotational and translational submodels, is that it is possible to describe various rotations and translations using only dual quaternion products, thus providing computational and mathematical simplicity to the model and control algorithms.

2.4.2.1 Quadcopter Dual Quaternion Kinematics

As is done when modeling a vehicle with unitary quaternions, the quadrotor is considered a rigid body with 6 DOF and under-actuated. Thus, the state vector of the quadrotor can be expressed in terms of its orientation, position, and twist as

$$\mathbf{x}_{quad} := \begin{bmatrix} \hat{q} \\ \hat{\zeta} \end{bmatrix} = \begin{bmatrix} q + \frac{q \otimes \mathbf{x}_B}{2}\epsilon \\ \Omega + [\Omega \times \mathbf{x}_B + \dot{\mathbf{x}}_B]\epsilon \end{bmatrix}. \tag{2.38}$$

2.4.2.2 Quadcopter Dual Quaternion Dynamic Model

The dynamic model of a quadcopter using the Newton–Euler approach with dual quaternions is inspired by [11]. Differentiating Eq. (2.36), it follows that

$$\dot{\hat{\zeta}} = \dot{\Omega} + [\dot{\Omega} \times \mathbf{x}_B + \Omega \times \dot{\mathbf{x}}_B + \ddot{\mathbf{x}}_B]\epsilon. \tag{2.39}$$

Thus, using (2.7), (2.8), and (2.39), the dynamic model is obtained as

$$\dot{\mathbf{x}}_{quad} = \begin{bmatrix} \dot{\hat{q}} \\ \dot{\hat{\zeta}} \end{bmatrix} = \begin{bmatrix} \frac{1}{2}\hat{q} \otimes \hat{\zeta} \\ \hat{F} + \hat{u} \end{bmatrix}, \tag{2.40}$$

where

$$\begin{aligned} \hat{F} &= a + (a \times \mathbf{x}_B + \Omega \times \dot{\mathbf{x}}_B)\epsilon, \\ \hat{u} &= \mathrm{I}^{-1}\tau + [\mathrm{I}^{-1}\tau \times \mathbf{x}_B + m^{-1}F_\xi]\epsilon, \\ a &= -\mathrm{I}^{-1}(\Omega \times \mathrm{I}\Omega). \end{aligned} \tag{2.41}$$

Equilibrium points. In order to find the equilibrium points, it is necessary to find the states and control inputs $\mathbf{x}^*_{quad}$, $\hat{u}^*$ where $\dot{\mathbf{x}}_{quad} = 0$. Taking this into account, the equilibrium points are

$$\mathbf{x}^*_{quad} = \begin{bmatrix} \hat{q}^* \\ \hat{0} \end{bmatrix} = \begin{bmatrix} q^* + \frac{q^* \otimes \mathbf{x}^*_B}{2}\epsilon \\ \hat{0} \end{bmatrix}, \; \hat{u}^* = \hat{0} \begin{cases} \Rightarrow F_\xi = 0 \Rightarrow \hat{F} = -F_{ext}, \\ \Rightarrow \tau = 0 \Rightarrow \tau_u = -\tau_{ext}. \end{cases} \tag{2.42}$$

This corroborates the intuitive approach for finding the equilibrium points. The platform will stay still at any position and orientation in space as long as its angular velocity and translational velocity equal to zero, and also as long as the control forces and torques counteract the external ones.

2.5 DISCUSSION

The classical approaches to mathematically represent an aerial vehicle are the Euler–Lagrange and Newton–Euler formalisms. A new tendency for modeling a UAV is the quaternion approach which gives no singularity and yields an easy design and implementation of control algorithms; nevertheless, it is less intuitive to understand, and sometimes the reader could get lost. In this chapter our aim was to introduce these three approaches to clarify the mathematical representation of the dynamics of an aerial vehicle, especially a quadcopter. The Euler–Lagrange model by design is given without perturbation, the idea is to slowly introduce beginners to the UAV dynamics. The next Newton–Euler approach includes aerodynamic and drag effects, the goal here is to give more basic facts to the readers or experts in the UAV community needed for nonlinear control design. Finally, the quaternion methodology is introduced to the UAV expert community working with aggressive (singularity free) trajectories. This, by no means, implies that this approach cannot be used by other readers.

REFERENCES

1. P. Pounds, R. Mahony, P. Corke, Modelling and control of a quad-rotor robot, in: Proceedings of the Australasian Conference on Robotics and Automation, Auckland, New Zealand, 2006.
2. R.W. Prouty, Helicopter Performance, Stability, and Control, Krieger Publishing Company, 2001.
3. P. Pounds, J. Gresham, R. Mahony, J. Robert, P. Corke, Towards dynamically favourable quadrotor aerial robots, in: Proceedings of the Australasian Conference on Robotics and Automation, Canberra, Australia, 2004.
4. G. Hoffmann, H. Huang, S. Waslander, C. Tomlin, Quadrotor helicopter flight dynamics and control: Theory and Experiment, in: Proceedings of the AIAA Guidance, Navigation, and Control Conference, USA, 2007.
5. H. Goldstein, Classical Mechanics, Addison Wesley Series in Physics, Addison–Wesley, USA, 1980.
6. P. Castillo, R. Lozano, A.E. Dzul, Modelling and Control of Mini-Flying Machines, Advances in Industrial Control, Springer-Verlag, London, UK, 2005.
7. B. Etkin, L. Duff Reid, Dynamics of Flight, John Wiley and Sons, New York, USA, 1959.
8. R. Lozano, B. Brogliato, O. Egeland, B. Maschke, Passivity-Based Control System Analysis and Design, 2nd edition, Communications and Control Engineering Series, Springer-Verlag, 2006.
9. G.M. Hoffmann, H. Huang, S.L. Waslander, C.J. Tomlin, Precision flight control for a multi-vehicle quadrotor helicopter testbed, Control Engineering Practice 19 (9) (2011) 1023–1036.
10. J.B. Kuipers, Quaternions and Rotation Sequences, vol. 66, Princeton University Press, 1999.
11. X. Wang, C. Yu, Feedback linearization regulator with coupled attitude and translation dynamics based on unit dual quaternion, in: International Symposium on Intelligent Control (ISIC), Yokohama, Japan, IEEE, 2010, pp. 2380–2384.

PART 2

Improving Sensor Signals for Control Purposes

To perform autonomous or semi-autonomous flights, that is, when an UAV is being controlled remotely, the UAV needs an adequate estimate of the attitude in semi-autonomous mode, and similarly, it needs estimates of the attitude and position in the 3D space in the case of an autonomous flight. There are many sensors in the world market that provide good estimates of the orientation and position. The most common sensor to measure the orientation and rate is the Inertial Measurement Unit (IMU), and for the position one uses the Global Position System (GPS). However, these commercial sensors are generally expensive, or, in the case of the GPS, not useful when the UAV operates indoors.

In the following chapters we present some approaches to improve the performance of the different sensors used in the autonomous or semi-autonomous flights. Chapter 3 provides a comparative evaluation of attitude estimation algorithms using low-cost gyroscopes and accelerometers which normally do not include good estimation algorithms. Chapter 4 presents a predictor scheme to improve the precision of the attitude estimation algorithm and also includes a scheme to compensate delays in the control input. The main source of delays in the UAV is the different sampling periods coming from different measurements or when computing the control law(s). Therefore it is important to take into account the delays to avoid the system becoming unstable. Finally, Chap. 5 deals with the UAV position problem. UAV localization is a currently interesting problem which has not been solved, yet. An efficient solution using common sensors in UAVs is proposed. These sensors include inertial sensors, ultrasound sensors, and a camera. An Extended Kalman Filter (EKF) is used to fuse the sensors' measurements to estimate the position of a quadcopter.

CHAPTER 3

Inertial Sensors Data Fusion for Orientation Estimation*

The problem of attitude estimation consists in recovering the true attitude using the signals provided by the gyroscopes and accelerometers. The gyroscopes measure the angular velocity of the body, and thus they can be integrated to obtain the attitude. This approach quickly produces a drift due to the errors introduced by the integration of bias and noise, as will be discussed below [1]. The accelerometers sense the orientation of the gravity acceleration, from which the attitude can be obtained directly. However, the accelerometer signal is highly corrupted by noise due to vibrations, and this approach yields an estimate too noisy to be used in practice. A simple approach that is widely used and provides good results is complementary filtering where the accelerometers are low-pass filtered and the gyroscopes are high-pass filtered [2].

This chapter aims at providing a comparative evaluation of attitude estimation algorithms using low-cost sensors (gyroscopes and accelerometers). Several aspects must be taken into account while choosing the most suitable approach for a given application: singularity existence, convergence guarantee, computational time, bias estimation, etc. The comparison is focused on Kalman filtering methods as they provide a suitable framework for an easy integration in higher level localization techniques based on laser range finders, cameras, or GPS [3]. The results show that it is possible to obtain a performance with a hobbyist-grade IMU similar to that of an industrial-grade IMU. The chapter is ended by explaining in detail the simplified Kalman filtering algorithm, and it is shown that such a proposal is very easy-to-code, computationally efficient, and yet has good performance.

3.1 ATTITUDE REPRESENTATION

Let us denote by $\{\hat{e}_1, \hat{e}_2, \hat{e}_3\}$ the unit basis vectors of the Earth-centered Earth-fixed (ECEF) reference frame, $\mathcal{E}$, which is assumed to be inertial. Let

* The results in this chapter were developed in collaboration with R. Sanz Diaz and P. Albertos from the Universidad Politécnica de Valencia, Spain.

Indoor Navigation Strategies for Aerial Autonomous Systems.
DOI: http://dx.doi.org/10.1016/B978-0-12-805189-4.00005-6

$\boldsymbol{\Omega}_{\mathcal{B}/\mathcal{E}}^{\mathcal{B}} = [\omega_x, \omega_y, \omega_z]^T$, denoted simply by $\boldsymbol{\Omega}$ in what follows, be the angular velocity of the aircraft with respect to $\mathcal{E}$ expressed in the body frame $\mathcal{B}$. Thus the rotational kinematics relating these angular velocities to the Euler angles are expressed as

$$\dot{\boldsymbol{\eta}}(t) = \begin{bmatrix} 1 & \sin\phi\tan\theta & \cos\phi\tan\theta \\ 0 & \cos\phi & -\sin\phi \\ 0 & \frac{\sin\phi}{\cos\theta} & \frac{\cos\phi}{\cos\theta} \end{bmatrix} \boldsymbol{\Omega}(t). \tag{3.1}$$

It is a well-known result that the rate of change of the basis vectors is given by

$$\frac{d\hat{\mathbf{e}}_i}{dt} = \boldsymbol{\Omega}_{\mathcal{E}/\mathcal{B}} \times \hat{\mathbf{e}}_i = -\boldsymbol{\Omega} \times \hat{\mathbf{e}}_i = [\boldsymbol{\Omega}]_\times^T \hat{\mathbf{e}}_i, \tag{3.2}$$

where the skew-symmetric operator is defined as

$$[\boldsymbol{\Omega}]_\times = \begin{bmatrix} 0 & -\omega_z & \omega_y \\ \omega_z & 0 & -\omega_x \\ -\omega_y & \omega_x & 0 \end{bmatrix}. \tag{3.3}$$

According to the roll–pitch–yaw sequence of Euler angles, the rotation matrix is expressed as

$${}^{\mathcal{B}}\mathbf{R}_{\mathcal{E}} = \begin{bmatrix} c_\theta c_\psi & c_\theta s_\psi & -s_\theta \\ s_\phi s_\theta c_\psi - c_\phi s_\psi & s_\phi s_\theta s_\psi + c_\phi c_\psi & s_\phi c_\theta \\ c_\phi s_\theta c_\psi + s_\phi s_\psi & c_\phi s_\theta s_\psi - c_\phi c_\psi & c_\phi c_\theta \end{bmatrix}, \tag{3.4}$$

where ${}^{\mathcal{B}}\mathbf{R}_{\mathcal{E}}$ maps vectors from $\mathcal{E}$ onto $\mathcal{B}$. Noticing that ${}^{\mathcal{B}}\mathbf{R}_{\mathcal{E}} = [\hat{\mathbf{e}}_1\, \hat{\mathbf{e}}_2\, \hat{\mathbf{e}}_3]$ and the time derivative of this rotation matrix can be derived by generalizing (3.2), it follows that the kinematics in terms of the rotation matrix, also referred to as the Direct Cosine Matrix (DCM), is given by

$${}^{\mathcal{B}}\dot{\mathbf{R}}_{\mathcal{E}}(t) = [\boldsymbol{\Omega}]_\times^T {}^{\mathcal{B}}\mathbf{R}_{\mathcal{E}}(t). \tag{3.5}$$

Rotations are also represented using quaternions, which are an extension to the complex numbers. Quaternions are mainly used because they provide a singularity-free representation, and also because quaternion algebra is computationally efficient, see Sect. 2.4 of Chap. 2. The unitary quaternion could be also represented by $\mathbf{q} = [\bar{q}\, \mathbf{q}_0]^T$ where $\bar{q}$ defines the scalar part of the quaternion and $\mathbf{q}_0 = [q_1\, q_2\, q_3]^T$ represents the vector part. From (2.30) the rigid body angular motion can be written using quaternion notation as

$$\frac{\mathrm{d}\mathbf{q}(t)}{\mathrm{d}t} = \breve{\boldsymbol{\Omega}}\mathbf{q}(t), \tag{3.6}$$

where

$$\breve{\mathbf{\Omega}} = \frac{1}{2}\begin{bmatrix} [\mathbf{\Omega}]_\times & \mathbf{\Omega} \\ -\mathbf{\Omega}^T & 0 \end{bmatrix}. \tag{3.7}$$

The rotation matrix (3.4) expressed in terms of the quaternion components is defined as follows:

$${}^{\mathcal{B}}\mathbf{R}_{\mathcal{E}} = \begin{bmatrix} q_1^2 - q_2^2 - q_3^2 + q_4^2 & 2(q_1q_2 - q_3q_4) & 2(q_1q_3 + q_2q_4) \\ 2(q_1q_2 + q_3q_4) & q_1^2 + q_2^2 - q_3^2 + q_4^2 & 2(q_2q_3 - q_1q_4) \\ 2(q_1q_3 - q_2q_4) & 2(q_2q_3 - q_1q_4) & q_1^2 - q_2^2 + q_3^2 + q_4^2 \end{bmatrix}. \tag{3.8}$$

3.2 SENSOR CHARACTERIZATION

The output of an MEMS (Micro-Electro-Mechanical System) sensor is corrupted by noise and an offset, and is usually referred to as bias [4]. The bias can be calibrated prior to each flight. However, it is dependent on temperature, causing the bias to drift. This effect is especially noticeable within the first few minutes of operation because of the internal warm-up of the electronic components [5].

For the attitude estimation problem, the biases of the gyroscopes are much more crucial than of the other components because they are time-forward integrated, originating an error that grows linearly with time. Even when considering an ideal scenario without bias, integration of the gyro output corrupted by white noise would give rise to an error growing with the square root of time. As a consequence, a method is needed to correct the errors introduced by the integration of bias and white noise. Regarding the biases of the accelerometers, they are not so important because after calibration they will result only in a small offset with respect to the actual horizontal plane (perpendicular to the gravity vector).

Let us consider the following error models for the sensors:

$$\begin{aligned} \bar{\mathbf{\Omega}} &= \tilde{\mathbf{\Omega}} + \boldsymbol{\beta}_\Omega + \boldsymbol{\lambda}_\Omega, \\ \bar{a} &= \tilde{\mathbf{a}} + \boldsymbol{\lambda}_a, \end{aligned} \tag{3.9}$$

where the velocity measurement $\bar{\mathbf{\Omega}}$ is composed of its actual value $\tilde{\mathbf{\Omega}}$, bias $\boldsymbol{\beta}_\Omega$, and noise $\boldsymbol{\lambda}_\Omega$. The same applies to the acceleration measurements but the biases are not included. As mentioned above, these errors are not so critical as they are not integrated over time. The measurement noises are considered subject to a Gaussian representation as follows:

$$\begin{aligned} \mathbb{E}[\boldsymbol{\lambda}_\Omega] &= 0, & \mathbb{E}[\boldsymbol{\lambda}_\Omega\boldsymbol{\lambda}_\Omega^T] &= \sigma_\Omega^2\mathbf{I}_3, \\ \mathbb{E}[\boldsymbol{\lambda}_a] &= 0, & \mathbb{E}[\boldsymbol{\lambda}_a\boldsymbol{\lambda}_a^T] &= \sigma_a^2\mathbf{I}_3, \end{aligned}$$

where σ_i^2 defines the variance, $\mathbf{I}_3$ represents the 3×3 identity matrix, and $\boldsymbol{\Sigma}_i$ denotes the diagonal covariance matrix. A random walk process,

$$\begin{aligned} \dot{\boldsymbol{\beta}}_{\Omega} &= \boldsymbol{\lambda}_{\beta}, \\ \mathbb{E}[\boldsymbol{\lambda}_{\beta}] = 0, \qquad \mathbb{E}[\boldsymbol{\lambda}_{\beta}\boldsymbol{\lambda}_{\beta}^T] &= \boldsymbol{\Sigma}_{\beta} = \sigma_{\beta}^2 \mathbf{I}_3, \end{aligned} \tag{3.10}$$

is used to model the "slowly-varying" biases of the gyros. The variance σ_{β}^2 determines how much the bias drifts.

Now, it is necessary to know how the measurements are related to the variables to be estimated. The relation of the gyroscope measurements with the different representations of the kinematic model of the vehicle have been already presented in (3.1)–(3.8). Regarding the accelerometers, they measure the specific force acting on the vehicle expressed in $\mathcal{B}$, as they are aligned with the body-fixed reference frame. The measurement can be expressed as

$$\tilde{\boldsymbol{a}}^{\mathcal{B}} = \frac{1}{m}\left(\boldsymbol{f}^{\mathcal{B}} - {}^{\mathcal{B}}\mathbf{R}_{\mathcal{E}}\,(mg)\hat{\boldsymbol{e}}_3\right) = \dot{\boldsymbol{v}}^{\mathcal{B}} - {}^{\mathcal{B}}\mathbf{R}_{\mathcal{E}}(g\hat{\boldsymbol{e}}_3), \tag{3.11}$$

where $\dot{\boldsymbol{v}}^{\mathcal{B}}$ is the acceleration vector due to the external forces vector $\boldsymbol{f}^{\mathcal{B}}$, both of them are expressed in $\mathcal{B}$. In case the vehicle is not expected to perform aggressive maneuvers, the linear acceleration ($\dot{\boldsymbol{v}}^{\mathcal{B}} \approx 0$) can be neglected. If the vector of acceleration measurements is normalized, it can be simply related to the roll and pitch angles as

$$\tilde{\boldsymbol{a}} = \frac{\tilde{\boldsymbol{a}}^{\mathcal{B}}}{|\tilde{\boldsymbol{a}}^{\mathcal{B}}|} \approx -{}^{\mathcal{B}}\mathbf{R}_{\mathcal{E}}\hat{\boldsymbol{e}}_3 = \begin{bmatrix} \sin\theta \\ -\sin\phi\cos\theta \\ -\cos\phi\cos\theta \end{bmatrix}. \tag{3.12}$$

3.3 ATTITUDE ESTIMATION ALGORITHMS

Several attitude estimation algorithms are compared and presented in the following. The Kalman Filter (KF) approach is used in all of them, regardless of the formulation of the system. If the resulting system is nonlinear then the Extended Kalman Filter (EKF) is applied. Recall that the Kalman Filter is derived under the assumption of Gaussian errors for both the system and the measurement models. Therefore, it is necessary to include the biases in the vector of estimated variables, as they are not removed in the filtering process. The equations of the discrete EKF for a general nonlinear system[1]

$$\dot{\boldsymbol{x}}(t) = \boldsymbol{f}(\boldsymbol{x}, \boldsymbol{u}) + \boldsymbol{\lambda}_\xi, \qquad \mathbb{E}[\boldsymbol{\lambda}_\xi \boldsymbol{\lambda}_\xi^T] = \mathbf{Q},$$
$$\boldsymbol{y}(t) = \boldsymbol{h}(\boldsymbol{x}) + \boldsymbol{\lambda}_v, \qquad \mathbb{E}[\boldsymbol{\lambda}_v \boldsymbol{\lambda}_v^T] = \mathbf{R}$$

can be found in [6] and are summarized as

$$\begin{aligned} \hat{\boldsymbol{x}}_k^- &= \hat{\boldsymbol{x}}_{k-1} + T_s \boldsymbol{f}(\boldsymbol{x}_k, \boldsymbol{u}_k), \\ \boldsymbol{P}_k^- &= \mathbf{A}_{k-1} \boldsymbol{P}_{k-1} \mathbf{A}_{k-1}^T + \mathbf{Q}_k, \end{aligned} \tag{3.13}$$

$$\begin{aligned} \boldsymbol{S}_k &= \mathbf{H}_k \boldsymbol{P}_k^- \mathbf{H}_k^T + \mathbf{R}_k, \\ \boldsymbol{K}_k &= \boldsymbol{P}_k^- \mathbf{H}_k^T \boldsymbol{S}_k^{-1}, \\ \hat{\boldsymbol{x}}_k^+ &= \hat{\boldsymbol{x}}_k^- + \boldsymbol{K}_k \left(\boldsymbol{y}_k - \boldsymbol{h}(\hat{\boldsymbol{x}}_k^-)\right) \\ \boldsymbol{P}_k^+ &= (\mathbf{I} - \boldsymbol{K}_k \mathbf{H}_k) \boldsymbol{P}_k^-, \end{aligned} \tag{3.14}$$

where

$$\mathbf{A}_k = e^{\mathbf{A} T_s} \approx \mathbf{I} + T_s \left.\frac{\partial \boldsymbol{f}}{\partial \boldsymbol{x}}\right|_{\hat{\boldsymbol{x}}^-} \qquad \text{and} \qquad \mathbf{H}_k = \left.\frac{\partial \boldsymbol{h}}{\partial \boldsymbol{x}}\right|_{\hat{\boldsymbol{x}}^-}.$$

The KF equations are simply obtained by applying (3.13)–(3.14) to a linear system.

All of the algorithms presented in the next section, except for the first one, lead to nonlinear formulations. Some of them are referred to as DCM-based because the elements of the DCM are manipulated directly. Also, formulations using both Euler angles and quaternions are explored.

3.3.1 DCM1-Based

Algorithms based on the DCM directly update the components of the matrix, avoiding the representation by means of Euler angles. The time variation of the elements of the DCM is given by (3.5). Although the whole matrix could be updated, the last column (denoted by $\boldsymbol{r}_3$) suffices when providing information about ϕ and θ, as can be seen in (3.4). Using the common notation in control theory, the state vector is denoted by $\boldsymbol{x} = \boldsymbol{r}_3$, and the system is described by

$$\begin{aligned} \dot{\boldsymbol{x}} &= [\bar{\boldsymbol{\Omega}}]_\times \boldsymbol{x} + \boldsymbol{\lambda}_\xi, & \mathbb{E}[\boldsymbol{\lambda}_\xi \boldsymbol{\lambda}_\xi^T] &= \mathbf{Q} = q_r \mathbf{I}, \\ \boldsymbol{y} &= -\mathbf{I}_3 \boldsymbol{x} + \boldsymbol{\lambda}_v, & \mathbb{E}[\boldsymbol{\lambda}_v \boldsymbol{\lambda}_v^T] &= \mathbf{R} = \sigma_a^2 \mathbf{I}_3. \end{aligned} \tag{3.15}$$

Assuming that σ_a^2 can be identified by analyzing the static output of the accelerometers, there is only one tuning parameter, q_r, which determines the

[1] In Chap. 5 the EKF is also used, and the equations are also described given more details.

relative importance of the state propagation with respect to the information recovered from the accelerometers. Notice that this methodology results in a very simple filter, but it would imply the need of zero-velocity updates in practice as the bias of the gyroscopes is not explicitly estimated. For more details about this formulation see [7].

3.3.2 DCM2-Based

The previous filter can be modified to include the bias estimation [8]. According to the measurement model in (3.9), the bias-free angular velocity is given by $\tilde{\mathbf{\Omega}} = \bar{\mathbf{\Omega}} - \boldsymbol{\beta}_\Omega$. Denoting by $\boldsymbol{x} = [\boldsymbol{r}_3^T, \boldsymbol{\beta}_\Omega^T]^T$ the state vector, the resulting system can be expressed as

$$\dot{x} = \mathbf{A}(x)x + \boldsymbol{\lambda}_\xi, \qquad \mathbb{E}[\boldsymbol{\lambda}_\xi \boldsymbol{\lambda}_\xi^T] = \mathbf{Q},$$
$$y = \mathbf{C}x + \boldsymbol{\lambda}_v, \qquad \mathbb{E}[\boldsymbol{\lambda}_v \boldsymbol{\lambda}_v^T] = \mathbf{R},$$

where

$$\mathbf{A}(x) = \begin{bmatrix} [\bar{\mathbf{\Omega}} - \boldsymbol{\beta}_\Omega]_\times & 0 \\ 0 & 0 \end{bmatrix}, \qquad \mathbf{C} = \begin{bmatrix} -\mathbf{I}_3 & 0 \end{bmatrix},$$
$$\mathbf{Q} = \begin{bmatrix} q_r \mathbf{I}_3 & 0 \\ 0 & q_\beta \mathbf{I}_3 \end{bmatrix}, \qquad \mathbf{R} = \sigma_a^2 \mathbf{I}_3.$$

Notice that there is an extra parameter, q_β, accounting for the bias noise. It has a physical meaning as it describes the random walk model for the bias. Although used as a tuning parameter, it has to be small enough to allow only slow variations of the bias, which is known to happen in reality.

3.3.3 New DCM3-Based

In order to improve the angular velocity estimation, the above algorithm can be extended by including the angular velocity in the state vector, $\boldsymbol{x} = [\boldsymbol{r}_3^T, \boldsymbol{\beta}_\Omega^T, \mathbf{\Omega}^T]^T$. The process model for the angular velocities is assumed to be constant, so they are only modified in the updating phase of the Kalman filter. In practice, this will work as a low-pass filter, controlled by the parameter q_Ω. In this case, the matrices describing the system are

$$\mathbf{A}(x) = \begin{bmatrix} 0 & 0 & [\boldsymbol{r}_3]_\times \\ 0 & 0 & 0 \\ 0 & 0 & 0 \end{bmatrix}, \qquad \mathbf{C} = \begin{bmatrix} -\mathbf{I}_3 & 0 & 0 \\ 0 & -\mathbf{I}_3 & \mathbf{I}_3 \end{bmatrix},$$

$$\mathbf{Q} = \begin{bmatrix} q_r\mathbf{I}_3 & \mathbf{0} & \mathbf{0} \\ \mathbf{0} & q_\beta\mathbf{I}_3 & \mathbf{0} \\ \mathbf{0} & \mathbf{0} & q_\Omega\mathbf{I}_3 \end{bmatrix}, \qquad \mathbf{R} = \begin{bmatrix} \sigma_a^2\mathbf{I}_3 & \mathbf{0} \\ \mathbf{0} & \sigma_\Omega^2\mathbf{I}_3 \end{bmatrix}.$$

3.3.4 Euler EKF

In this algorithm six variables are estimated: three Euler angles and three biases of the gyroscopes. The state vector can be written as $\boldsymbol{x} = [\boldsymbol{\eta}^T\ \boldsymbol{\beta}^T]^T$. The nonlinear equations are described in a state-space form as

$$\begin{aligned} \dot{\boldsymbol{x}} &= \boldsymbol{f}(\boldsymbol{x}) + \boldsymbol{\lambda}_\xi, & \mathbb{E}[\boldsymbol{\lambda}_\xi\boldsymbol{\lambda}_\xi^T] &= \mathbf{q}, \\ \boldsymbol{y} &= \boldsymbol{h}(\boldsymbol{x}) + \boldsymbol{\lambda}_v, & \mathbb{E}[\boldsymbol{\lambda}_v\boldsymbol{\lambda}_v^T] &= \mathbf{R}, \end{aligned}$$

where

$$\mathbf{Q} = \begin{bmatrix} q_\eta\mathbf{I}_3 & \mathbf{0} \\ \mathbf{0} & q_\beta\mathbf{I}_3 \end{bmatrix}, \qquad \mathbf{R} = \sigma_a^2\mathbf{I}_3.$$

The process equation is derived from (3.1) by directly incorporating the measurements from the gyroscopes as follows:

$$\boldsymbol{f}(\boldsymbol{x}) = \begin{bmatrix} (\bar{\omega}_x - \beta_{\omega_x}) + (\bar{\omega}_y - \beta_{\omega_y})\sin\phi + (\bar{\omega}_z - \beta_{\omega_z})\tan\theta \\ (\bar{\omega}_y - \beta_{\omega_y})\cos\phi - (\bar{\omega}_z - \beta_{\omega_z}) - \sin\phi \\ (\bar{\omega}_y - \beta_{\omega_y})\frac{\sin\phi}{\cos\theta} + (\bar{\omega}_z - \beta_{\omega_z})\frac{\cos\phi}{\cos\theta} \\ 0 \\ 0 \\ 0 \end{bmatrix}, \tag{3.16}$$

while the measurement model is given by (3.12), i.e.,

$$\boldsymbol{h}(\boldsymbol{x}) = \begin{bmatrix} \sin\theta \\ -\sin\phi\cos\theta \\ -\cos\phi\cos\theta \end{bmatrix}. \tag{3.17}$$

In this case the parameter q_η determines how reliable the state propagation (based on gyroscopes measurements) is with respect to the observations made by the accelerometers.

3.3.5 Quaternions

Four quaternions and two biases of the gyroscope variables are estimated in this algorithm. The state vector can be written as $\boldsymbol{x} = [\mathbf{q}^T,\ \boldsymbol{\beta}^T]^T$. The nonlinear equations are described as

$$\begin{aligned} \dot{x} &= \mathbf{A}(x)x + \boldsymbol{\lambda}_{\xi}, & \mathbb{E}[\boldsymbol{\lambda}_{\xi}\boldsymbol{\lambda}_{\xi}^{T}] &= \mathbf{Q}, \\ y &= \boldsymbol{h}(x) + \boldsymbol{\lambda}_{v}, & \mathbb{E}[\boldsymbol{\lambda}_{v}\boldsymbol{\lambda}_{v}^{T}] &= \mathbf{R}, \end{aligned} \tag{3.18}$$

where

$$\mathbf{A}(x) = \begin{bmatrix} \breve{\boldsymbol{\Omega}} & 0 \\ 0 & 0 \end{bmatrix}, \qquad \mathbf{C} = \begin{bmatrix} -\mathbf{I}_3 & 0 \end{bmatrix},$$

$$\mathbf{Q} = \begin{bmatrix} q_q\mathbf{I}_4 & 0 \\ 0 & q_\beta\mathbf{I}_3 \end{bmatrix}, \qquad \mathbf{R} = \sigma_a^2\mathbf{I}_3.$$

The measurement model is given by the third column of (3.8), namely,

$$\boldsymbol{h}(x) = \begin{bmatrix} 2(q_1q_3 + q_2q_4) \\ 2(q_2q_3 - q_1q_4) \\ q_1^2 - q_2^2 + q_3^2 + q_4^2 \end{bmatrix}. \tag{3.19}$$

3.3.6 Comparative Real-Time Analysis in the Quanser Platform

The algorithms presented in the previous section have different levels of complexity, make use of different orientation representations, and even estimate different variables. A comparison was carried out in order to see if the different formulations have a significant impact on the estimation accuracy.

3.3.6.1 *Experimental Platform and Devices*

The experimental platform used for the comparative analysis is shown in Fig. 3.1. It is the 3D HOVER platform manufactured by Quanser and thought of as a test-bed platform of control algorithms for vertical take-off and landing (VTOL) vehicles, so that the translational degrees of freedom are clamped for convenience [9]. The Euler angles are measured by means of three optical encoders with an accuracy of 0.04 degrees, which can be sampled with a frequency up to 1000 Hz. These encoders provide a reliable pattern for the evaluation and comparison of the algorithms.

For the sake of comparison, two Inertial Measurement Units (IMU) have been used, the MicroStrain 3DM-GX2 and the Sparkfun Sensor Stick. The 3DM-GX2 has typical inertial sensor sets which are internally processed [10]. The angle estimations can be delivered in either Euler angles or rotational matrix at up to 250 Hz. The Sensor Stick is a hobbyist-level

Figure 3.1 Experimental platform and devices.

unit which consists of only a plain circuit board that gathers inertial sensors connected to a unique bus [11]. Some of the relevant information about these devices is shown in Table 3.1.

3.3.6.2 Results

In order to evaluate the performance of the different algorithms, the platform described in Sect. 3.3.6.1 was used to follow alternating step references. These references were tracked using a PD controller. The angular estimates of the encoders and the 3DM-GX2 were stored, along with the information of gyroscopes and accelerometers of the Sensor Stick. The whole sampling procedure was implemented in a single real-time C++ code running at 100 Hz. This frequency was selected since it allows for a reasonable time to run the algorithms in a low-cost microcontroller. The estimation algorithms are executed offline using Matlab and the sampled data.

One of the experiments comparing the attitude measured by the encoders with that of the 3DM-GX2 and the one computed with the Euler EKF algorithm during an experiment with ± 2 degree step references is presented in Fig. 3.2. At first sight the angular estimate obtained with the Sensor Stick is fast and noise-free, very similar to the estimate given by the 3DM-GX2.

Since it is very difficult to reach any conclusion from the naked eye observations, a set of five experiments like the one in Fig. 3.2 were carried out for different step amplitudes. The root mean square value (rmse)

Table 3.1 IMUs specifications

	Microstrain 3DM-GX2	Sensor stick	MPU-6050
Size	41 × 63 × 32	35 × 10 × 2	21 × 17 × 2
Weight	50 g	1.2 g	6 g
Gyro range	±75 to ±1200 degree/s	±2000 degree/s	±250 to ±2000 degree/s
Gyro bias	±0.2 degree/s	Not defined	±20 degree/s
Gyro nonlinearity	0.2%	0.2%	0.2%
Gyro noise performance	0.17 degree/s (rms)[a]	0.38 degree/s (rms)	0.025 degree/s (rms)
Acc. bias	±5 mg (for ±5 g)	Not defined	±50 mg
Acc. nonlinearity	0.2%	0.5%	0.5%
Acc. noise performance	0.6 mg (rms)[a]	2.9 mg (rms)	1.3 mg (rms)

[a] Measured from static output of the sensor.

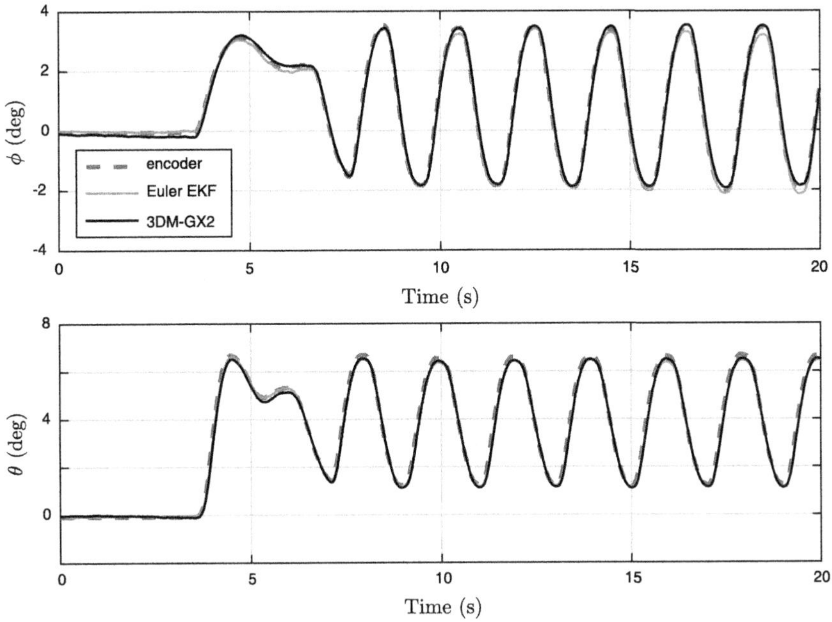

Figure 3.2 Comparison of the angle measured by the encoders with the one given by the 3DM-GX2 and the estimation of the Euler EKF algorithm during an experiment tracking a ± 2 degree input command.

was chosen as a performance index, taking the encoder measurement as the ideal one. This index was averaged over the whole set of experiments for each step amplitude and for every algorithm. For each algorithm an optimization by trial and error was carried out to find the tuning of the parameters leading to the best performance.

The results of these experiments are depicted in Fig. 3.3 where the mean value of the rmse is represented along with its standard deviation. As can be seen, the accuracy of the estimation gets worse as the amplitude of the steps increases. All the algorithms perform very well for low angle oscillations.

However, for higher angles the 3DM-GX2 performs slightly better. This is sensible given that higher outputs of the sensors will manifest their imperfections, e.g., scale factor inaccuracy, nonlinearity, or cross-axis sensitivity. In any case, performance is very similar, and the comparison points out that a cheap board like the Sensor Stick and a simple linear KF provides sufficient accuracy for small angles.

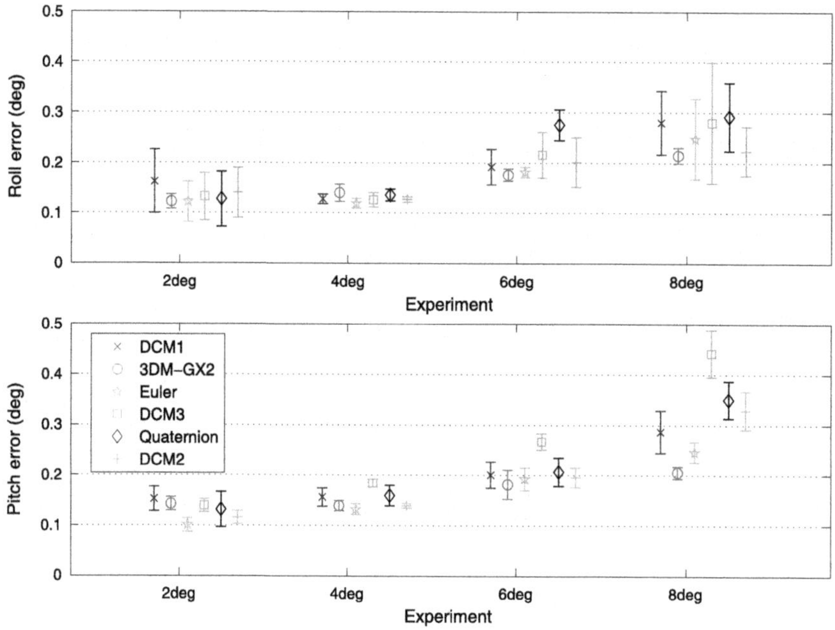

Figure 3.3 Root mean square errors and standard deviations.

Fig. 3.4 shows a situation in which the algorithms did not perform so well. It is possible to observe that one of the main sources of error is the deviation in a stationary position, far from the horizontal. One of the reasons for that is due to small nonlinearities of the accelerometers as they are calibrated using a linear model. There is also another effect, which is not so obvious. The estimation is slightly worse in one of the axes when there is velocity in the other. This is a consequence of the cross-sensitivity of the gyroscopes.

Among the evaluated algorithms, the DMC3-based is the only one which includes the angular velocities as estimated variables. Fig. 3.5 shows an example of how the estimation looks like, compared to the angular velocity extracted from the 3DM-GX2. The estimation exhibits less noise, and no delay is appreciated.

3.4 A COMPUTATIONALLY-EFFICIENT KALMAN FILTER

Some of the aforementioned algorithms may need a large amount of computational resources. Computationally efficient algorithms are crucial be-

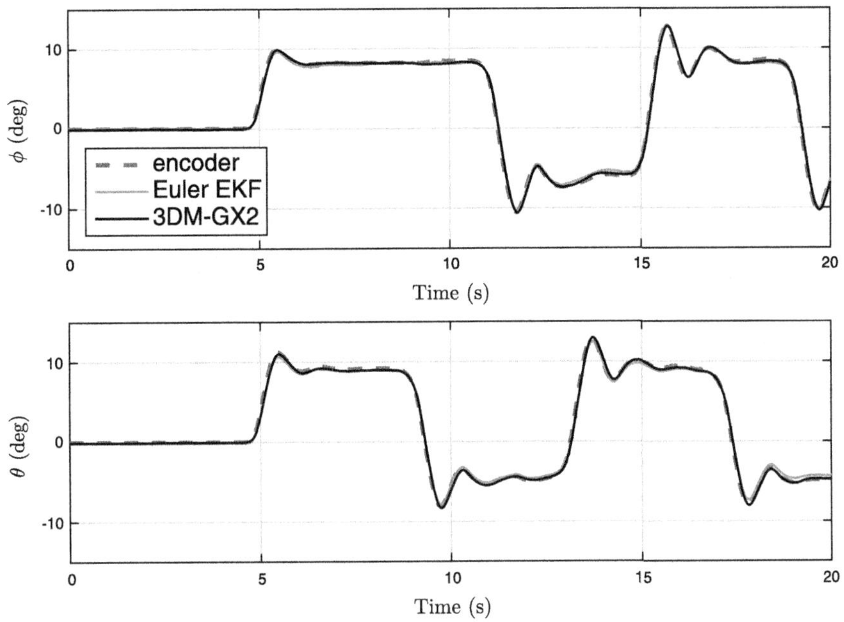

Figure 3.4 Comparison of the angles estimation for ±8 degrees.

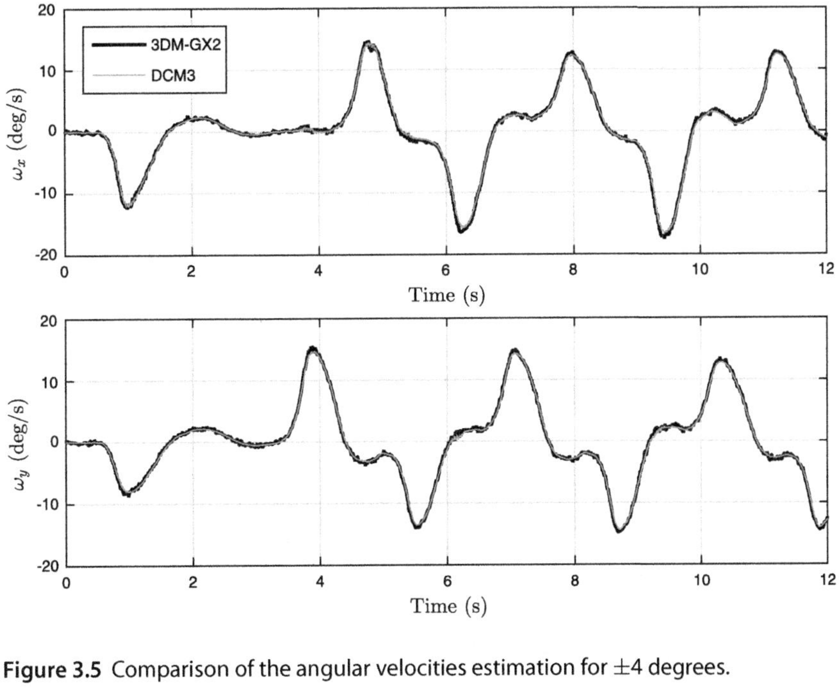

Figure 3.5 Comparison of the angular velocities estimation for ±4 degrees.

cause of two reasons. First of all, the computational resources onboard are limited, especially in small aerial vehicles. Second, the fast rotational dynamics of these vehicles requires high frequency controllers, and thus the measurements should be updated as fast as possible.

After several tests, we have arrived at a computationally efficient algorithm to estimate the attitude of a rotorcraft. The algorithm uses a Kalman filter applied to the kinematics of the vehicle represented by means of the Euler formulation and includes the biases of the gyroscopes into the vector of estimated variables. It was validated in flight tests, and the experimental results were compared with the data coming from a commercial IMU showing good behavior of the estimation algorithm which is presented next; for more details see [12].

3.4.1 Simplified Algorithm

An advantage of the Euler formulation is that the yaw angle can be removed from the equations if the rotation matrix is well defined. Let us denote the vector of variables to be estimated by $\boldsymbol{x} = [\phi,\ \beta_x,\ \theta,\ \beta_y]$. It is reasonable to simplify the equations by assuming non-aggressive maneuvers, $\sin\alpha \approx \alpha$ and $\cos\alpha \approx 1$, with $\alpha = \{\phi,\ \theta\}$. From previous considerations, the kinematics of an aerial vehicle and the measurement model can be expressed by (3.1) and (3.12), respectively, which leads to the following linear equations:

$$\begin{aligned} \dot{\boldsymbol{x}} &= \underbrace{\begin{bmatrix} 0 & -1 & 0 & 0 \\ 0 & 0 & 0 & 0 \\ 0 & 0 & 0 & -1 \\ 0 & 0 & 0 & 0 \end{bmatrix}}_{\mathbf{A}} \boldsymbol{x} + \underbrace{\begin{bmatrix} 1 & 0 \\ 0 & 0 \\ 0 & 1 \\ 0 & 0 \end{bmatrix}}_{\mathbf{B}} \boldsymbol{u} + \boldsymbol{\lambda}_w, \\ \hat{\boldsymbol{y}} &= \underbrace{\begin{bmatrix} 0 & 0 & 1 & 0 \\ -1 & 0 & 0 & 0 \end{bmatrix}}_{\mathbf{H}} \boldsymbol{x} + \boldsymbol{\lambda}_v, \end{aligned} \tag{3.20}$$

where the input vector consists of the angular velocity measurements $\boldsymbol{u} = [\bar{\omega}_x,\ \bar{\omega}_y]$, and $\boldsymbol{y} = [\bar{a}_x,\ \bar{a}_y]$ contains the acceleration measurements. The third equation of the measurement model which involves the third accelerometer axis has been removed, as it has very low sensitivity with respect to the roll–pitch orientation for small angles.

These simplifications result in a smaller-size linear system, thus reducing the computational load substantially. Furthermore, it will be shown next how the structure of the matrices can also be exploited to reduce the Kalman filter to a set of simple equations. The continuous-time system (3.20) can be discretized with sample time T_s, assuming zero-order hold of the input, as follows:

$$\begin{aligned} \hat{\boldsymbol{x}}_{k+1} &= \mathbf{A}_k\hat{\boldsymbol{x}}_k + \mathbf{B}_k\boldsymbol{u}_k + \boldsymbol{\lambda}_{wk}, & \mathbb{E}[\boldsymbol{\lambda}_{wk}\boldsymbol{\lambda}_{w_k}^T] &= \mathbf{q}_k, \\ \boldsymbol{y}_k &= \mathbf{H}_k\hat{\boldsymbol{x}}_k + \boldsymbol{\lambda}_{vk}, & \mathbb{E}[\boldsymbol{\lambda}_{vk}\boldsymbol{\lambda}_{v_k}^T] &= \mathbf{R}_k, \end{aligned} \tag{3.21}$$

where

$$\mathbf{A}_k = e^{\mathbf{A}T_s} = \begin{bmatrix} 1 & -T_s & 0 & 0 \\ 0 & 1 & 0 & 0 \\ 0 & 0 & 1 & -T_s \\ 0 & 0 & 0 & 1 \end{bmatrix}, \qquad \mathbf{B}_k = \left(\int_0^{T_s} e^{\mathbf{A}s}\mathrm{d}s\right)\mathbf{B} = \begin{bmatrix} T_s & 0 \\ 0 & 0 \\ 0 & T_s \\ 0 & 0 \end{bmatrix},$$

$$\mathbf{H}_k = \begin{bmatrix} 0 & 0 & 1 & 0 \\ -1 & 0 & 0 & 0 \end{bmatrix}.$$

Notice that, as a consequence of the simplifications, the resulting system is decoupled. Therefore, two different Kalman filters could be separately implemented for roll and pitch dynamics. Furthermore, one can take advantage of the fact that matrix $\boldsymbol{P}$ is symmetric, so that only three of its entries need to be stored in the memory. Using (3.13) and (3.14), the following equations for the Kalman filter of the roll angle can be derived:

$$\begin{aligned} \hat{\phi}_k^- &= \hat{\phi}_{k-1} + T_s(\bar{\omega}_x - \beta_{x_k}), \\ \beta_{x_k}^- &= \beta_{x_{k-1}}, \\ p_{11_k}^- &= p_{11_{k-1}} - 2T_s p_{12_{k-1}} + T_s^2 p_{22_{k-1}} + q_{11_k}, \\ p_{12_k}^- &= p_{12_{k-1}} - T_s p_{22_{k-1}}, \\ p_{22_k}^- &= p_{22_{k-1}} + q_{22_k}, \end{aligned} \tag{3.22}$$

$$\begin{aligned} \hat{\phi}_k^+ &= (1-\alpha_\phi)\hat{\phi}_k^- - \alpha_\phi \bar{a}_y \\ \beta_{x_k}^+ &= \beta_{x_k}^- - \gamma_\phi(\bar{a}_y + \hat{\phi}_k^-), \\ p_{11_k}^+ &= (1-\alpha_\phi)p_{11_k}^-, \\ p_{12_k}^+ &= (1-\alpha_\phi)p_{12_k}^-, \\ p_{22_k}^+ &= -\gamma_\phi p_{12_k}^- + p_{22_k}^-, \end{aligned} \tag{3.23}$$

where

$$\alpha_\phi = \frac{p_{11_k}^+}{p_{11_k}^+ + r_{11}}, \qquad \gamma_\phi = \frac{p_{12_k}^+}{p_{11_k}^+ + r_{11_k}}.$$

In a similar way, the derivation of the Kalman filter equations for the pitch angle leads to the following set of equations:

$$
\begin{aligned}
\hat{\theta}_k^- &= \hat{\theta}_{k-1} + T_s(\bar{\omega}_x - \beta_{x_k}),\\
\beta_{y_k}^- &= \beta_{y_{k-1}},\\
p_{11_k}^- &= p_{11_{k-1}} - 2T_s p_{12_{k-1}} + T_s^2 p_{22_{k-1}} + q_{33_k},\\
p_{12_k}^- &= p_{12_{k-1}} - T_s p_{22_{k-1}},\\
p_{22_k}^- &= p_{22_{k-1}} + q_{44_k},
\end{aligned}
\tag{3.24}
$$

$$
\begin{aligned}
\hat{\theta}_k^+ &= (1-\alpha_\phi)\hat{\theta}_k^- + \alpha_\phi \bar{a}_y,\\
\beta_{y_k}^+ &= \beta_{y_k}^- + \gamma_\phi(\bar{a}_x - \hat{\theta}_k^-),\\
p_{11_k}^+ &= (1-\alpha_\theta)p_{11_k}^-,\\
p_{12_k}^+ &= (1-\alpha_\theta)p_{12_k}^-,\\
p_{22_k}^+ &= -\gamma_\theta p_{12_k}^- + p_{22_k}^-,
\end{aligned}
\tag{3.25}
$$

where

$$
\alpha_\theta = \frac{p_{11_k}^+}{p_{11_k}^+ + r_{22_k}}, \qquad \gamma_\theta = \frac{p_{12_k}^+}{p_{11_k}^+ + r_{22_k}}.
$$

3.4.2 Numerical Validation

The Kalman filter algorithm for estimating the roll and pitch angle (Eqs. (3.22)–(3.25)) has been first validated using the encoders of the Quanser platform. A commercial IMU (3DM-GX2) and a low cost inertial sensor (MPU6050) were also included in the platform in order to validate and compare the measurements. The MPU6050 is composed of a 3-axis gyroscope and a 3-axis accelerometer. It does not provide the angles of the rigid body, only the raw measurements of the sensors. The characteristics of the both devices are given in Table 3.1.

Since the angles of the experimental platform are unstable, a simple PD controller was used to stabilize the system to constant references of pitch and roll. The system was then perturbed by applying manual disturbances. All data was collected at 333 Hz, and the algorithm equations (3.22)–(3.25) were computed offline using Matlab. A trial and error tuning process re-

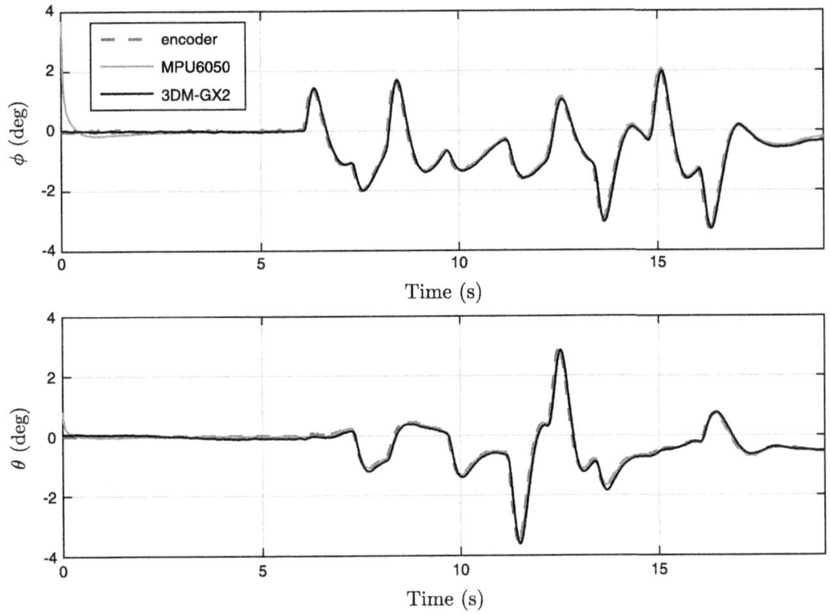

Figure 3.6 Attitude estimation.

sulted in the following covariance matrices:

$$\mathbf{Q}_k = \begin{bmatrix} 0.94 \cdot 10^{-6} & 0 & 0 & 0 \\ 0 & 0.91 \cdot 10^{-6} & 0 & 0 \\ 0 & 0 & 0 & 0 \\ 0 & 0 & 0 & 0 \end{bmatrix}, \qquad \mathbf{R}_k = \begin{bmatrix} 0.37 & 0 \\ 0 & 0.39 \end{bmatrix}.$$

The estimation obtained by means of this procedure can be seen in Fig. 3.6. At first sight, it can be noticed that the proposed algorithm performs fairly well, thus validating the simplifications made in its derivation.

As it is difficult to visually evaluate the quality of both estimations, three performance indexes were chosen, namely, the root mean squared error, the maximum absolute error, and the delay, all of them computed with respect to the estimate given by the encoders (which are assumed to provide a "real" measurement). Table 3.2 gathers the information of these indices for an experiment of several minutes where the system was again stabilized around zero and perturbed manually. Notice that the proposed algorithm performs even better than the 3DM-GX2.

The evolution of the bias estimate is shown in Fig. 3.7. The real bias of the gyroscopes was computed by averaging the first few seconds during

Table 3.2 Performance indices

		rmse (degrees)	$\|error\|_{max}$ (degrees)	Delay (ms)
3DM-GX2	Roll	0.3	1.56	25
	Pitch	0.27	1.46	
MPU 6050	Roll	0.14	0.72	15
	Pitch	0.19	0.91	

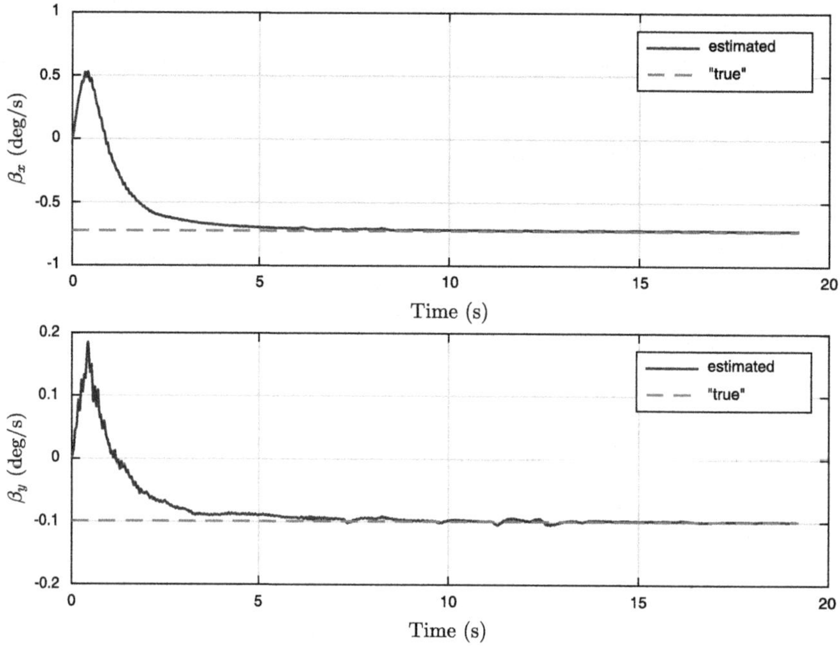

Figure 3.7 Bias estimation.

which the system remains steady. It is possible to see how the estimated bias converges to the real value within a few seconds.

3.4.3 Flight Tests

Although the platform described above is a suitable scenario for numerical comparison, there are still some handicaps to overcome in real flight, like vibrations and linear accelerations. The estimation algorithm is finally tested online and onboard to compute the roll and pitch angles and to stabilize a quadrotor prototype.

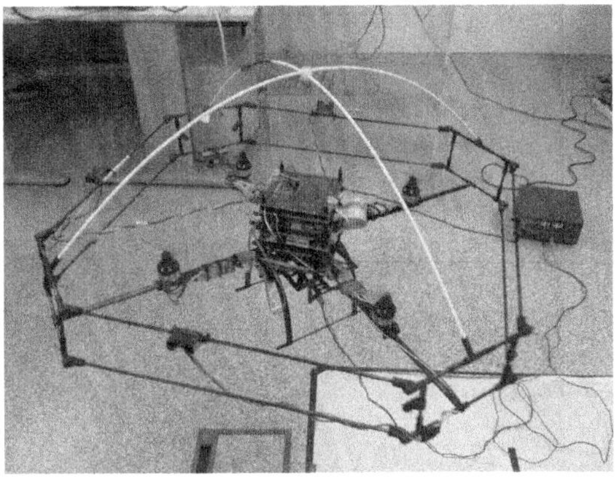

Figure 3.8 UPV quadrotor prototype.

3.4.3.1 UPV Quadrotor Prototype

The quadrotor prototype developed at the UPV (Universidad Politécnica de Valencia, Spain) is depicted in Fig. 3.8. It has a distance of 41 cm between rotors, weighs around 1.3 kg without battery, and is outfitted with the IMU MicroStrain 3DM-GX2 and MPU6050, among other sensors. The basic hardware consists of a MikroKopter frame, YGE 25i electronic speed controllers, RobbeRoxxy 2827-35 brushless motors, and 10×4.5 plastic propellers. All the computations are made onboard using an Arduino Due which is based on an Atmel SAM3X8E ARM Cortex-M3 microcontroller running at 84 MHz, and an Igep v2 board running Xenomai real-time operating system at 1 GHz.

The Arduino Due is in charge of reading every sensor, computing the Kalman filter and attitude control algorithms, controlling the motor's speed, and sending the data to the Igep board. The control algorithm consists of a PD controller with nested saturations. The Igep board is only used for data storage and also serves as a WiFi bridge. Although the proposed Kalman filter takes only 2 ms to run, the main loop in the microcontroller runs at 333 Hz restricted by the communications with the 3DM-GX2 and Igep board.

3.4.3.2 Experiments

The quadrotor was controlled by adjusting roll and pitch angles using PD controllers computed with the estimated values $\hat{\phi}$, $\hat{\theta}$, $\dot{\hat{\phi}}$, and $\dot{\hat{\theta}}$. Furthermore, angular references were sent to the vehicle using a joystick. The yaw

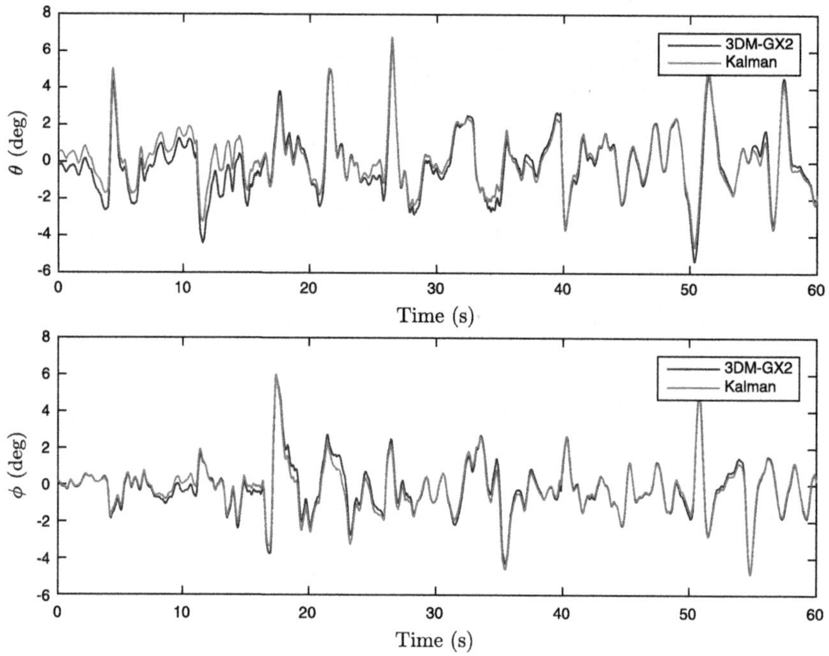

Figure 3.9 Comparison of the attitude estimations in-flight.

angle was stabilized separately with the measurement of the Microstrain sensor, also with a PD controller.

Due to the lack of a motion capture system, the in-flight attitude estimate of the proposed algorithm is compared to that of the 3DM-GX2. Fig. 3.9 shows the attitude estimate during one minute of flight. One can see that both estimates are very similar (an rmse of 2.7 degrees and 1.6 degrees in pitch and roll axes, respectively). Although the 3DM-GX2 is not a fully reliable pattern, it can be seen that the proposed algorithm provides a fast, noise-free, and drift-free estimate. Furthermore, it should be also noticed that the control is computed with the estimated attitude. The small oscillations around the equilibrium point evidence that the attitude and velocity estimates lead to a very good control performance.

The angular velocities are shown in Fig. 3.10. The estimated angular velocities consist of the raw gyroscope measurements corrected with the estimated biases. The estimation of the bias avoids the need of correcting the offset of the gyroscopes prior to each flight and allows operating over long time periods. This comparison shows that, in spite of the simplifica-

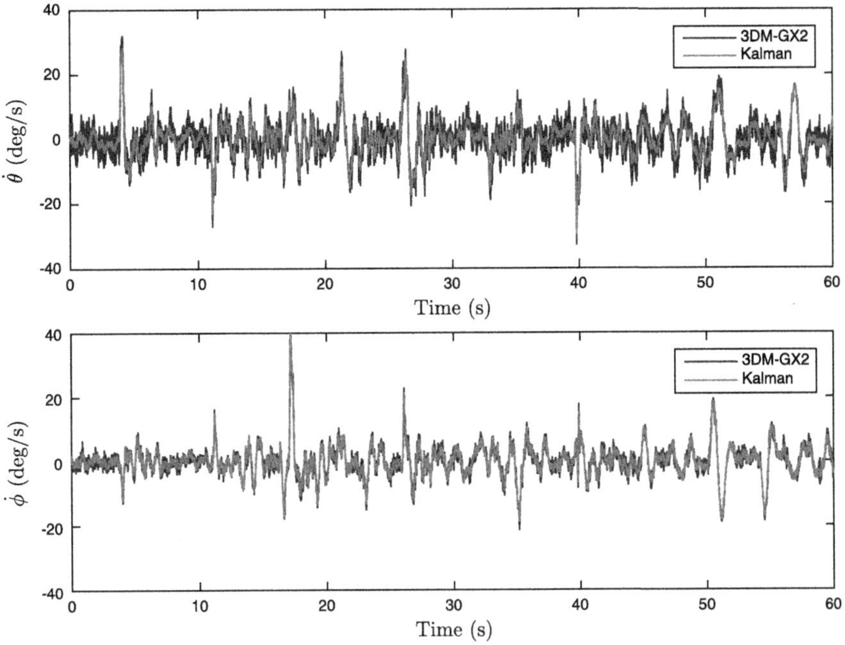

Figure 3.10 Comparison of the angular velocity estimations in-flight.

tions, the accuracy of the estimates is fairly good compared to that of a commercial device, at least for non-aggressive flights. Because of the low computational cost of the algorithm, the angular estimates exhibit a very small delay, which also improves control performance.

3.5 DISCUSSION

Attitude estimation is a serious problem for many research teams working in UAVs. The best solution for some teams is to buy classical IMUs such as MicroStrain; nevertheless, their price (around 2000 USD) demotivates small research groups (also particular researchers). Low-priced arrays of MEMS (accelerometers and gyrometers) can be found on the market; nevertheless, in some cases an estimation algorithm needs to be used. In this chapter the most common estimation algorithms were described and analyzed. When evaluating performance, it is important to take into account the computational load of the algorithms and the available resources onboard, as the controllers must run at very high frequencies. In our results, Kalman filter exhibited a performance that could be compared with a commercial IMU, the Microstrain 3DM-GX2. The performance of these algorithms can be further improved by using other control techniques; for example, in the next chapter, we propose a predictor scheme to restore all the states and improve the precision of the estimation algorithm.

REFERENCES

1. K. Nonami, F. Kendoul, S. Suzuki, W. Wang, D. Nakazawa, Autonomous Flying Robots: Unmanned Aerial Vehicles and Micro Aerial Vehicles, Springer, 2010.
2. A. Tayebi, S. McGilvray, Attitude stabilization of a VTOL quadrotor aircraft, IEEE Transactions on Control Systems Technology 14 (2006) 562–571.
3. S. Bouabdallah, R. Siegwart, Full control of a quadrotor, in: International Conference on Intelligent Robots and Systems, IROS, IEEE/RSJ, San Diego, CA, USA, 2007, pp. 153–158.
4. J. Thienel, R.M. Sanner, A coupled nonlinear spacecraft attitude controller and observer with an unknown constant gyro bias and gyro noise, IEEE Transactions on Automatic Control 48 (2003) 2011–2015.
5. S. Bonnet, C. Bassompierre, C. Godin, S. Lesecq, A. Barraud, Calibration methods for inertial and magnetic sensors, Sensors and Actuators A, Physical 156 (2009) 302–311.
6. J.L. Crassidis, J.L. Junkins, Optimal Estimation of Dynamic Systems, vol. 24, CRC Press, 2011.
7. H. Rehbinder, X. Hu, Drift-free attitude estimation for accelerated rigid bodies, Automatica 40 (2004) 653–659.
8. C. Liu, Z. Zhou, X. Fu, Attitude determination for MAVs using a Kalman filter, Tsinghua Science and Technology 13 (2008) 593–597.
9. http://www.quanser.com/products/3dof_hover.
10. http://www.microstrain.com/inertial/3dm-gx2.
11. https://www.sparkfun.com/products/10724.
12. R. Sanz, L. Rodenas, P. Garcia, P. Castillo, Improving attitude estimation using inertial sensors for quadrotor control systems, in: International Conference on Unmanned Aircraft Systems, ICUAS, Orlando, FL, USA, 2014.

CHAPTER 4

Delay Signals & Predictors*

Several control algorithms aiming at stabilization of aerial vehicles can be found in the literature [1,2]. Some of these control/navigation strategies have been applied in real time using efficient but expensive sensing systems [3]. Other research teams use commercial sensors combined with observer/predictor algorithms that improve measurements [4]. Whatever the case, a common problem when applying control strategies in real time is the measurement delay represented as input delays. When these delays are significant, the performance of the closed-loop system suffers and a crash may occur.

In the past, the control of linear processes with time delay in either the input or output has been addressed with strategies such as the Smith Predictor (SP) [5] and the Finite Spectrum Assignment (FSA) [6]. The Smith Predictor [5] can be considered as the first predictor-based control for SISO (Single-Input Single-Output) open-loop stable linear systems [7]. Later the same concept was extended for MIMO (Multiple-Input Multiple-Output) open-loop unstable systems by introducing an h units of time ahead state predictor [6,8]. The original proposals to deal with time-delay systems suffer from robustness problems and implementation issues (a good survey can be found in [9]). In [10] a discrete predictor for continuous-time plants with time delay is proposed, and the closed-loop stability is proved. More recently, in [11], the robustness of the proposed predictor was proved with respect to a delay mismatch or model uncertainties [12], and also to perform well in combination with a state observer [13]; the issue of time-varying delays was also addressed in [14]. Provided that nowadays almost any control system application is implemented using a computer, discrete-time control schemes are more interesting in practical applications [15].

This chapter presents an observer–predictor algorithm (OP-A) which is used to counteract the inherent delays in the attitude estimates. The

* The results in this chapter were developed in collaboration with R. Sanz Diaz from the Universidad Politecnica de Valencia, Spain, and Angel G. Alatorre and Sabine Mondié from CINVESTAV-IPN, Mexico.

Indoor Navigation Strategies for Aerial Autonomous Systems.
DOI: http://dx.doi.org/10.1016/B978-0-12-805189-4.00006-8

origin of such delays is discussed and illustrated by comparing the measurements of a commercial IMU (Inertial Measurement Unit) with those coming from optical encoders. The OP-A is used to improve the estimation of the pitch and roll angles, and it is based on a Kalman Filter (KF) and a discrete-time predictor to fuse the measurements coming from gyroscopes and accelerometers. The topic of time delay systems is reviewed before introducing a discrete-time predictor and a state predictor which is based on the Fundamental Calculus Theorem and the Backstepping approach. The advantages of using predicted measurements is demonstrated in simulations and in real-time experiments.

4.1 OBSERVER–PREDICTOR ALGORITHM FOR COMPENSATION OF MEASUREMENT DELAYS

As aforementioned, the IMUs are the core of lightweight aerial robotic applications as they provide attitude estimates and angular velocity measurements. In practice it is observed that the attitude estimated by an IMU exhibits a delay with respect to its real value. This fact is more significant when using low-cost sensors. Since low-cost devices have typically worse noise performance, their signals must be filtered at lower frequencies, thus introducing larger delays. In any case, even high-performance IMUs provide delayed measurements. In order to illustrate this fact, the experimental platform described in Sect. 3.3.6.1 of Chap. 3 is used to compare the angular measurements against a commercial IMU (Mircrostrain 3DM-GX2) with those of coming from a set of encoders which are faster and more accurate than any IMU. A zoom of the comparison is represented in Fig. 4.1 where a delay of approximately 40 ms is appreciated.

One of the unavoidable sources of delay is the low-pass filtering before sampling. During the data acquisition process in an IMU, the signals are low-pass filtered to remove noise and avoid aliasing effects. Another issue is the computational time required to run the estimation algorithm which is often carried out in an onboard microcontroller. The incorporation of delayed measurements into the Kalman filter while preserving optimality is far from being trivial. When the delay consists of only a few sample periods, the problem can be handled optimally by augmenting the state vector [16]. However, for larger delays the computational burden of this approach becomes too large. This topic has been investigated in [17]. More recent work on this topic has been done in [18] where a general delayed KF framework is derived for linear time-invariant systems.

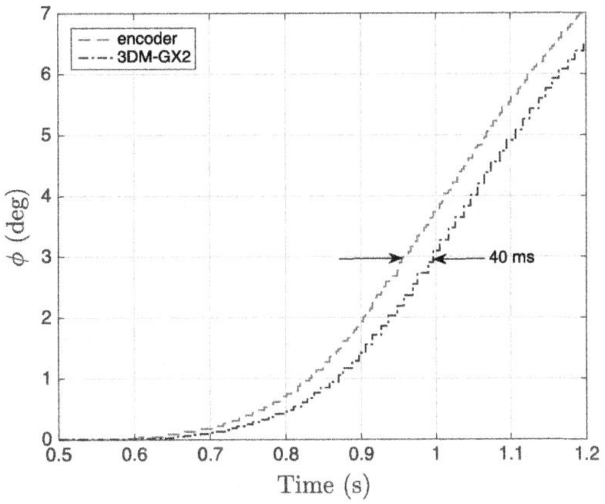

Figure 4.1 Measurement delay in the ϕ angle estimated by a 3DM-GX2 IMU with respect to an optical encoder.

4.1.1 The Finite Spectrum Assignment Problem

Let us consider the following time-delay systems:

$$\begin{aligned} \dot{\boldsymbol{x}}(t) &= \mathbf{A}\boldsymbol{x}(t) + \mathbf{B}\boldsymbol{u}(t-\tau), & \qquad \dot{\boldsymbol{x}}(t) &= \mathbf{A}\boldsymbol{x}(t) + \mathbf{B}\boldsymbol{u}(t), \\ \boldsymbol{y}(t) &= \boldsymbol{x}(t), & \qquad \boldsymbol{y}(t) &= \boldsymbol{x}(t-\tau), \end{aligned} \tag{4.1}$$

where the time delay τ affects either the input or the output. Notice that, under the feedback control law $u(t) = Ky(t)$, both systems are equivalent in terms of stability. To be more specific, such a control law acting on the systems in (4.1) leads to

$$\dot{\boldsymbol{x}}(t) = \mathbf{A}\boldsymbol{x}(t) + \mathbf{BK}\boldsymbol{x}(t-\tau) \tag{4.2}$$

which has an infinite-dimensional characteristic equation $|s\mathbf{I} - \mathbf{A} - \mathbf{BK}e^{-\tau s}| = 0$. This equation (quasipolynomial) is transcendental, which makes it very difficult to solve. Moreover, if the system has uncertainties, the solution becomes even more complicated. Therefore, it would be convenient to get rid of the delay in the characteristic equation, so that conventional analysis and design techniques could be applied.

It is well known that for the systems in (4.1), the feedback control law

$$\boldsymbol{u}(t) = e^{\mathbf{A}\tau}\boldsymbol{x}(t) + \int_0^{\tau} e^{\mathbf{A}\theta}\mathbf{B}\boldsymbol{u}(t-\tau)\,\mathrm{d}\theta \tag{4.3}$$

yields the closed-loop characteristic equation $|s\mathbf{I} - (\mathbf{A} + \mathbf{BK})| = 0$; see [6]. The finite spectrum assignment (FSA) feature of this control law is a significant advantage from a design point of view. However, the practical implementation of the distributed control law (4.3) has raised several problems. A reasonable approach consists in using a numerical quadrature rule to approximate the integral term. It has been shown in [19] that the control law (4.3) with a large discretization step size may not stabilize the systems in (4.1). A safe implementation of the controller (4.3) was introduced in [20], based on including a low-pass filter in the control loop. In the next section, the discrete-time predictor presented in [10] is developed and proved to be suitable for open-loop unstable time-delay systems, robust with respect to a delay mismatch [11], or model uncertainties [12].

4.1.2 An h-Step Ahead Predictor

As discussed in the introduction of the chapter, in a quadrotor vehicle, the estimated variables are unavoidably affected by small time delays which can be counteracted by using a predictive feedback. As it can be seen in (4.3), for the implementation of the predictive control law, a model of the system is needed. The rotational dynamics of the quadrotor using the Newton–Euler approach are given by (2.8). It has been showed in several works (and corroborated in flight tests) that, under some circumstances, the rotational dynamics of the quadrotor can be reduced to a double integrator on each axis

$$\ddot{\boldsymbol{\eta}} = \tilde{\boldsymbol{\tau}}, \tag{4.4}$$

where $\boldsymbol{\eta} = [\phi, \theta, \psi]^T$ is the orientation of the vehicle, $\tilde{\boldsymbol{\tau}} = \mathbb{J}^{-1}\tau = [\tilde{\tau}_\phi, \tilde{\tau}_\theta, \tilde{\tau}_\psi]^T$ is a control transformation, τ is the vector of external torques, and $\mathbb{J}$ is the inertia matrix. In the simplified model (4.4), the axes are decoupled, and each of them can be represented in a discrete-time state-space form

$$\boldsymbol{x}_{k+1} = \underbrace{\begin{bmatrix} 1 & T_s \\ 0 & 1 \end{bmatrix}}_{\mathbf{A}_k} \boldsymbol{x}_k + \underbrace{\begin{bmatrix} \frac{T_s^2}{2} \\ T_s \end{bmatrix}}_{\mathbf{B}_k} u_k, \tag{4.5}$$

where $\boldsymbol{x}_k = [s_k, \dot{s}_k]^T$, $s = \{\phi, \theta, \psi\}$, and T_s is the sampling period. The discretized matrices have been obtained using

$$\mathbf{A}_k = e^{\mathbf{A}T_s} = \mathbf{I} + T_s\mathbf{A}, \qquad \mathbf{B}_k = \left(\int_0^{T_s} e^{\mathbf{A}s}\, ds \right) \mathbf{B}. \tag{4.6}$$

Notice that for the assumed simplified model (4.4), the matrix $\mathbf{A} = \begin{bmatrix} 0 & 1 \\ 0 & 0 \end{bmatrix}$ is nilpotent with $\mathbf{A}^j = 0$ for any $j \geq 2$, and thus there is no discretization error because the truncated expansion of the exponential matrix is exact.

Let us as assume now that the state of the plant is fully accessible, but there is a known constant transmission delay τ which is approximated[1] to be a multiple of the sampling period T_s, that is, $\tau = T_s d$ with $d \in \mathbb{Z}^+$. The measured delayed state is denoted by

$$\bar{\boldsymbol{x}}_k \triangleq \boldsymbol{x}_{k-d}. \tag{4.7}$$

The source of the measurement $\bar{\boldsymbol{x}}_k$ can be the output of some of the estimation algorithms presented in Chap. 3. If a conventional state feedback $u_k = \mathbf{K}\bar{\boldsymbol{x}}_k$ is used to stabilize (4.5), the resulting closed-loop system will be $\boldsymbol{x}_{k+1} = \mathbf{A}_k \boldsymbol{x}_k + \mathbf{B}_k \mathbf{K} \boldsymbol{x}_{k-d}$. Such a system may have poor performance or even become unstable if d is sufficiently large. This is especially critical for open-loop unstable systems, as is the case for quadrotor vehicles.

In order to counteract the effect of the measurement delay, an h-step ahead predicted state is obtained by using (4.5) recursively to propagate the state of the system. Notice that

$$\begin{aligned}
\boldsymbol{x}_{(k-d)+1} &= \mathbf{A}_k \boldsymbol{x}_{k-d} + \mathbf{B}_k u_{k-d}, \\
\boldsymbol{x}_{(k-d)+2} &= \mathbf{A}_k \boldsymbol{x}_{(k-d)+1} + \mathbf{B}_k u_{(k-d)+1}, \\
&= \mathbf{A}_k^2 \boldsymbol{x}_{k-d} + \mathbf{A}_k \mathbf{B}_k u_{k-d} + \mathbf{B}_k u_{(k-d)+1}, \\
&\cdots \\
\boldsymbol{x}_{(k-d)+h} &= \mathbf{A}_k^h \boldsymbol{x}_{k-d} + \sum_{i=0}^{h-1} \mathbf{A}_k^{h-1-i} \mathbf{B}_k u_{k-d+i}.
\end{aligned}$$

Taking into account the definition of the delayed measurement in (4.7), the predicted state is defined by

$$\tilde{\boldsymbol{x}}_{k+h} \triangleq \mathbf{A}_k^h \bar{\boldsymbol{x}}_k + \sum_{i=0}^{h-1} \mathbf{A}_k^{h-1-i} \mathbf{B}_k u_{k-d+i}, \tag{4.8}$$

with the prediction horizon $h \in \mathbb{Z}^+$ being a design parameter. Notice that choosing $h = d$ (assuming perfect model matching), the prediction (4.8)

[1] As the control structure should be robust under model parameters uncertainty, the round-off of the fractional delay will not be a problem [10].

will be exact and $\tilde{\boldsymbol{x}}_{k+h} \equiv \boldsymbol{x}_{(k-d)+d} = \boldsymbol{x}_k$. Therefore, with the predictive feedback $u_k = \mathbf{K}\tilde{\boldsymbol{x}}_{k+h}$, a delay-free closed-loop system $\boldsymbol{x}_{k+1} = (\mathbf{A}_k + \mathbf{B}_k\mathbf{K})\boldsymbol{x}_k$ is recovered.

Digital implementation of (4.8) is straightforward. Using the $\mathcal{Z}$-transform, the predicted state can be written as

$$\tilde{\boldsymbol{x}}(z) = \mathbf{F}_1(z)\boldsymbol{x}(z) + \mathbf{F}_2(z)u(z), \tag{4.9}$$

with

$$\mathbf{F}_1(z) = \mathbf{A}_k^h \qquad \text{and} \qquad \mathbf{F}_1(z) = \sum_{i=0}^{h-1} \mathbf{H}_i z^{-(d-i)}. \tag{4.10}$$

A buffer of size $h - 1$ is needed to store the history of control inputs. The coefficients $\mathbf{H}_i = \mathbf{A}_k^{h-1-i}\mathbf{B}_k$ and matrix $\mathbf{A}_k^h$ can be pre-computed to save some computational time in slow microcontrollers.

Regarding the instability issues related to the implementation of feedback controllers for open-loop unstable systems, notice that $\mathbf{F}_1(z)$ is a static gain and $\mathbf{F}_2(z)$ is a finite impulse response filter, which means it is BIBO stable. Then, with this approach, the instability of the predicted state for open-loop unstable plants is avoided, and if the roots of $|z\mathbf{I} - (\mathbf{A}_k + \mathbf{B}_k\mathbf{K})| = 0$ are inside the unit circle, the closed-loop system is guaranteed to be internally stable.

The algorithm consists in applying the predictor to the Kalman filter output, that is, taking $\tilde{\boldsymbol{x}}_k \leftarrow \hat{\boldsymbol{x}}_k$. In the simulations and experiments that follow, the measurement $\hat{\boldsymbol{x}}_k$ is computed using the simplified Kalman filter presented in (3.13)–(3.14) or (3.22)–(3.23). The resulting algorithm can be considered as a self-contained predictor-based observer which is depicted in Fig. 4.2.

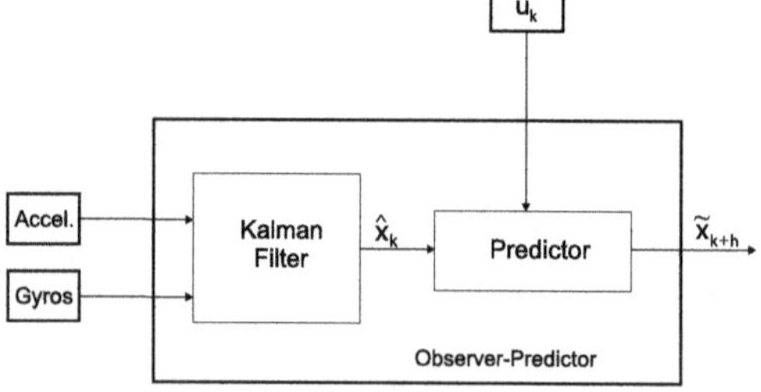

Figure 4.2 Observer–predictor scheme diagram.

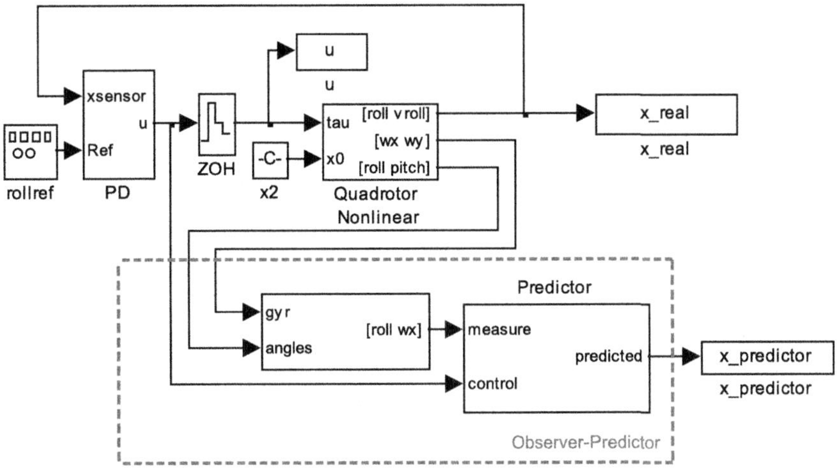

Figure 4.3 Simulink model of the open-loop Observer–Predictor algorithm.

4.1.3 Simulations

For the sake of clarity, only the roll axis of the quadrotor is considered in what follows. Therefore, the state of the plant is given by $\boldsymbol{x}_k = [\phi_k, \dot{\phi}_k]^T$ while the dynamic model is given by (4.5). Simulations were carried out using the Simulink model depicted in Fig. 4.3. The nonlinear quadrotor model in (3.16) is used to represent the plant, and an artificial delay of 40 ms was added to the Kalman filter output. References in the roll angle are tracked by using a simple state-feedback controller

$$u_k = \mathbf{K}\big([\phi_k^{ref}, 0]^T - \boldsymbol{x}_k\big) = k_p(\phi_k^{ref} - \phi_k) - k_d\dot{\phi}_k \tag{4.11}$$

while the pitch and yaw are driven to zero using the same approach. The predictor is applied to the delayed measurements given by the Kalman filter. The parameter h is chosen to be equal to the number of delayed sample periods d. In simulations, d is known, whereas in the experiments it has to be measured.

In the first simulation, the control loop is closed using the ideal (non-delayed) measurements. The Kalman filter and the predictor are run in parallel and do not affect the behavior of the system. The results are shown in Fig. 4.4. One can see how the Kalman estimates are delayed and the predicted state matches the ideal one.

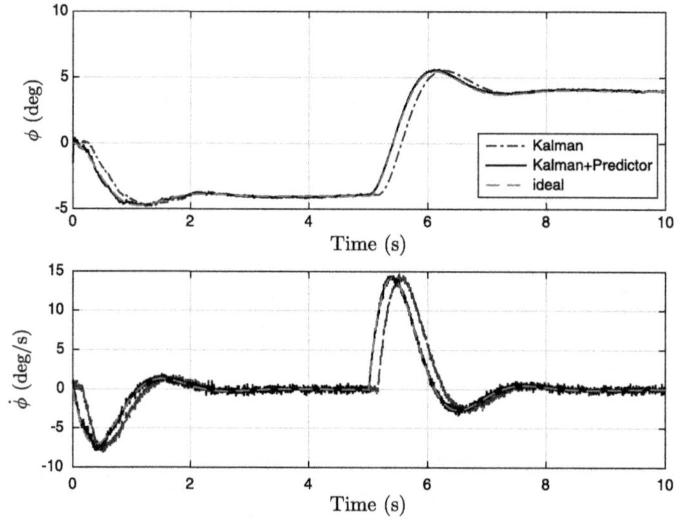

Figure 4.4 Delay-free closed-loop evolution (dashed line) and estimation algorithms running in parallel: Kalman output (dashed-dotted line) and predicted state (solid line).

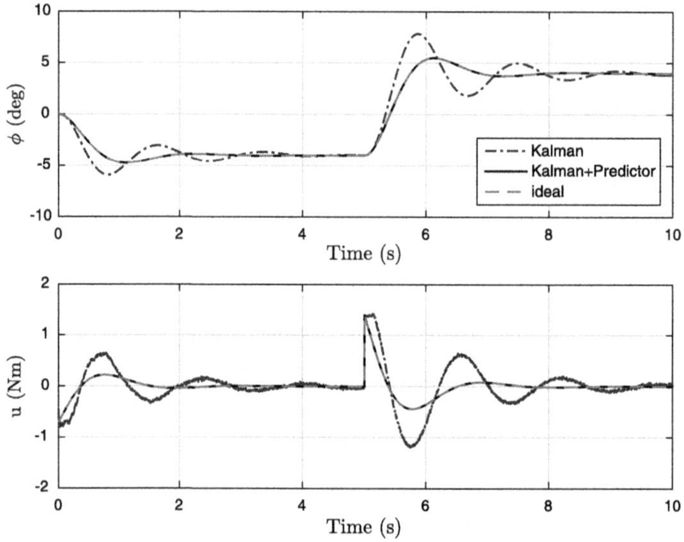

Figure 4.5 Comparison of closed-loop responses when the system is controlled with: ideal (dashed line), Kalman (dashed-dotted line), and predicted (solid line) measurements.

As mentioned above, a delayed measurement decreases the performance of a given controller. Fig. 4.5 shows the output of the closed-loop system and the control action when different state measurements are fed to the controller. Notice how oscillations arise when the system is controlled with the delayed measurements from the Kalman filter. However, the use of the predictor improves the performance substantially, and the response gets very close to that of the system when controlled using a non-delayed measurement.

4.1.4 Experiments

The purpose of the experiments is twofold. On the one hand, open-loop experiments are accomplished to show how the predictor works over experimental data. The output of the predictor should look like the actual state measurement but shifted backwards in time. On the other hand, closed-loop experiments illustrate the improvement in stability when the predictor is used.

4.1.4.1 Quanser Platform

Some experiments were carried out using the Quanser platform described in Sect. 3.3.6.1 of Chap. 3. The measurements coming from the optical encoders are considered to be ideal (non-delayed), and they are only used to evaluate the delay of the other state estimates, not for control purposes. The non-delayed angular rate was computed offline from the encoder measurements by using central difference approximation and filtering.

In the first experiment, the system was controlled via state feedback, according to (4.11), using the measurements coming from the 3DM-GX2. The proposed algorithm was computed offline, discretized with $T_s = 4$ ms, and choosing $h = 5$, that is, a predictive horizon of 20 ms. The resulting state estimates are shown in Fig. 4.6. A detail of the rising phase of the response can be seen in this figure (below). One can see how the delay of the predicted state is almost negligible, while the delay of the 3DM-GX2 is about 40 ms.

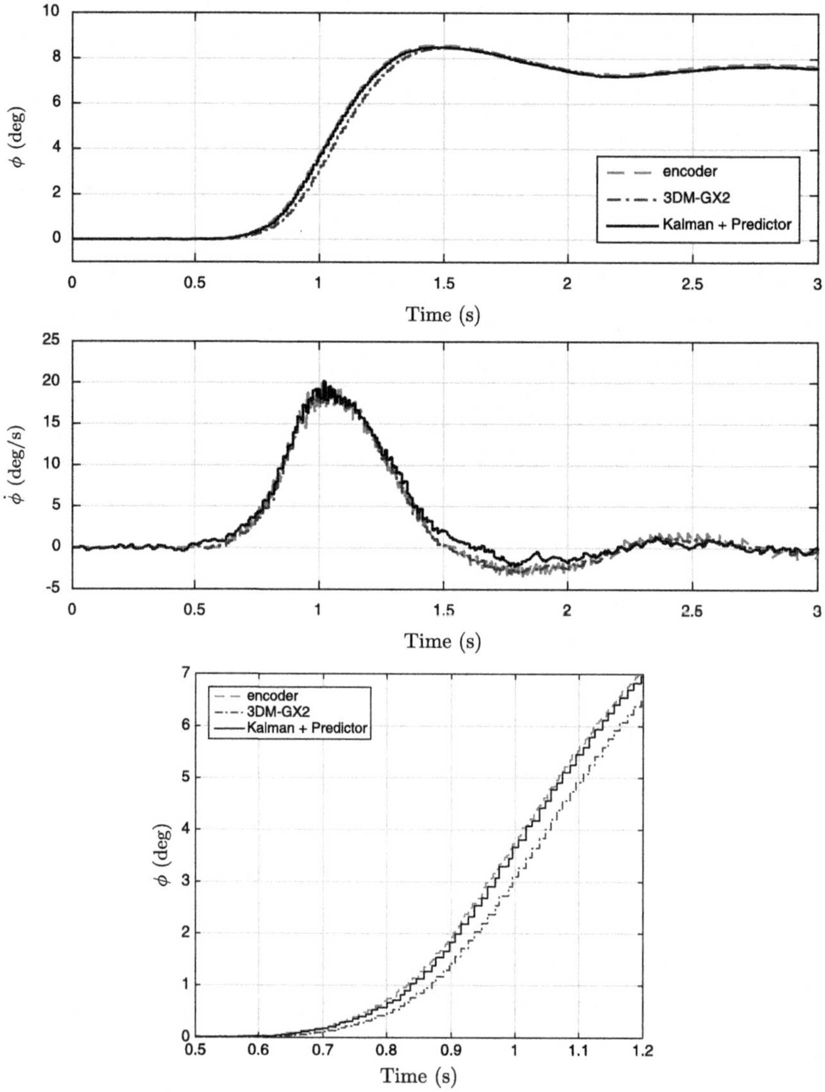

Figure 4.6 Open-loop experiments comparing estimates of the ϕ angle: 3DM-GX2 (dashed-dotted line), proposed OP algorithm (solid line), and optical encoders (dashed line). Bottom figure represents a zoom of top figure.

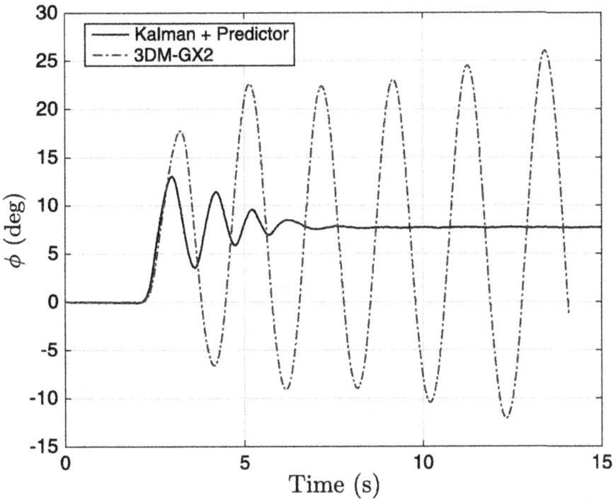

Figure 4.7 Closed-loop step responses in the ϕ angle when the system is controlled with the measurement given by 3DM-GX2 (dashed-dotted line) or with the one given by the proposed algorithm (solid line).

The influence of delayed measurements in the closed-loop stability was also analyzed. For this purpose the predictive scheme was implemented in real time. The gain of the controller was increased until the system controlled with 3DM-GX2 became unstable. Fig. 4.7 shows two overlapped experiments: one using 3DM-GX2 and the other using the predicted state. In both of them a step reference of 8 degrees was applied. Notice that, for a given controller, the system became unstable when the measurement of 3DM-GX2 was used, while it remained stable with the predicted measurement. This experiment shows that the observer–predictor algorithm increases the stability of the system in the presence of measurement delays.

4.1.5 Flight Tests

Recently the feasibility of this proposal has also been validated in flight tests with a commercial device, the Parrot AR Drone 2.0, at the Heudiasyc laboratory. For these experiments, the delay in the measurements coming for the IMU was too small to clearly show the effect of the predicted algorithm, because of that artificial delays were introduced.

The prototype AR Drone has Linux 2.6.32 onboard, the embedded processor is an ARM Cortex A8 at 1 GHz, it also has a DDR2 RAM of

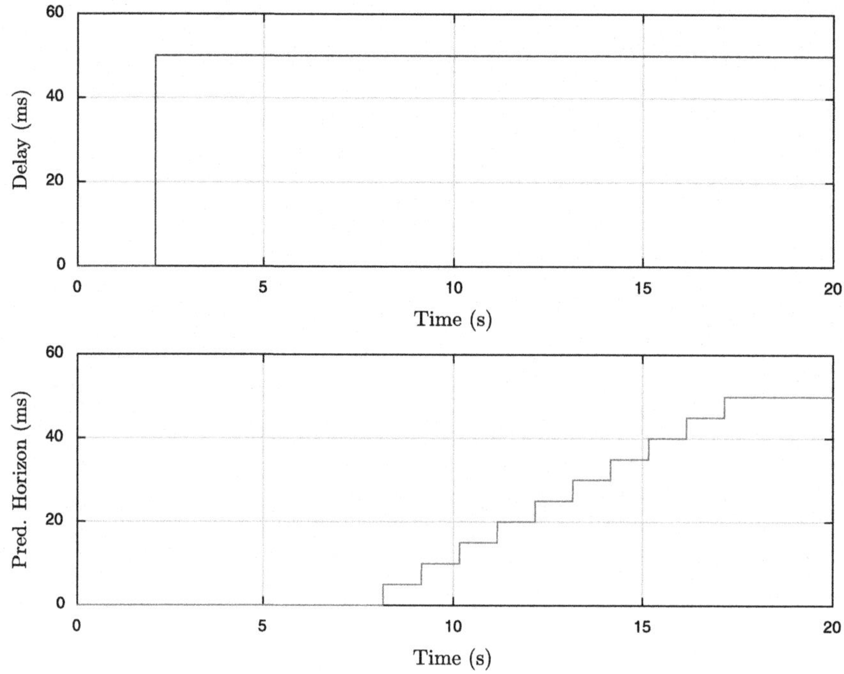

Figure 4.8 Artificial delay and prediction horizon evolution during the flight test.

1 GB at 200 MHz, the sampling frequency is 200 Hz. The sensors for the navigation are a three axis gyroscope with 2000°/s, an accelerometer which accuracy is ±50 mg. The hardware of the Parrot vehicle has been used; nevertheless, several changes have been made to the software in order to access all states and variables that usually cannot be modified in the original software and firmware of Parrot.

During the hovering experiments the measurements from an IMU where artificially delayed by 50 ms (or 10 sampling periods). The predictor was initially disabled, and then activated by sequentially increasing the prediction horizon from 0 to 50 ms (or, from 0 to 10 sampling periods). The history of both variables is depicted in Fig. 4.8. The results of the experiment (variables ϕ and $\dot{\phi}$) are shown in Fig. 4.9. Notice how the system starts oscillating when the delay is introduced, while the performance is recovered when the predictor kicks in (the oscillations have already disappeared with a prediction horizon of 25 ms).

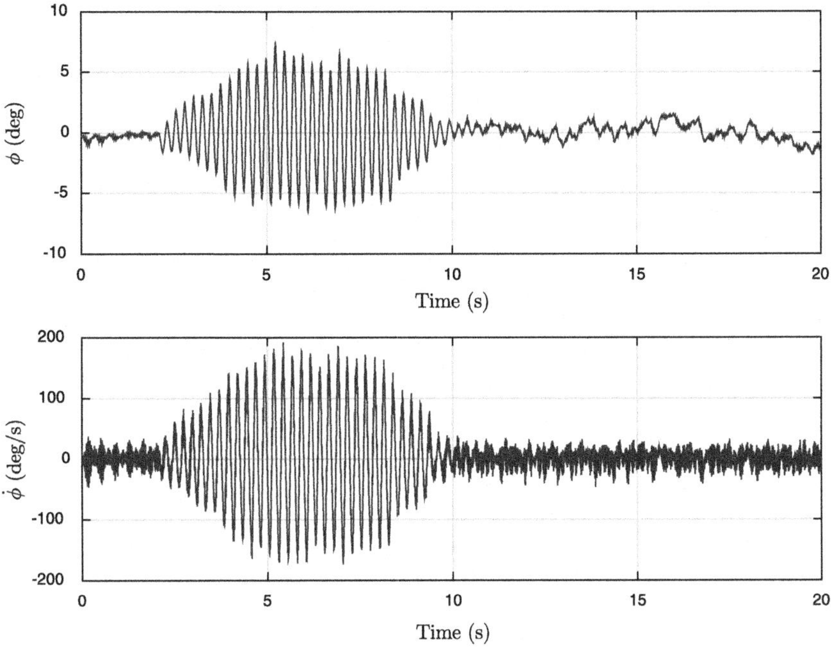

Figure 4.9 Recorded flight data of the ϕ and $\dot{\phi}$ variables showing the influence of measurement delays in stability and ability of the predictor to counteract them.

4.2 STATE PREDICTOR–CONTROL SCHEME

When working with UAVs or robotic systems with fast dynamics, it is crucial to have good closed-loop performance in order to measure the states very quickly. In general, autonomous vehicles (ground, aerial, or underwater) use slow sensors to auto-locate, this is crucial when they are performing quick tasks or when moving in unstructured environments (with obstacles). Improving the performance of the closed-loop system has been a challenge for researchers. Several works can be found in the literature; nevertheless, only a few results have been applied in real time. In this section, we introduce a state predictor–control scheme applied to a quadcopter in order to improve its behavior when the control inputs are delayed. Our result is based on the Fundamental Calculus Theorem and the demonstration made by Krstic [21]. Nevertheless, the proof is modified to better illustrate our predictive-control scheme.

Theorem 4.1 (Fundamental Theorem of Calculus). *Let f be a continuous function in the closed interval $[a, b]$, and let g be a function which satisfies*

$$g'(x) = f(x) \quad \forall x \in [a, b]. \tag{4.12}$$

Then

$$\int_a^b f(t)dt = g(b) - g(a). \tag{4.13}$$

Note that if $x = a$, the derivative in (4.12) is the right-hand side derivative, and if $x = b$, the derivative in (4.12) is the left-hand side derivative.

The main result can be stated in the following theorem.

Theorem 4.2. *Consider a delayed chain of integrators represented as*

$$\dot{\mathbf{x}}(t) = \mathbf{A}\mathbf{x}(t) + \mathbf{B}u(t-\tau), \tag{4.14}$$

where $u(t-\tau)$ is the control input delayed by τ units of time. Then the controller

$$u(t) = \mathbf{K}\hat{\mathbf{x}}(t) \tag{4.15}$$

stabilizes system (4.14). Here $\mathbf{K}$ is a vector gain that stabilizes the delay-free system, and $\hat{\mathbf{x}}(t)$ is the predicted state defined as

$$\hat{\mathbf{x}}(t) = e^{\tau\mathbf{A}}\mathbf{x}(t-\tau) + \varphi(u(t)), \tag{4.16}$$

where

$$e^{\tau A} = \begin{pmatrix} 1 & \tau & \frac{\tau^2}{2} & \dots & \frac{\tau^{n-1}}{(n-1)!} \\ 0 & 1 & \tau & \dots & \frac{\tau^{n-2}}{(n-2)!} \\ 0 & 0 & 1 & \dots & \frac{\tau^{n-3}}{(n-3)!} \\ \vdots & \vdots & \vdots & \ddots & \vdots \\ 0 & 0 & 0 & \dots & \tau \\ 0 & 0 & 0 & \dots & 1 \end{pmatrix};$$

$$\varphi(u(t)) = \begin{pmatrix} \int_{t-\tau}^{t}\int_{t-\tau}^{t_1}\dots\int_{t-\tau}^{t_{n-1}} u(t_n)dt_n dt_{n-1}\cdots dt_1 \\ \int_{t-\tau}^{t}\int_{t-\tau}^{t_1}\dots\int_{t-\tau}^{t_{n-2}} u(t_{n-1})dt_{n-1} dt_{n-2}\cdots dt_1 \\ \vdots \\ \int_{t-\tau}^{t}\int_{t-\tau}^{t_1} u(t_2)dt_2 dt_1 \\ \int_{t-\tau}^{t} u(t_1)dt_1 \end{pmatrix}.$$

Proof. Consider the following system

$$\dot{x}(t) = Ax(t) + Bu(t - \tau), \tag{4.17}$$

where $x(t) \in \mathbb{R}^n$, $A \in \mathbb{R}^{n\times n}$, $B \in \mathbb{R}^{n\times p}$, and $u(t-\tau) \in \mathbb{R}^p$. Using the results in [21], the delay in the control input can be modeled by a first-order hyperbolic partial differential equation, also referred to as transport PDE. Thus,

$$\begin{aligned} u_t(s,t) &= u_s(s,t), \\ u(\tau,t) &= U(t). \end{aligned} \tag{4.18}$$

The solution to this equation is

$$u(s,t) = u(t + s - \tau), \tag{4.19}$$

and therefore the output $u(0,t) = u(t-\tau)$ gives the delayed input. Hence system (4.17) can be written as

$$\dot{x}(t) = Ax(t) + Bu(0,t). \tag{4.20}$$

The backstepping transformation

$$w(s,t) = u(s,t) - \varphi(u(s,t)) - \Gamma(x(t)) \tag{4.21}$$

maps the system (4.18)–(4.20) into the target system

$$\begin{aligned} \dot{x}(t) &= (A + BK)x(t) + Bw(0,t), \\ w_t(s,t) &= w_s(s,t), \\ w(\tau,t) &= 0. \end{aligned} \tag{4.22}$$

This transformation which is defined as $(x,u) \longmapsto (x,w)$ has the lower-triangular form

$$\begin{pmatrix} x \\ w \end{pmatrix} = \begin{pmatrix} I_{n\times n} & 0_{n\times(0,\tau)} \\ \Gamma & \varphi + I_{(0,\tau)\times(0,\tau)} \end{pmatrix} \begin{pmatrix} x \\ u \end{pmatrix}. \tag{4.23}$$

Here Γ denotes the operator $\Gamma : x(t) \longmapsto \Theta(s)x(t)$, where $\Theta(s)$ is defined as

$$\Theta(s) = \begin{pmatrix} 1 & s & s^2 & \dots & s^{n-1} \\ 0 & 1 & s & \dots & s^{n-2} \\ 0 & 0 & 1 & \dots & s^{n-3} \\ \vdots & \vdots & \vdots & \ddots & \vdots \\ 0 & 0 & 0 & \dots & s \\ 0 & 0 & 0 & \dots & 1 \end{pmatrix}$$

and φ denotes the operator

$$\varphi : u(s,t) \longmapsto \begin{pmatrix} \Lambda \int_0^s \int_0^{s_1} \ldots \int_0^{s_{n-1}} u(s_{n,t}) ds_n ds_{n-1} \cdots ds_1 \\ \Lambda \int_0^s \int_0^{s_1} \ldots \int_0^{s_{n-2}} u(s_{n-1}, t) ds_{n-1} ds_{n-2} \cdots ds_1 \\ \vdots \\ \Lambda \int_0^s \int_0^{s_1} u(s_2, t) ds_2 ds_1 \\ \Lambda \int_0^s u(s_1, t) ds_1 \end{pmatrix},$$

where Λ is a scalar that will be defined later.

Computing the time and spatial derivatives of the backstepping transformation (4.21) yields

$$\begin{aligned} w_s(s,t) &= u_s(s,t) - \Theta'(s)x(t) - \Lambda \int_0^s u(s_n, t) ds_n, \\ w_t(s,t) &= u_s(s,t) - \Theta(s)Ax(t) - \Theta(s)Bu(0,t) \\ &\quad - \Lambda \int_0^s u(s_n, t) ds_n + u(0,t)\zeta(s), \end{aligned}$$

where

$$\zeta(s) := \begin{pmatrix} \frac{s^{n-1}}{(n-1)!} \\ \frac{s^{n-2}}{(n-2)!} \\ \vdots \\ s \\ 1 \end{pmatrix}. \tag{4.24}$$

Subtracting $w_t(s,t) - w_s(s,t) = 0$ gives

$$\begin{aligned} 0 &= \Theta(s)\zeta(s)u(0,t) - \Theta(s)Ax(t) - \Theta(s)Bu(0,t) + \Theta'(s)x(t), \\ 0 &= (\Lambda\zeta(s) - \Theta(s)B)\, u(0,t) + (\Theta'(s) - \Theta(s)A)x(t), \\ 0 &= \Lambda\zeta(s) - \Theta(s)B, && (4.25) \\ 0 &= \Theta'(s) - \Theta(s)A. && (4.26) \end{aligned}$$

Expression (4.25) is an ordinary differential equation. To find its initial condition, we set $s = 0$ in (4.21). Hence

$$w(0,t) = u(0,t) - \Theta(0)x(t). \tag{4.27}$$

Introducing (4.27) into (4.22) gives

$$\dot{x}(t) = Ax(t) + Bu(0, t) + B(K - \Theta(0))x(t).$$

Comparing this equation with (4.20) implies $\Theta(0) = K$. Therefore, the solution of (4.25) is of the form

$$\Theta(s) = Ke^{As}. \tag{4.28}$$

From (4.26), $\Lambda = \Theta(s)B\zeta^{-1}(s) = Ke^{As}B\zeta^{-1}(s)$, where $\zeta^{-1}(s)$ denotes the inverse operator. Therefore, we have that

$$e^{As}B = \zeta(s)$$

and then $\Lambda = K$. Substituting the gains Λ and $\Theta(s)$ into the transformation (4.21) and setting $s = \tau$, the controller is then given by

$$u(\tau, t) = Ke^{\tau A} + K\varphi(u(\tau, t)). \tag{4.29}$$

We now prove that the closed-loop system composed of the plant (6.5) and (4.20) with the controller (4.29) is exponentially stable at the origin in the sense of the norm

$$\left(\|x(t)^2\| + \int_0^\tau u(s, t)^2 ds \right)^{1/2}.$$

We first demonstrate that the origin of the target system (4.22) is exponentially stable. Consider the following Lyapunov–Krasovskii functional

$$V(t) = x(t)^T Px(t) + \frac{a}{2} \int_0^\tau (1 + s)w(s, t)^2 ds, \tag{4.30}$$

where $P = P^T > 0$ is the solution of the Lyapunov equation for some $Q = Q^T > 0$ and $a > 0$. The time derivative of (4.30) is

$$\dot{V}(t) \le -x(t)^T Qx(t) + \frac{2}{a}\|x(t)^T PB\|^2 - \frac{a}{2} \int_0^\tau w(s, t)^2 ds.$$

Let us choose

$$a = \frac{4\lambda_{max}(PBB^T P)}{\lambda_{min}(Q)},$$

where $\lambda_{\max}$ and $\lambda_{\min}$ are the maximum and minimum eigenvalues of the corresponding matrices. It follows that

$$\dot{V}(x) \le -\frac{\lambda_{min}(Q)}{2}\|x(t)\|^2 - \frac{a}{2(1+\tau)} \int_0^\tau (1 + s)w(s, t)^2 ds,$$

hence

$$\dot{V}(t) \leq -\mu V(t), \tag{4.31}$$

where

$$\mu := -\min\left\{\frac{\lambda_{min}(Q)}{2\lambda_{max}(P)}, \frac{1}{1+\tau}\right\}. \tag{4.32}$$

Using the backstepping transformation and its inverse form, we get

$$w(s,t) = u(s,t) - \Lambda \int_0^s u(v,t)dv - \Theta(s)x(t), \tag{4.33}$$

$$u(s,t) = w(s,t) + \Omega \int_0^s w(v,t)dv + \Phi(s)x(t). \tag{4.34}$$

Expression (4.30) implies that

$$\begin{aligned}
\omega_1 \left(||x(t)||^2 + \int_0^\tau w(s,t)^2 ds \right) &\leq V(t), \\
\omega_2 \left(||x(t)||^2 + \int_0^\tau w(s,t)^2 ds \right) &\geq V(t),
\end{aligned}$$

where

$$\begin{aligned}
\omega_1 &= \min\left\{\lambda_{min}(P), \frac{a}{2}\right\}, \\
\omega_2 &= \max\left\{\lambda_{max}(P), \frac{a(1+\tau)}{2}\right\}.
\end{aligned}$$

Furthermore (4.33) and (4.34) give

$$\int_0^\tau w(s,t)^2 ds \leq \alpha_1 \int_0^\tau u(s,t)^2 ds + \alpha_2 ||x(t)||^2, \tag{4.35}$$

$$\int_0^\tau u(s,t)^2 ds \leq \beta_1 \int_0^\tau w(s,t)^2 ds + \beta_2 ||x(t)||^2, \tag{4.36}$$

where

$$\begin{aligned}
\alpha_1 &:= 3(1+\tau\,||\Lambda||), \quad & \beta_1 &:= 3(1+\tau\,||\Omega||), \\
\alpha_2 &:= 3\,||K\Theta||^2, & \beta_2 &:= 3\,||K\Phi||^2.
\end{aligned}$$

Hence we obtain

$$\rho_1 \left(||x(t)||^2 + \int_0^\tau u(s,t)^2 ds \right) \le ||x(t)||^2 + \int_0^\tau w(s,t)^2 ds, \quad (4.37)$$

$$\rho_2 \left(||x(t)||^2 + \int_0^\tau u(s,t)^2 ds \right) \ge ||x(t)||^2 + \int_0^\tau w(s,t)^2 ds, \quad (4.38)$$

where

$$\rho_1 := \frac{1}{\max\{\beta_1, \beta_2 + 1\}}, \qquad \rho_2 := \max\{\alpha_1, \alpha_2 + 1\}.$$

Notice that the inequalities for ρ_i, ω_i for $i = 1, 2$ can be combined. Then it follows that

$$\rho_1\omega_1 \left(||x(t)||^2 + \int_0^\tau u(x,t)^2 dx \right) \le V(t),$$

$$\rho_2\omega_2 \left(||x(t)||^2 + \int_0^\tau u(x,t)^2 dx \right) \ge V(t).$$

Finally, combining the above with (4.31) gives

$$||x(t)||^2 + \int_0^\tau u(x,t)^2 dx \le \xi(t) \left(||x(0)||^2 + \int_0^\tau w(x,0)^2 dx \right),$$

where $\xi(t) = \frac{\rho_2\omega_2}{\rho_1\omega_1} e^{-\mu t}$, which completes the proof of exponential stability. □

4.2.1 Aerial Vehicle Validation

The predictor algorithm in Theorem 4.2 is obtained for linear systems and delayed linear controllers; however, we will show in the following that it could be applied with delayed nonlinear algorithms forcing its nonlinear dynamics to have a linear behavior, and then guarantee the closed-loop stability. To better illustrate, we apply our approach to a quadcopter vehicle.

4.2.1.1 Mathematical Representation of a Quadrotor with Delayed Inputs

Aerial vehicles generally move in translational flight (or classical maneuvers) with angles between ± 10 grad, which in the aerial control community and for stability analysis is considered as small angles and substantially simplifies the nonlinear equations (2.4)–(2.5) because the Coriolis effects can be neglected given the following equations[2]:

$$\ddot{\xi} = \mathbf{f}_1(\theta, \phi, u_z), \tag{4.39}$$

$$\ddot{\eta} = \mathbf{f}_2(u_\psi, u_\theta, u_\phi), \tag{4.40}$$

with $\xi^T = \begin{pmatrix} x & y & z \end{pmatrix}$ and $\eta^T = \begin{pmatrix} \psi & \theta & \phi \end{pmatrix}$ respectively defining the position and orientation of the vehicle. It is well known that variable τ denotes the delay in the control community. Thus, to avoid confusion in this chapter, the torques $\tau_\psi, \tau_\phi, \tau_\theta$ will be represented as u_ψ, u_ϕ, u_θ, respectively.

Consider now that the control inputs u_i are delayed, then functions $\mathbf{f}_1(\cdot)$, $\mathbf{f}_2(\cdot)$ can be defined as

$$\mathbf{f}_1(\theta, \phi, \mathbf{u_z}) := \begin{pmatrix} -u_{z(t,d_1)} \sin(\theta) \\ u_{z(t,d_1)} \cos(\theta) \sin(\phi) \\ u_{z(t,d_{1,})} \cos(\theta) \cos(\phi) - mg \end{pmatrix},$$

$$\mathbf{f}_2(\mathbf{u}_\psi, \mathbf{u}_\theta, \mathbf{u}_\phi) := \begin{pmatrix} u_{\psi(t,d_1)} \\ u_{\theta(t,d_1,d_2)} \\ u_{\phi(t,d_1,d_2)} \end{pmatrix}.$$

Remember that θ, ϕ, and ψ represent the pitch, roll, and yaw angles, respectively. In the previous system, it is considered that the control inputs are delayed d_i sampling periods.

4.2.1.2 Altitude and Yaw Control

From (4.40), notice that the yaw angle is a linear system and could be considered decoupled from the others, i.e., $\ddot{\psi} \approx u_{\psi(t-\tau)}$. Therefore, applying Theorem 4.2, it follows that

$$\begin{aligned} u_\psi(t) &:= \mathbf{K}_\psi \hat{\psi} \\ &= k_{d_\psi} \dot{\hat{\psi}} + k_{p_\psi}(\hat{\psi} - \psi_d). \end{aligned}$$

[2] For further analysis and without loss of generality, we set the values of the mass to one and the inertial matrix to the identity matrix.

For the altitude dynamics, $\ddot{z} = \cos\theta \cos\phi \; u_{z(t-\tau)} - g$, the following controller is given by

$$u_z(t-\tau) := \frac{u_{\gamma(t-\tau)} + g}{\cos\hat{\theta}\cos\hat{\phi}}, \tag{4.41}$$

and the closed-loop system is

$$\ddot{z} := \epsilon_{e_{\hat{\theta},\hat{\phi}}} u_\gamma(t-\tau), \tag{4.42}$$

where $u_\gamma(t-\tau)$ is the virtual delayed input and $\epsilon_{e_{\hat{\theta},\hat{\phi}}}$ represents the error in the prediction. Hence, if the pair $(\hat{\theta}, \hat{\phi}) \rightarrow (\theta, \phi)$, then $\epsilon_{e_{\hat{\theta},\hat{\phi}}} = 1$. However, applying the predictor–control scheme to (4.42), it follows that

$$\begin{aligned} u_\gamma(t) &= \mathbf{K}_z\hat{\mathbf{z}} \\ &= k_{d_z}\dot{\hat{z}} + k_{p_z}(\hat{z} - z_d). \end{aligned}$$

The predicted states for ψ and z have the form

$$\begin{aligned} \dot{\hat{\beta}}(t) &= \dot{\beta}(t-d_i) + \int_{t-d_i}^{t} u_\beta(l)dl, \\ \hat{\beta}(t) &= \beta(t-d_i) + d_i\dot{\beta}(t-d_i) + \int_{t-d_i}^{t}\int_{t-d_i}^{s} u_\beta(l)dlds, \end{aligned}$$

where β is z or ψ.[3] The equations for $\hat{\theta}$ and $\hat{\phi}$ will be defined later. Introducing (4.41) into (4.39), we have

$$\begin{aligned} \ddot{x} &\approx -g\epsilon_{e_{\hat{\theta},\hat{\phi}}}\frac{\tan\theta}{\cos\hat{\phi}}, \\ \ddot{y} &\approx g\epsilon_{e_{\hat{\theta},\hat{\phi}}}\tan\phi. \end{aligned}$$

4.2.1.3 Modified Nested Saturation Algorithm for Translational Movements

The previous equations with the roll and pitch dynamics mathematically represent the lateral and longitudinal movements of the quadrotor. These

[3] d_i denotes the delayed sampling period: for two integrators in cascade there is only one d_1, for four integrators there are d_1 and d_2. It does not signify that d_i needs to be the same for the yaw or altitude state, or that $d_1 \neq d_2$ strictly.

dynamics are the strongest and define the stability of the flying aerial vehicle. Following the ideas proposed in [22], we will prove that nonlinear controllers based on nested saturations can be used with the predictor algorithm to stabilize the quadrotor even in the presence of delayed inputs.

It has been also proved in several works that controllers based on saturation functions impose a linear behavior in nonlinear systems when bounding mainly the angular position. Thus in our analysis we will take this fact into account, and it will signify that the quadcopter will move with small angles, i.e., $\cos\phi \approx 1$ and $\sin\phi \approx \phi$. This implies that the lateral and longitudinal dynamics can be represented by four integrators in cascade. In the following, the procedure to fully obtain the controller for the longitudinal dynamics is described step by step, a similar procedure is used to acquire the controller for the lateral dynamics.

Let us define the linear longitudinal dynamics of the quadrotor as

$$\begin{aligned} \ddot{x}(t) &\approx -g_\epsilon \theta_1(t), \\ \ddot{\theta}(t) &= u_\theta(t, d_1, d_2), \end{aligned}$$

where $g_\epsilon = g\epsilon_{e_{\hat{\theta},\hat{\phi}}}$ and $u_{\theta(t,d_1,d_2)} = u_\theta(x_{(t,d_1)}, \dot{x}_{(t,d_1)}, \theta_{(t,d_2)}, \dot{\theta}_{(t,d_2)})$ is the delayed input.

Define the observation error as $e_{\hat{\theta}_i} = \hat{\theta}_i - \theta_i$. Observe from Theorem 4.1 that if $e_{\hat{\theta}_i} \to 0$, then $\hat{\theta}_i \to \theta_i$. Thus we can consider that after a time $T_{\hat{\theta}_i}$, $e_{\hat{\theta}_i} \ll 1$, so that $|e_{\hat{\theta}_i}| \leq \delta_{\hat{\theta}_i}$ with $\delta_{\hat{\theta}_i} > 0$.

We propose the following control using the predicted states:

$$u_\theta(t, d_1, d_2) := -\sigma_1(\hat{\theta}_2 + \varepsilon_1), \tag{4.43}$$

where $\sigma_i(s)$ is a saturation function such that $|\sigma_i(s)| \leq M_i$ and ε_i is a bounded function, $|\varepsilon_i| \leq M_{\varepsilon_i}$, that will be defined later to prove convergence. We propose the following positive definite function $V_1(t) = \frac{1}{2}\theta_2^2(t)$, then $\dot{V}_1(t) = -\theta_2\sigma_1(\hat{\theta}_2 + \varepsilon_1)$.

If $e_{\hat{\theta}_2} \to 0$ then $\text{sign}(\hat{\theta}_2) = \text{sign}(\theta_2)$, and if $|\hat{\theta}_2(t)| > M_{\varepsilon_1}$, this implies that $\dot{V}_1(t) \leq 0$. Then after a time T_1, $|\hat{\theta}_2| \leq M_{\varepsilon_1}$. If we choose $M_1 \geq 2M_{\varepsilon_1}$, then $\forall\, t > T_1$

$$u_\theta(t, d_1, d_2) := -\hat{\theta}_2 - \varepsilon_1. \tag{4.44}$$

Define $z_1(t) \equiv \theta_1(t) + \theta_2(t)$, then

$$\dot{z}_1(t) = \theta_2 - \hat{\theta}_2 - \varepsilon_1 \leq \delta_{\hat{\theta}_2} - \varepsilon_1.$$

Define $\varepsilon_1 := \sigma_2(z_1(t) + \varepsilon_2)$, then $M_{\varepsilon_1} = M_2$. We propose $V_2(t) = \frac{1}{2}z_1^2(t)$, then

$$\dot{V}_2(t) \leq -z_1\left(\sigma_2(z_1(t) + \varepsilon_2) - \delta_{\hat{\theta}_2}\right).$$

Notice that $\delta_{\hat{\theta}_2}$ is arbitrarily small and $M_2 \gg \delta_{\hat{\theta}_2}$, thus, if $|z_1| > M_{\varepsilon_2}$, this implies that $\dot{V}_2(t) \leq 0$ and then after a time T_2, $|z_1| \leq M_{\varepsilon_2}$. If we choose $M_2 \geq 2M_{\varepsilon_2}$, then $\forall\, t > T_2$

$$u_\theta(t, d_1, d_2) := -2\hat{\theta}_2 - \hat{\theta}_1 + e_{\hat{\theta}_2} + e_{\hat{\theta}_1} - \varepsilon_2. \tag{4.45}$$

Define $z_2(t) = z_1(t) + \theta_1(t) - \frac{x_2(t)}{g_\varepsilon}$, then

$$\dot{z}_2(t) = -e_{\hat{\theta}_2} - \varepsilon_2 \leq \delta_{\hat{\theta}_2} - \varepsilon_2.$$

Take $\varepsilon_2 := \sigma_3(z_2(t) + \varepsilon_3)$, then $M_{\varepsilon_2} = M_3$. Propose $V_3(t) = \frac{1}{2}z_2^2(t)$, then

$$\dot{V}_3(t) \leq -z_2\left(\sigma_3(z_2(t)\varepsilon_3) - \delta_{\hat{\theta}_2}\right).$$

Observe again that $\delta_{\hat{\theta}_2}$ is arbitrarily small and $M_3 \gg \delta_{\hat{\theta}_2}$, thus, if $|z_2| > M_{\varepsilon_2}$, this implies that $\dot{V}_2(t) \leq 0$ and then after a time T_3, $|z_2| \leq M_{\varepsilon_3}$. If we choose $M_3 \geq 2M_{\varepsilon_3}$, then $\forall\, t > T_3$

$$u_\theta(t, h_1, h_2) = -3\hat{\theta}_2 - 3\hat{\theta}_1 + 2e_{\hat{\theta}_2} + 3e_{\hat{\theta}_1} + \frac{\hat{x}_2}{g_\varepsilon} - \frac{e_{\hat{x}_2}}{g_\varepsilon} - \varepsilon_3.$$

Finally, define $z_3(t) \equiv z_2(t) - 2\frac{x_2(t)}{g_\varepsilon} + \theta_1 - \frac{x_1(t)}{g_\varepsilon}$, then

$$\dot{z}_3(t) = -e_{\hat{\theta}_2} - \varepsilon_3 \leq \delta_{\hat{\theta}_2} - \varepsilon_3.$$

Take $\varepsilon_3 := \sigma_4(z_3(t))$, then $M_{\varepsilon_3} = M_4$. Propose $V_4 = \frac{1}{2}z_3^2$, then $\dot{V}_4 = -z_3(\sigma_4(z_3(t)) + \delta_{\hat{\theta}_2})$. Notice that $M_4 \gg \delta_{\hat{\theta}_2}$ and this implies that z_3 is bounded and decreases, so after a time T_4 the controller can be rewritten as

$$u_\theta(t, d_1, d_2) \;=\; -4\hat{\theta}_2 - 6\hat{\theta}_1 + \frac{4\hat{x}_2}{g_\varepsilon} + \frac{\hat{x}_1}{g_\varepsilon} + 3e_{\hat{\theta}_2} + 6e_{\hat{\theta}_1} - \frac{4e_{\hat{x}_2}}{g_\varepsilon} - \frac{e_{\hat{x}_1}}{g_\varepsilon}.$$

From Theorem 4.2, $u_\theta(t, d_1, d_2)$ could be expressed as $u_\theta(t, \hat{\mathbf{x}})$, and, in addition, remembering that $\hat{\mathbf{x}}_i \to \mathbf{x_i}$, this implies that $\delta_{(\hat{\theta}_i, \hat{x}_i)} \to 0$. Hence the previous controller becomes

$$\begin{aligned} u_\theta(t, \hat{\mathbf{x}}) &= -4\hat{\theta}_2 - 6\hat{\theta}_1 + \frac{4\hat{x}_2}{g_\varepsilon} + \frac{\hat{x}_1}{g_\varepsilon} \\ &= \mathbf{K}\hat{\mathbf{x}}(t), \end{aligned}$$

where $\mathbf{K} = [1/g \;\; 4/g \;\; -6 \;\; -4]$ and $\hat{\mathbf{x}}^T = [\hat{x}_1 \;\; \hat{x}_2 \;\; \hat{\theta}_1 \;\; \hat{\theta}_2]$. The convergence of the states is proved by applying Theorem 4.2.

The predicted states are defined as

$$
\begin{aligned}
\dot{\hat{\beta}}_1(t) &= \dot{\beta}_1(t-d_2) + \int_{t-d_2}^{t} u_{\beta_1}(l)dl, \\
\hat{\beta}_1(t) &= \beta_1(t-d_2) + d_2\,\dot{\beta}_1(t-d_2) + \int_{t-d_2}^{t}\int_{t-d_2}^{s} u_{\beta_1}(l)dlds, \\
\dot{\hat{\beta}}_2(t) &= \dot{\beta}_2(t-d_1) + \int_{t-d_1}^{t} \hat{\beta}_1(s)ds, \\
\hat{\beta}_2(t) &= \beta_2(t-d_1) + d_1\,\dot{\beta}_2(t-d_1) + \int_{t-d_1}^{t}\int_{t-d_1}^{s} \hat{\beta}_1(l)dlds,
\end{aligned}
$$

where β_1 stands for θ and ϕ while β_2 denotes x or y.

4.2.2 Simulation Results

Several simulations are done to validate the proposed predictive-control scheme in the quadrotor model (4.39)–(4.40). Two delays are introduced to test the performance, d_1 for the translational motion (x, y, z) and d_2 for the attitude (ψ, θ, ϕ). The sampling period, T_s, used in simulation is $T_s = 0.001$, the desired values are $x_d = 10$, $y_d = 5$, and $z_d = 1$, and the initial conditions for all states are set to zero except for $z_1(0) = 10$. First, simulations are made to illustrate good performance of the closed-loop system (4.39)–(4.40) with the control inputs without considering the delays $(d_1 = d_2 = 0)$, see Fig. 4.10. For clarity, only the positions (x, y, z) are depicted in this figure.

Simulations were conducted adding and increasing the delays, (d_1, d_2), until instability is presented. The Critical Delay Values (CDV) that induce system instability are $d_1^* = 90T_s$ and $d_2^* = 240T_s$. The critical values are increased until divergence in the states is observed. Two graphs are presented here to illustrate the performance. On the one hand, Fig. 4.11 depicts the behavior for the position with $d_1 = 100T_s$, here the x position diverges at time $t = 5$ s. On the other hand, in Fig. 4.12 d_2 is increased to $250T_s$, and the system diverges in just 4 seconds.

The predictor control scheme is applied with $d_1 = 100T_s$ and $d_2 = 250T_s$, and the closed-loop system is restored. Nevertheless, we continue to

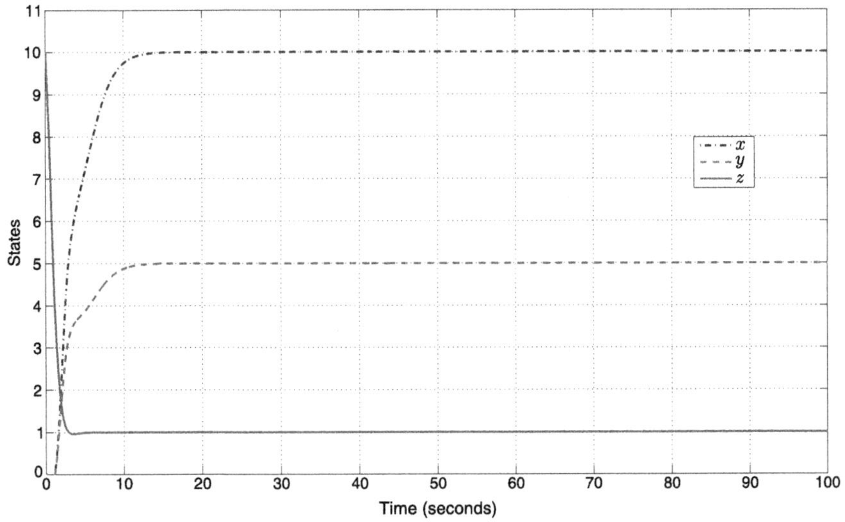

Figure 4.10 State responses without delay in the input, $d_1 = d_2 = 0$.

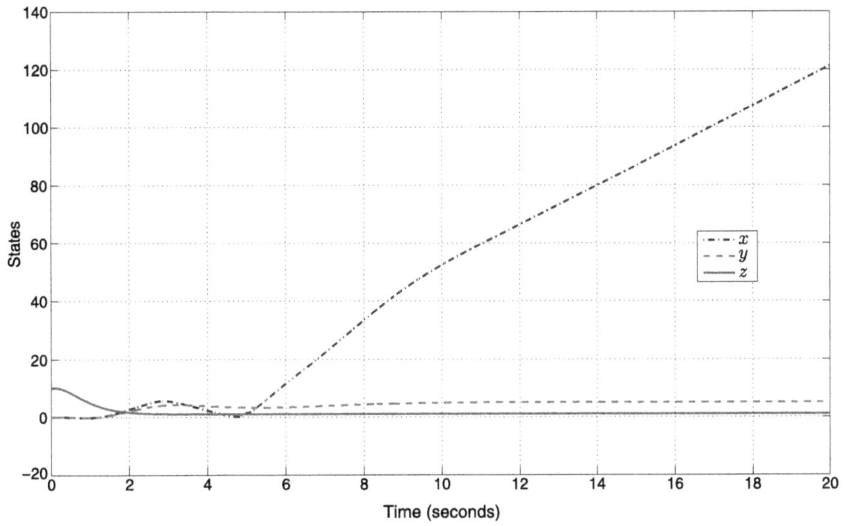

Figure 4.11 Position response with delay $d_1 = 0.1$ s.

increase these delays to observe the robustness of the predictor. The critical values were again increasing until $d_1 = 900T_s$ and $d_2 = 300T_s$; the behavior when applying these delays and the PC-S is shown in Fig. 4.13. Observe

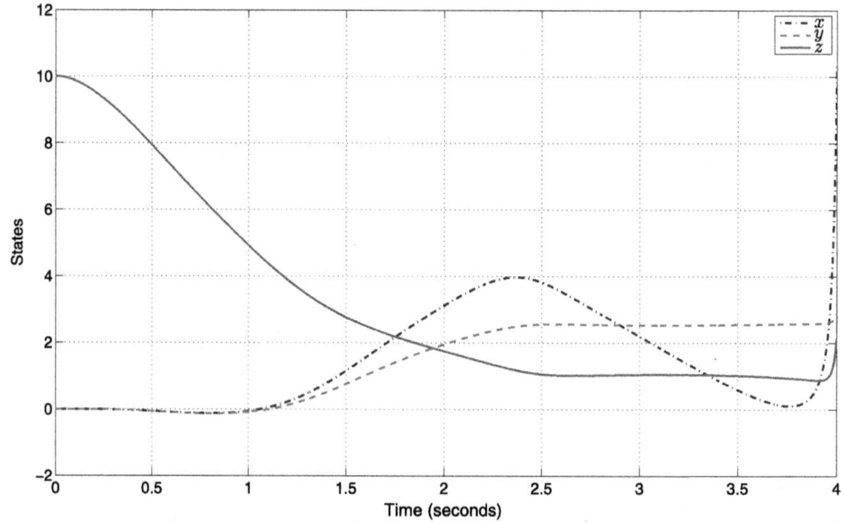

Figure 4.12 x, y, z responses with delay $d_2 = 0.25$ s.

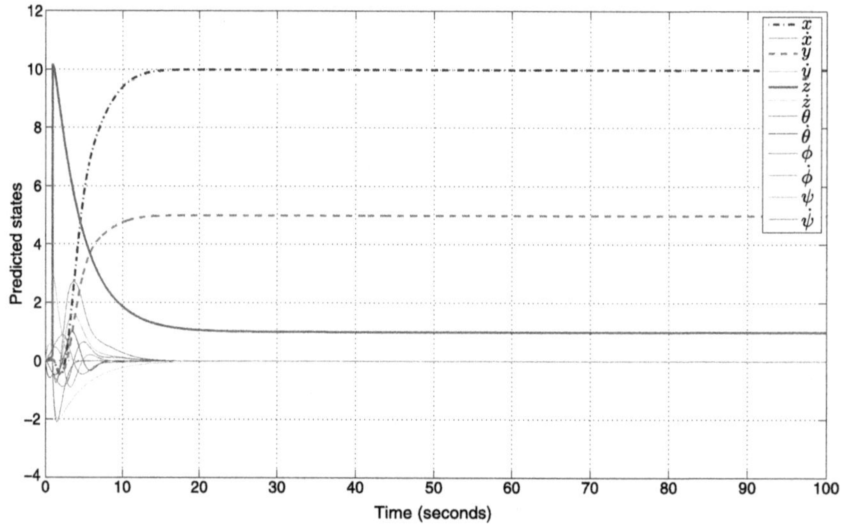

Figure 4.13 System response when applying the predictor–control scheme and with delay values $d_1 = 900T_s$, $d_2 = 300T_s$.

here that all the states converge to the desired values even though the delays are bigger than the CDV.

4.2.3 Experimental Results

Several experiments are carried out to accomplish the practical goal, i.e., validate the proposed scheme with real-data or in real-time experiments. Here only three cases using different sensors will be presented. First, the predictive algorithm is tested offline and in open-loop with data collected from manual tests. A virtual delay is induced to the data collection, and then the predictive algorithm is applied to the delayed data. Next the resulting array is compared with the original data to verify the performance of the predictive scheme.

In the second case, a flight test is realized to follow a square trajectory without any delay in the input. With the collected data virtual delays are induced in all the states to validate the predictive algorithm and verify its behavior with the dynamics of the quadrotor. Finally, for an UAV in a real flight mission, the yaw dynamics (ψ, $\dot{\psi}$) was virtually delayed until signs of instability appeared, then the predictor control scheme is used in closed-loop to compensate the delay.

Open-Loop with GPS Measurements

A rectangular trajectory was made in the manual mode with a LEA-6s GPS sensor, produced by *ublox*. Collected data were used to test the predictive algorithm offline. The sampling period of the GPS is $T_s = 200$ ms. The position data is virtually delayed by $9T_s = 1.8$ s while the velocity data is delayed by just $3T_s = 0.6$ s. Figs. 4.14–4.15 show the obtained results and illustrate the delayed measure (red dotted line), the actual state (blue solid line), and the predicted state (black dashed line). Data are depicted in the coordinate system ECEF, and for the length of the chapter only results in the x-axis are presented. In Fig. 4.14 the x-displacement is presented, and to better illustrate the result a zoom of this figure is also shown.

In Fig. 4.15 the velocity response and a zoom of this figure in the x-axis is presented. Recall that for aerial vehicles the velocity responses need to be very fast in order to stabilize the vehicle. Delaying this state by 1.8 s is not realistic for the sensor characteristics; therefore, it was delayed by only three sampling periods to improve the prediction. Recall also that the sampling period is 200 ms, so the delay for the velocity is 600 ms.

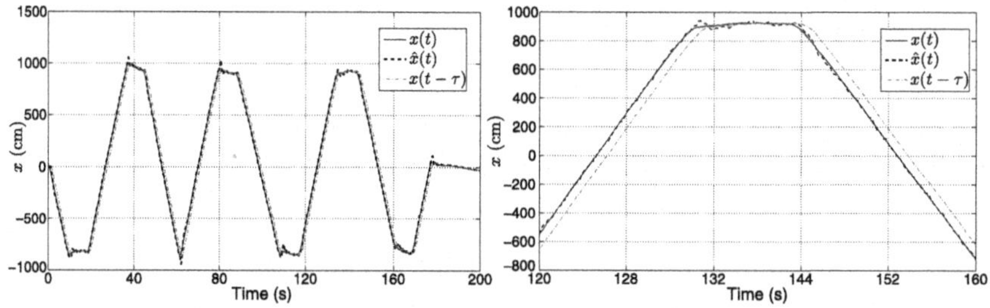

Figure 4.14 State response in the x-axis (and a zoom of it) in the ECEF coordinate system. (For interpretation of the colors in this figure legend, the reader is referred to the web version of this chapter.)

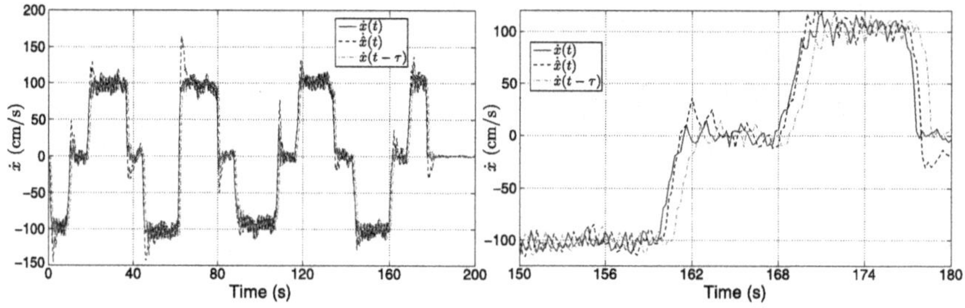

Figure 4.15 Velocity response in the x-direction (and a zoom of it) of the ECEF coordinate system. (For interpretation of the colors in this figure legend, the reader is referred to the web version of this chapter.)

Open-Loop with OptiTrack Measures

Another flight was made to follow a rectangle trajectory in automatic mode without delay. The collected data was delayed in order to apply our predictor control scheme. The height z was imposed to be constant. In the following figures, and due to chapter length limitations, only the position in the horizontal plane is presented (longitudinal and lateral dynamics).

First, the delays d_1 and d_2 were set to $15T_s$. The predictor responses show that the real and predicted values of the states are similar and their difference is almost zero, see Fig. 4.16. Next delays d_1 and d_2 were increased to $40T_s$. The horizontal position responses are illustrated in Fig. 4.17. Notice that even when the delay is big, the predictor scheme is able to correctly predict the values of the states. From the obtained result, we carried on real-time experiments to implement the predictor algorithm in the closed-loop system of an octarotor vehicle.

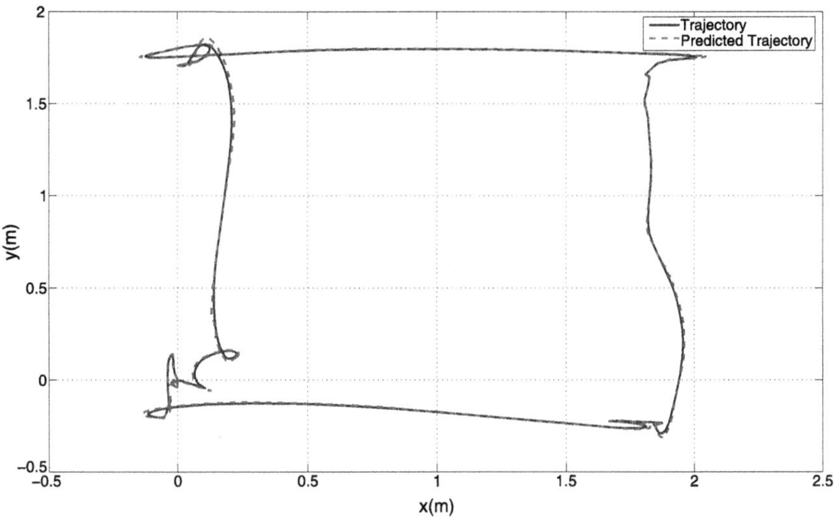

Figure 4.16 x–y plane trajectory vs $\hat{x}$–$\hat{y}$ plane trajectory with delays $d_1 = d_2 = 15T_s$.

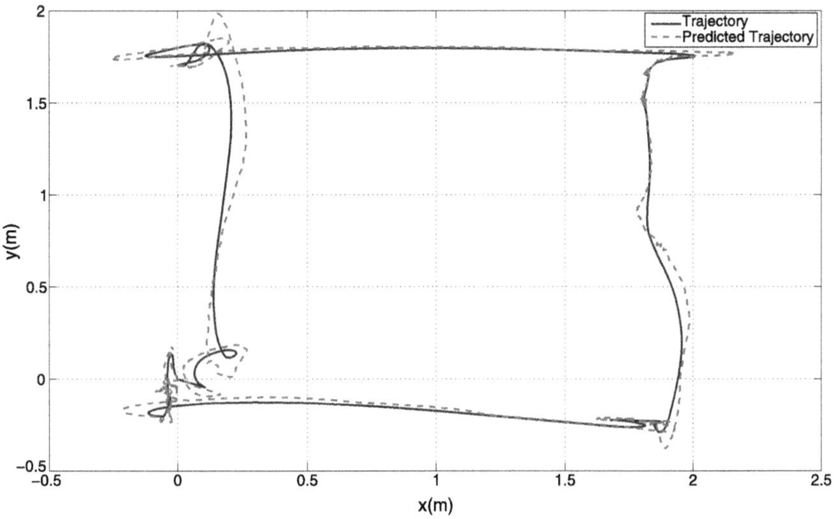

Figure 4.17 x–y plane trajectory vs $\hat{x}$–$\hat{y}$ plane trajectory with delays $d_1 = d_2 = 40T_s$.

4.2.3.1 CoQua Vehicle

Multirotor vehicles are the most popular aerial vehicles that can hover. In this part, we propose to validate the predictor–control scheme in a closed-loop system for the yaw dynamics. This test was made with our

Figure 4.18 CoQua vehicle.

CoQua (Coaxial-quadrotor) vehicle, or octarotor, shown in Fig. 4.18. The aerial vehicle is controlled with an *IGEP* that controls eight drivers *mikrokopter blctrlv*2, and consequently, the eight motors *roxxy bl* 2827 − 35 which gives the thrust and torques of the vehicle. To measure the attitude (ψ, θ, ϕ), an inertial measurement unit model $3dmgx3-25$ of the *Microstrain* brand is used. The weight of this UAV is 1.6 kg.

In this experiment virtual delays are introduced in the yaw dynamics in order to destabilize the yaw dynamics. First, the critical delay value (CDV, delay value capable to destabilize the system) was found using a control scheme without the predictive algorithm. This value was 15 sampling periods, see Fig. 4.19. Notice the unstable ψ response in this figure when applying the delayed controller without predictor, similarly for the $\dot{\psi}$ state illustrated in Fig. 4.20.

The second part consisted of a repeated experiment with the CDV and the predictive algorithm in closed loop. When the predictor is applied to the unstable system, the closed-loop system becomes stable, see Figs. 4.21–4.22. Also in this experiment we added manual changes of reference in order to analyze the system's performance. A zoom of the yaw response can be also appreciated in Fig. 4.21. Fig. 4.22 illustrates the yaw rate response. Notice here that the stability is also recovered for this state.

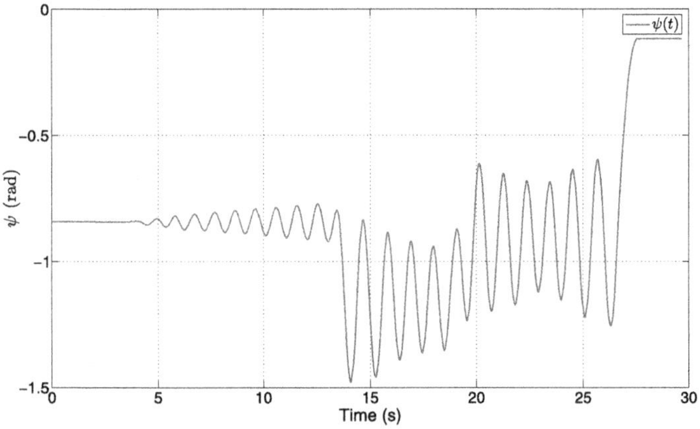

Figure 4.19 ψ response with a delayed input of $15T_S$.

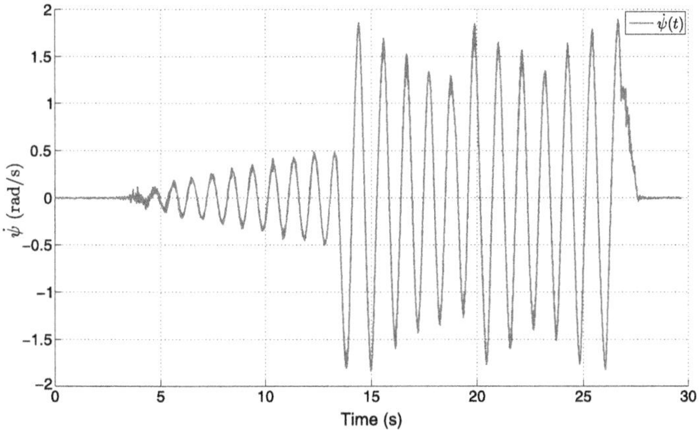

Figure 4.20 $\dot{\psi}$ response with a delayed input of $15T_S$.

4.3 DISCUSSION

Improving system's behavior is a challenge to many researchers. Several teams work directly on proposing robust controllers that can reject uncertainties, perturbations, or sometimes delays. In this chapter two predictor schemes to improve the closed-loop behavior in the presence of delays in the inputs were proposed. The algorithms considered real cases as delayed measurements on sensors, and it was proved that using these predictors the states are recovered.

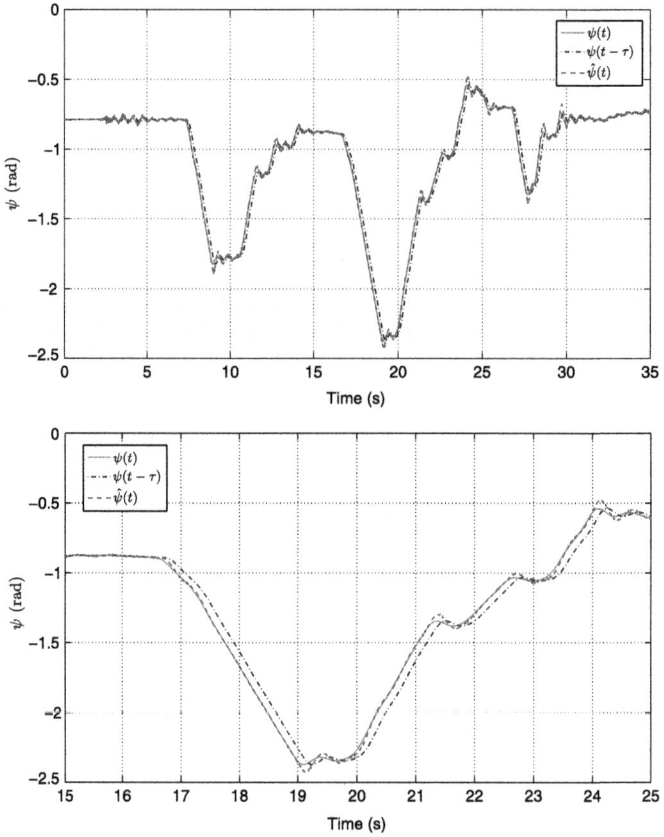

Figure 4.21 ψ response (and a zoom of it) with the predictor control scheme with the delay of $15T_S$.

The first scheme was based on an Observer–Predictor Algorithm that compensates delays in the control inputs. The algorithm includes an *h*-step ahead predictor to estimate the delayed state. Simulation and experimental results have demonstrated good performance of the closed loop even when it had been degraded. The second algorithm used the Fundamental Theorem of Calculus to design a predictor scheme to compensate the delays in the system (sensors, etc.). Stability analysis was proved using the ideas in [21]. Simulation and experimental results have corroborated good performance of the algorithm.

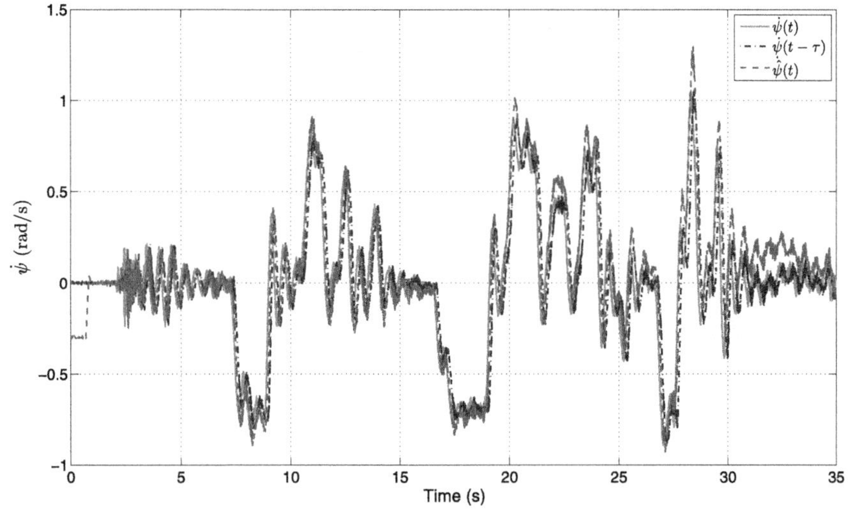

Figure 4.22 $\dot{\psi}$ response with the predictor control scheme with the delay of $15T_S$.

REFERENCES

1. K.U. Lee, H.S. Kim, J.B. Park, Y.H. Choi, Hovering control of a quadrotor, in: International Conference on Control, Automation and Systems – ICCAS, Jeju Island, Korea, 2012, pp. 162–167.
2. T. Madani, A. Benallegue, Backstepping control for a quadrotor helicopter, in: International Conference on Intelligent Robots and Systems, Beijing, China, 2006.
3. D. Mellinger, N. Michael, V. Kumar, Trajectory generation and control for precise aggressive maneuvers with quadrotors, The International Journal of Robotics Research 31 (5) (2012) 664–674.
4. A. Chan, S. Tan, C. Kwek, Sensor data fusion for attitude stabilization in a low cost quadrotor system, in: International Symposium on Consumer Electronics, Singapore, Singapore, 2011.
5. O.J. Smith, Closer control of loops with dead time, Chemical Engineering Progress 53 (1957) 217–219.
6. A. Manitius, A.W. Olbrot, Finite spectrum assignment problem for systems with delays, IEEE Transactions on Automatic Control 24 (1979) 541–552.
7. J. Normey-Rico, E.F. Camacho, Dead-time compensators: a survey, Control Engineering Practice 16 (2008) 407–428.
8. Z. Artstein, Linear systems with delayed controls: a reduction, IEEE Transactions on Automatic Control 27 (1982) 869–879.
9. J. Richard, Time-delay systems: an overview of some recent advances and open problems, Automatica 39 (2003) 1667–1694.
10. R. Lozano, P. Castillo, P. Garcia, A. Dzul, Robust prediction-based control for unstable delay systems: application to the yaw control of a mini-helicopter, Automatica 40 (2004) 603–612.

11. P. Garcia, P. Castillo, R. Lozano, P. Albertos, Robustness with respect to delay uncertainties of a predictor–observer based discrete-time controller, in: Proceedings of the 45th Conference on Decision and Control (CDC), San Diego, CA, USA, IEEE, 2006, pp. 199–204.
12. A. Gonzalez, A. Sala, P. Garcia, P. Albertos, Robustness analysis of discrete predictor-based controllers for input-delay systems, International Journal of Systems Science 44 (2) (2013) 232–239.
13. P. Albertos, P. García, Predictor–observer-based control of systems with multiple input/output delays, Journal of Process Control 22 (2012) 1350–1357.
14. A. Gonzalez, P. Garcia, P. Albertos, P. Castillo, R. Lozano, Robustness of a discrete-time predictor-based controller for time-varying measurement delay, Control Engineering Practice 20 (2012) 102–110.
15. J.E. Normey-Rico, E. Camacho, Control of Dead-Time Processes, Springer, 2007.
16. E. Kaszkurewicz, A. Bhaya, Discrete-time state estimation with two counters and measurement delay, in: Proceedings of the 35th Conference on Decision and Control, vol. 2, Kobe, Japan, IEEE, 1996, pp. 1472–1476.
17. T.D. Larsen, N.A. Andersen, O. Ravn, N.K. Poulsen, Incorporation of time delayed measurements in a discrete-time Kalman filter, in: Proceedings of the 37th Conference on Decision and Control, vol. 4, Tampa, Florida, USA, IEEE, 1998, pp. 3972–3977.
18. Q. Ge, T. Xu, X. Feng, C. Wen, Universal delayed Kalman filter with measurement weighted summation for the linear time invariant system, Chinese Journal of Electronics 20 (2011) 67–72.
19. V. Van Assche, M. Dambrine, J. Lafay, J. Richard, Some problems arising in the implementation of distributed-delay control laws, in: Proceedings of the 38th Conference on Decision and Control, vol. 5, Phoenix, Arizona, USA, IEEE, 1999, pp. 4668–4672.
20. S. Mondié, W. Michiels, Finite spectrum assignment of unstable time-delay systems with a safe implementation, IEEE Transactions on Automatic Control 48 (2003) 2207–2212.
21. M. Krstic, Delay Compensation for Nonlinear Adaptive and PDE Systems, Birkhäuser, 2009.
22. P. Castillo, R. Lozano, A.E. Dzul, Modelling and Control of Mini-Flying Machines, Advances in Industrial Control, Springer-Verlag, London, UK, 2005.

CHAPTER 5

Data Fusion for UAV Localization

In recent years, there has been increasing interest in UAV (Unmanned Aerial Vehicle) civilian applications. UAVs have several potential applications such as search and rescue missions, surveillance, environmental exploration, inspection for maintenance, disaster monitoring, among others. To accomplish these tasks autonomously, UAVs must have precise feedback on the position in order to maintain a desired flight path.

In outdoor environments UAVs can obtain location information from a GPS (Global Positioning System) receiver. However, the GPS signal may be weak, imprecise, or not available in indoor environments and urban environments. Therefore, it is important to propose alternative options to obtain the UAVs position. Most of these alternatives employ a laser ranger and ultrasonic, infrared or visual sensors [1–4].

This chapter addresses an alternative low-cost localization system using classical sensors to perform autonomous trajectory tracking missions. The vehicle is outfitted only with the following sensors: an IMU (Inertial Measurement Unit) to measure orientation, angular velocities and accelerations, an ultrasonic sensor for the altitude, and a camera to obtain the translational velocities in the horizontal plane.

In order to prove the performance of the position estimation algorithm, several flight tests in real time were carried out in open and in closed loop. To close the loop, two control laws widely used in the literature were applied. The aim is to demonstrate that the localization system performs well in closed loop, regardless of the control law input.

5.1 SENSOR DATA FUSION

The proposed estimation algorithm is based on the Extended Falman Filter – EKF. The estimator fuses measurements from the IMU, ultrasonic sensor, and camera. With the camera the translational velocities in the plane are obtained using Optical Flow (OF) computation. The ultrasonic sensor gives us the altitude, and finally, the angular position and rate are provided by the IMU. Then, using a nonlinear model of the aerial vehicle, the EKF estimates the position $(\hat{x}, \hat{y})$ in the horizontal plane, see Fig. 5.1. In the following, the different system components will be described briefly.

Indoor Navigation Strategies for Aerial Autonomous Systems.
DOI: http://dx.doi.org/10.1016/B978-0-12-805189-4.00007-X

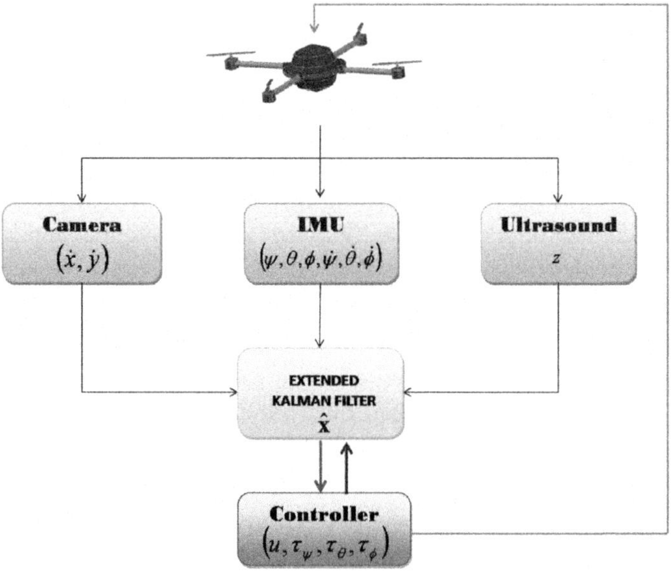

Figure 5.1 Position estimation schematic.

5.1.1 Inertial Measurement Unit

Attitude determination systems normally consist of gyroscopes, accelerometers, and magnetometers (see Fig. 5.2), whereat a device, containing gyroscopes and accelerometers, is commonly called Inertial Measurement Unit (IMU) [5]. In general, an IMU uses three gyroscopes to detect changes in rotational motion like yaw, pitch, and roll and three orthogonally mounted accelerometers to measure acceleration.

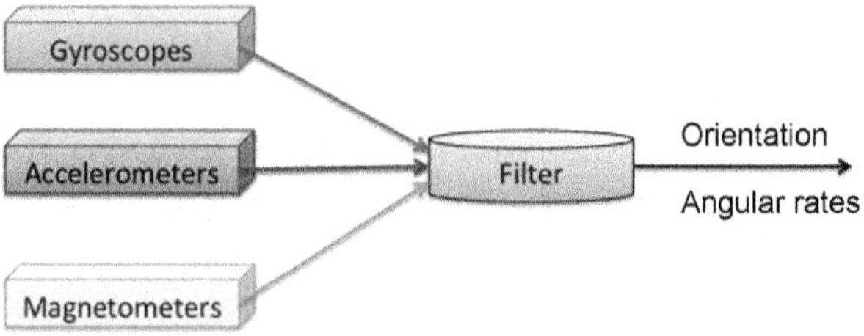

Figure 5.2 Inertial measurement unit scheme.

5.1.2 Ultrasonic Rangefinder

An Ultrasonic Rangefinder is basically a sensor that measures distance with sound waves. The sensor has one transmitter and one receiver. The transmitter could be considered as a speaker that sends out a sound wave while the receiver could be considered as a microphone. By emitting an ultrasonic pulse and timing how long it takes to hear an echo, the Ultrasonic Rangefinder can estimate the distance between the sensor and an object. The ultrasonic sensor has many advantages such as the strong directive property, slow energy consumption, easy processing of measured data, and low cost.

5.1.3 Optical Flow

As stated above, the optical flow was used to calculate the translational velocities in the horizontal plane. It is well known that optical flow is used to calculate the motion between two image frames. It takes advantage of gradients of the temporal information and spatial information to calculate approximate motion vectors; and in general, the optical flow is computed using the gradient and supposing that the intensity of a point in a scene is constant. The techniques to determine the optical flow can be categorized as differential, correlation, energy and frequency based methods.

The localization approach proposed in this chapter uses the Lucas and Kanade algorithm [6]. This algorithm is a differential method that takes advantage of spatiotemporal derivatives of image sequences [7], and it can be summarized as follows.

Let us define $I(x_{im}, y_{im}, t)$ as the grayscale density of the form

$$I(x_{im}, y_{im}, t) = I(x_{im} + \partial x_{im}, y_{im} + \partial y_{im}, t + \partial t),$$

where I denotes the intensity and (x_{im}, y_{im}) represents the position of a point in the image. Hereafter we will consider that $I = I(x_{im}, y_{im}, t)$, $I_x = \frac{\partial I(x_{im}, y_{im}, t)}{\partial x_{im}}$, $I_y = \frac{\partial I(x_{im}, y_{im}, t)}{\partial y_{im}}$, $I_t = \frac{\partial I(x_{im}, y_{im}, t)}{\partial t}$. Then, using the Taylor series,

$$I_x v_{x_{im}} + I_y v_{y_{im}} + I_t = 0$$

with $v_{x_{im}}$ and $v_{y_{im}}$ respectively defining the x_{im} and y_{im} components of the optical flow. Define $\Delta I = [I_x, I_y]^T$ and $V_{OF} = [v_{x_{im}}, v_{y_{im}}]^T$, thus it follows that

$$\Delta I \cdot V_{OF} + I_t = 0. \tag{5.1}$$

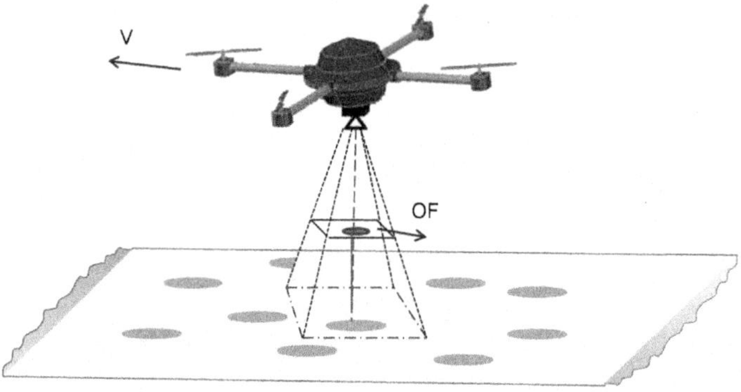

Figure 5.3 Optical flow scheme.

Observe from the previous equation that there are two unknowns and only one equation, therefore another condition is necessary to find a solution. The algorithm of Lucas and Kanade also suggests that the optical flow $(v_{x_{im}}, v_{y_{im}})$ is constant in a neighborhood (a window of $p \times p$, with $p > 1$) centered at the pixel which displacement we want to calculate. Then from previous considerations, p^2 equations with form

$$V = (A^T A)^{-1} A^T b \tag{5.2}$$

can be computed where

$$A = \begin{bmatrix} I_{x_1} & I_{y_1} \\ I_{x_2} & I_{y_2} \\ \vdots & \vdots \\ I_{x_n} & I_{y_n} \end{bmatrix}, \quad b = \begin{bmatrix} -I_{t_1} \\ -I_{t_2} \\ \vdots \\ -I_{t_n} \end{bmatrix},$$

and $n = p^2$. Finally, using (5.1) and (5.2), the optical flow V_{OF} can be computed, see Fig. 5.3.

Motion Parameters

Optical flow allows for the estimation of depth information relative to the speed and direction of the linear velocity, and rotational velocity for each of the axes in 3D [8]. Then optical flow is created by the translational and the rotational movements of a point $P(X_c, Y_c, Z_c)$ in the camera reference frame. Considering that the camera imaging geometry can be modeled as a perspective projection [9], the point $P(X_c, Y_c, Z_c)$ in the scene is projected onto (x_{im}, y_{im}) in the image plane with

$$x_{im} = F\frac{X_c}{Z_c}, \quad y_{im} = F\frac{Y_c}{Z_c},$$

where F is the focal length. The model equation for instantaneous or differential motion reads [8]:

$$\begin{bmatrix} \dot{x}_{im} \\ \dot{y}_{im} \end{bmatrix} = \frac{1}{Z_c}\begin{bmatrix} -F & 0 & x_{im} \\ 0 & -F & y_{im} \end{bmatrix}\begin{bmatrix} V_{c_x} \\ V_{c_y} \\ V_{c_z} \end{bmatrix} + \frac{1}{F}\begin{bmatrix} x_{im}y_{im} & -(F^2 + x_{im}^2) & F y_{im} \\ (F^2 + y_{im}^2) & -x_{im}y_{im} & -F x_{im} \end{bmatrix}\begin{bmatrix} \omega_{c_x} \\ \omega_{c_y} \\ \omega_{c_z} \end{bmatrix}, \quad (5.3)$$

where V_{c_i} is the camera's linear velocity and ω_{c_i} is the rotational velocity. This equation relates the motion parameters of the point to the measurable optical flow vector. Notice from Eq. (5.3) that only translational flow depends on the linear velocity and that rotational components do not depend on the depth of the point.

In our practical implementation, to determinate the linear velocity of the aerial vehicle from the optical flow, we take into account that the camera shares the movements of the UAV, i.e., the camera is fixed on the UAV so that it has no degrees of freedom. Therefore, the camera's linear velocity is equivalent to the UAV's linear velocity. Moreover, we consider that the UAV's angular velocity is small enough so that the rotational components of the OF can be neglected. Nevertheless, several tests of the influence of the rotational component were carried out, and they are described in Sect. 5.5 of this chapter.

5.1.4 Extended Kalman Filter – EKF

As mentioned in Chap. 3, the Kalman Filter algorithm is easy to implement, computationally efficient, and has very good performance. However, when either the system state dynamics or the observation dynamics are nonlinear, the Kalman Filter does not provide accuracy in the estimation because the conditional probability density functions are no longer Gaussian. A non-optimal approach to solve the problem is the Extended Kalman Filter (EKF). In the EKF the nonlinearities of the system dynamics are approximated by a linearized version of the nonlinear system model around the last estimated state. This approximation is valid, if the linearization is a good approximation of the nonlinear model in the entire uncertainty domain associated with the estimated state [10]. The EKF algorithm is basically composed of a prediction step, which involves propagating a state and finding the covariance estimate of the next time step based on the current estimates and system dynamics model; and an update step or filtered cycle, where the new measurement processing takes place and the prediction update is computed using the new information.

Let us consider the following nonlinear dynamics with external inputs:

$$\begin{aligned} \bar{\mathbf{x}}_{k+1} &= \mathbf{f}(\bar{\mathbf{x}}_k, \mathbf{u}_k) + \boldsymbol{\lambda}_{\xi k}, \\ \mathbf{y}_k &= \mathbf{h}(\bar{\mathbf{x}}_k) + \boldsymbol{\lambda}_{vk}, \end{aligned} \tag{5.4}$$

where $\bar{\mathbf{x}}_{k+1}$ represents the states vector, $\mathbf{f}(\bar{\mathbf{x}}_k, \mathbf{u}_k)$ denotes the nonlinear system dynamics, $\mathbf{y}_k$ signifies the output equation, $\mathbf{h}(\bar{\mathbf{x}}_k)$ is the desired vector output. Likewise $\boldsymbol{\lambda}_{\xi k}$ and $\boldsymbol{\lambda}_{vk}$ are assumed to be Gaussian noises with covariance matrices $\mathbf{Q}_k$ and $\mathbf{R}_k$, respectively. Then the equations for the EKF are given by

$$\hat{\bar{\mathbf{x}}}_{k+1|k} = \mathbf{f}(\hat{\bar{\mathbf{x}}}_{k|k}, \mathbf{u}_{k|k}), \tag{5.5a}$$

$$\hat{\mathbf{y}}_{k+1|k} = \mathbf{h}(\bar{\mathbf{x}}_{k+1|k}), \tag{5.5b}$$

$$\mathbf{P}_{k+1|k} = \mathbf{A}\mathbf{P}_{k|k}\mathbf{A}^T + \mathbf{Q}_k, \tag{5.5c}$$

$$\mathbf{K}_{k+1|k} = \mathbf{P}_{k+1|k}\mathbf{H}^T(\mathbf{H}\mathbf{P}_{k+1|k}\mathbf{H}^T + \mathbf{R}_{k+1})^{-1}, \tag{5.5d}$$

$$\hat{\bar{\mathbf{x}}}_{k+1|k+1} = \hat{\bar{\mathbf{x}}}_{k+1|k} + \mathbf{K}_{k+1}(\mathbf{y}_{k+1} - \hat{\mathbf{y}}_{k+1|k}), \tag{5.5e}$$

$$\mathbf{P}_{k+1|k+1} = \mathbf{P}_{k+1|k} - \mathbf{K}_{k+1|k}\mathbf{H}_{k+1}\mathbf{P}_{k+1|k}, \tag{5.5f}$$

where $\mathbf{A} = \left[\frac{\partial \mathbf{f}}{\partial \bar{\mathbf{x}}}(\hat{\bar{\mathbf{x}}}, \mathbf{u})\right]$, $\mathbf{H} = \left[\frac{\partial \mathbf{h}}{\partial \bar{\mathbf{x}}}(\hat{\bar{\mathbf{x}}})\right]$, and $\mathbf{K}_{k+1|k}$ defines the gain filter; the initial state $\hat{\bar{\mathbf{x}}}(0)$ and the initial covariance $\mathbf{P}(0)$ are assumed to be known. Let us remark that, contrary to the Kalman Filter, the EKF may diverge if the consecutive linearizations are not a good approximation of the linear model in the entire associated uncertainty domain. This fact is related to the constant parameters of the system.

On the other hand, when the EKF filter is implemented numerically, the most common problem is choosing the covariance matrices. If these matrices are not well conditioned in some cases, this filter diverges. However, from our experience, system matrices $\mathbf{A}$ and $\mathbf{H}$ need to be adjusted offline or online to improve the predictor's performance.

5.1.5 Quadrotor Simplified Nonlinear Model

From Chap. 2, the quadrotor translational dynamics in the presence of wind is expressed as $m\ddot{\xi} = R(\sum_{i=1}^{4} f_i + f_d) + f_g$ while the rotational dynamics is given by $\mathrm{I}\dot{\Omega} = -\Omega \times \mathrm{I}\Omega + \sum_{i=1}^{4}(\tau_{M_i} + \tau_{r_i}) + \tau_d$.[1] Remember that the angular velocity vector Ω can be related to the Euler velocities $\dot{\eta}$ by the following standard kinematic relationship: $\Omega = W_\eta \dot{\eta}$, see [11]. Thus $\dot{\Omega} = W_\eta \ddot{\eta} + \dot{W}_\eta W_\eta^{-1}\Omega$ and (2.18) yields

$$M(\eta)\ddot{\eta} = -C(\eta, \dot{\eta})\dot{\eta} + \sum_{i=1}^{4}(\tau_{M_i} + \tau_{r_i}) + \tau_d, \tag{5.6}$$

where $M(\eta)$ describes the full inertia matrix, W_η defines a matrix relating the body angular rates with the Euler velocities, and $C(\eta, \dot{\eta})\dot{\eta}$ represents the Coriolis matrix.

When the quadrotor is hovering, it can be assumed that the rotor thrust is proportional to the square of the propellers' rotation speed and that the rotor and body planes are aligned. As a consequence, the flapping angles are so small that $\sin(a_{1_{si}}) \approx a_{1_{si}}$, $\sin(b_{1_{si}}) \approx b_{1_{si}}$, and $\cos(a_{1_{si}})$, $\cos(b_{1_{si}}) \approx 1$.[2] Then the longitudinal and lateral rotor thrust can be neglected. Indeed, around hover (the quadrotors usually operates in an attitude range within $\pm 30°$), the motion of a VTOL (Vertical Take-Off and Landing) vehicle is largely decoupled in each axis, and then the Coriolis matrix becomes very small and can be neglected.

In summary, the dynamical model used for pose estimation takes into account the following assumptions: quasi-stationary maneuvers and small flapping angles. Then, by developing rotations in z, x, and y, i.e., ${}^{\mathcal{B}}R_{\mathcal{E}}(\psi, \theta, \phi)$, it follows that the simplified nonlinear model of a quadrotor is

$$m\ddot{x} = -(u + w_x)\sin(\theta), \tag{5.7a}$$

$$m\ddot{y} = (u + w_y)\cos(\theta)\sin(\phi), \tag{5.7b}$$

$$m\ddot{z} = (u + w_z)\cos(\theta)\cos(\phi) - mg, \tag{5.7c}$$

$$I_x\ddot{\theta} = \tau_\theta + w_\theta, \tag{5.7d}$$

$$I_y\ddot{\phi} = \tau_\phi + w_\phi, \tag{5.7e}$$

$$I_z\ddot{\psi} = \tau_\psi + w_\psi, \tag{5.7f}$$

[1] m defines the mass, ξ the position, R the rotation matrix; f_i, f_d, f_g denote the rotor, drag, and gravitational forces, respectively; I is the inertia, τ_{M_i} and τ_{r_i} stand for the main torques, and τ_d is the drag torque.

[2] For more details see Chap. 2.

where x, y, and z denote the position of the vehicle, ψ, θ, and ϕ are the yaw, pitch, and roll angles, respectively, I_j represents the inertia matrix in the j-axis, and g is the gravitational acceleration. Indeed, u defines the main thrust, τ_ψ, τ_θ, and τ_ϕ are the yaw, pitch, and roll moments, respectively. Similarly, w_k, where k takes values x, y, z, θ, ϕ or ψ, denotes the unknown disturbances due to the wind.

5.2 PROTOTYPE AND NUMERICAL IMPLEMENTATION

Our experimental platform consists of a quadrotor vehicle, an embedded navigation system, and a ground station shown in Fig. 5.4. The vehicle is based on a Mikrokopter structure and has four brushless motors controlled by i2c BlCtrl drivers. The electronic board is equipped with an IGEPv2 board based on the Texas Instruments system on Chip OMAP3530 with an ARM Cortex A8 processor running at 750 MHz and a Digital Signal Processor (DSP) C64x+ running at 520 MHz. The sensors employed are: an ultrasonic rangefinder SRF10, IMU 3DMGX3-25, and finally, PS3eye camera. The total mass of the prototype is 1.1 kg. A LiPO 11.1 V and 6000 mAh battery is employed to energize all the electronic systems. The

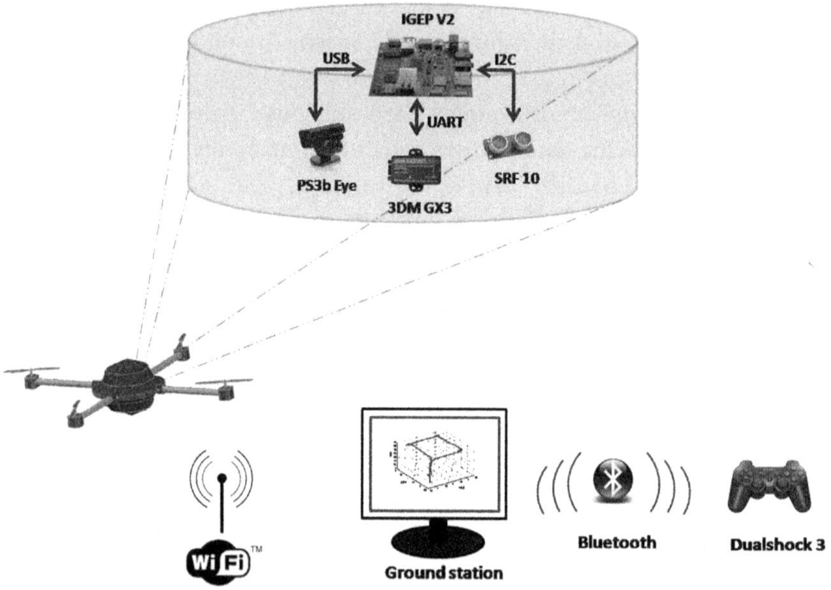

Figure 5.4 Quadrotor experimental platform.

electronic board, sensors and control algorithms compose the embedded navigation system.

All the information collected by the IGEPv2 is sent to a ground station using a Wi-Fi connection. Its objective is to graph the states to supervise the system responses, to tune the control parameters, and to redefine tasks or missions, all in real time. The base station is written with the QT library, making it a multi-platform. For flight manual phases a Playstation 3 joystick is used. The ground station and the joystick are connected using Bluetooth communication.

Rewriting (5.7) with $w_k = 0$ in the standard form $\dot{\bar{\mathbf{x}}} = \mathbf{f}(\bar{\mathbf{x}}, \mathbf{u})$ yields

$$\dot{\bar{\mathbf{x}}} = \begin{bmatrix} \dot{x} & -\frac{u}{m}\sin(\theta) & \dot{y} & \frac{u}{m}\cos(\theta)\sin(\phi) & \dot{z} & \frac{u}{m}\cos(\theta)\cos(\phi) - g & \dot{\theta} & \frac{\tau_\theta}{I_x} & \dot{\phi} & \frac{\tau_\phi}{I_y} & \frac{\tau_\psi}{I_z} \end{bmatrix}^T, \tag{5.8}$$

where $\bar{\mathbf{x}} = [x \quad \dot{x} \quad y \quad \dot{y} \quad z \quad \dot{z} \quad \theta \quad \dot{\theta} \quad \phi \quad \dot{\phi} \quad \psi \quad \dot{\psi}]$. Notice that since the EKF is applied in discrete time, thus considering a sampling period T_s small enough, the Euler approximation can be used to write system (5.8) in discrete variables [12]. Hence, without loss of generality, (5.8) becomes

$$\bar{\mathbf{x}}_{k+1} = \begin{bmatrix} \dot{x}_k T_s + x_k \\ -(\frac{u_k}{m}\sin(\theta_k))T_s + \dot{x}_k \\ \dot{y}_k T_s + y_k \\ (\frac{u_k}{m}\cos(\theta_k)\sin(\phi_k))T_s + \dot{y}_k \\ \dot{z}_k T_s + z_k \\ (\frac{u_k}{m}\cos(\theta_k)\cos(\phi_k) - g)T_s + \dot{z}_k \\ \dot{\theta}_k T_s + \theta_k \\ \frac{\tau_{\theta_k}}{I_x} T_s + \dot{\theta}_k \\ \dot{\phi}_k T_s + \phi_k \\ \frac{\tau_{\phi_k}}{I_y} T_s + \dot{\phi}_k \\ \dot{\psi}_k T_s + \psi_k \\ \frac{\tau_{\psi_k}}{I_z} T_s + \dot{\psi}_k \end{bmatrix} \tag{5.9}$$

with $Ts = t_{k+1} - t_k$ and t_k being the time t at instant k.

On the other hand, the state transition and measurement noises have been generated from Gaussian distributions with the following covariance matrices:

$$\mathbb{E}[\boldsymbol{\lambda}_{\xi_k}\boldsymbol{\lambda}_{\xi_k}^T] = \mathbf{Q}_k, \quad \mathbb{E}[\boldsymbol{\lambda}_{v_k}\boldsymbol{\lambda}_{v_k}^T] = \boldsymbol{R}_k,$$

where $\boldsymbol{\lambda}_{\xi k}$ represents the estimated random process noise and $\boldsymbol{\lambda}_{vk}$ defines the sensor resolution, see Eq. (5.4).

Our numerical implementation had three steps:

1. After a small identification of the quadrotor's altitude constant parameters, the implementation of the EKF algorithm in the helicopter's board was carried out. The goal in this part was to validate the algorithm onboard, online, and in an open-loop system. In this way, several flight tests were carried out manually to validate the algorithm. The behavior of the quadrotor was depicted in the ground station to analyze online; in addition, some small adjustments in the covariance matrices $\mathbf{Q}_k$ and $\boldsymbol{R}_k$ were realized to improve the convergence of the EKF.
2. In the second part, the EKF was validated in closed loop, online, and onboard, making sure that the flying vehicle autonomously follows desired trajectories.
3. Finally, some flights tests were realized using the OptiTrack system in order to corroborate good accuracy of the proposed algorithm.

The results of the numerical implementation are shown in the next section.

5.3 FLIGHT TESTS AND EXPERIMENTAL RESULTS

As mentioned above, before realizing autonomous control of the quadrotor vehicle, a number of flight tests in manual form (open loop) were conducted to validate the EKF implementation and heuristically verify the accuracy of the states' estimates.

5.3.1 Manual Flight Tests

In the most representative experiment, the pilot tried to realize a rectangle trajectory with $\ell_1 = 5$ m in the $\hat{x}$ axis and $\ell_2 = 6$ m in the $\hat{y}$ axis. The result is presented in Fig. 5.5 where small oscillations in the trajectory can be observed due the flight skills of the pilot. Nevertheless, a good estimate of $(\hat{x}, \hat{y})$ is also observed.

5.3.2 Autonomous Mode

In the case of autonomous mode, two common control techniques were used to close the loop, the PID controller and the saturation function approach. The PID control inputs used to stabilize the quadrotor are given by

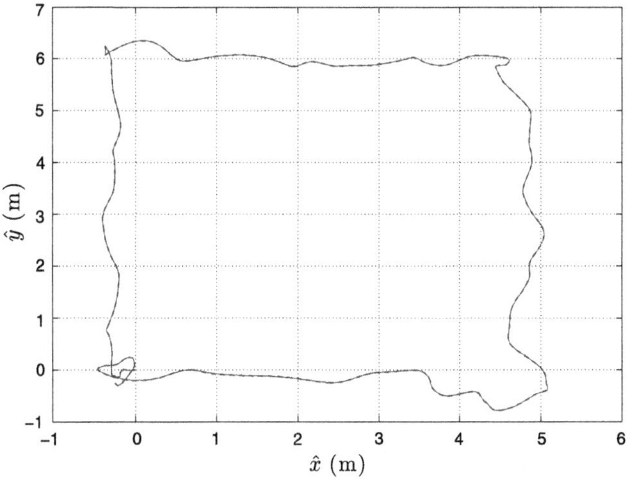

Figure 5.5 $\hat{x}$ and $\hat{y}$ position estimation in manual mode for a rectangle trajectory.

$$u = -K_{3_z}\dot{z} - K_{2_z}(z - z_d) - \int K_{1_z}(z - z_d)d\tau + g, \tag{5.10a}$$

$$\tau_\psi = -K_{3_\psi}\dot{\psi} - K_{2_\psi}(\psi - \psi_d) - \int K_{1_\psi}(\psi - \psi_d)d\tau, \tag{5.10b}$$

$$\tau_\theta = -K_{6_\theta}\dot{\theta} - K_{5\theta}(\theta - \theta_d) - \int K_{4_\theta}(\theta - \theta_d)d\tau + K_{3_\theta}\dot{x} + K_{2_\theta}(x - x_d) + \int K_{1_\theta}(x - x_d)d\tau, \tag{5.10c}$$

$$\tau_\phi = -K_{6_\phi}\dot{\phi} - K_{5_\phi}(\phi - \phi_d) - \int K_{4_\phi}(\phi - \phi_d)d\tau - K_{3_\phi}\dot{y} - K_{2_\phi}(y - y_d) - \int K_{1_\phi}(y - y_d)d\tau, \tag{5.10d}$$

where K_{i_j} is constant, z_d, ψ_d, θ_d, ϕ_d, x_d, and y_d are the desired values, and g is the gravity constant. For more details about the PID controller see Chap. 6.

The performance of the PID control law can be observed in Figs. 5.6–5.7. The control gains were first tuned to realize hover. The results were correct; however, when the mission for the quadrotor was to follow a line along the y-axis, the vehicle's behavior was different, and it was necessary to tune again the control gains to improve the performance. The mission was to take-off with coordinates $(x(0), y(0), z(0)) = (0, 0, 0.5)$ m, then move to $(x_d, y_d, z_d) = (0, 3, 0.5)$ m, and later return to the initial position and land. Observe in Fig. 5.6 (right) that only the roll angle changes

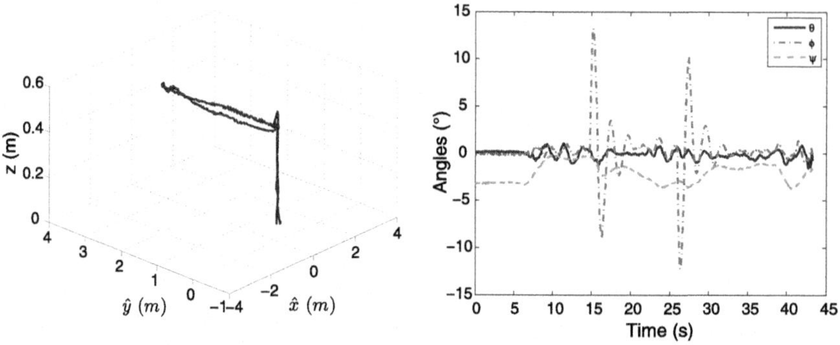

Figure 5.6 (Left) $\hat{x}$, $\hat{y}$, and z performance in autonomous flight following a line trajectory. (Right) θ, ϕ, and ψ responses.

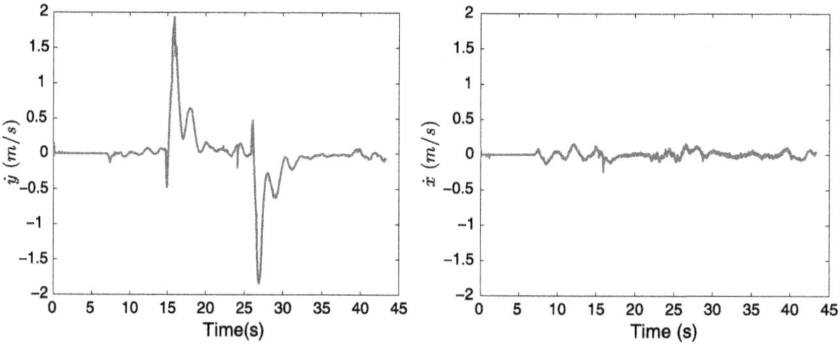

Figure 5.7 $\dot{x}$ and $\dot{y}$ performance in autonomous flight following a line trajectory.

to displace the aerial vehicle in the y coordinate, while the pitch angle is around 0°. This behavior can be also verified by Fig. 5.7, where the translational velocity $\dot{x}$ is almost zero and the velocity $\dot{y}$ is increased or decreased depending on the roll angle.

The second control approach adopted is based on the saturation functions. This type of control law is simple and computationally-efficient to implement. Let us consider a system of the form

$$\begin{aligned} \dot{r}_1 &= r_2, \\ &\vdots \\ \dot{r}_n &= u_r, \end{aligned} \tag{5.11}$$

and the saturation function $|\sigma_{b_i}(s)| \leqslant b_i$, with $b_i > 0$ being a constant. Then the control law

$$u_r = -\sum_{i=1}^{n} \sigma_{b_i}(k_i r_i) \tag{5.12}$$

stabilizes the system (5.11) for all constant $k_i > 0$ and bounds every state r_i, using the nonlinearity σ_{b_i}. It has been proved in [13] and [14] that the previous control law can be used to stabilize nonlinear systems given by Eqs. (5.7). Thus, Eq. (5.12) applied to the quadrotor dynamics takes the form

$$\bar{u} = -\sigma_{b_{2_z}}\left(K_{2_z}\dot{z}\right) - \sigma_{b_{1_z}}\left(K_{1_z}(z - z_d)\right), \tag{5.13a}$$

$$\tau_\psi = -\sigma_{b_{2_\psi}}(K_{2_\psi}\dot{\psi}) - \sigma_{b_{1_\psi}}\left(K_{1_\psi}(\psi - \psi_d)\right), \tag{5.13b}$$

$$\begin{aligned}\tau_\theta = & -\sigma_{b_{4_\theta}}(K_{4_\theta}\dot{\theta}) - \sigma_{b_{3_\theta}}(K_{3_\theta}(\theta - \theta_d)) \\ & + \sigma_{b_{2_\theta}}(K_{2_\theta}\dot{x}) + \sigma_{b_{1_\theta}}(K_{1_\theta}(x - x_d)),\end{aligned} \tag{5.13c}$$

$$\begin{aligned}\tau_\phi = & -\sigma_{b_{4_\phi}}(K_{4_\phi}\dot{\phi}) - \sigma_{b_{3_\phi}}(K_{3_\phi}(\phi - \phi_d)) \\ & - \sigma_{b_{2_\phi}}(K_{2_\phi}\dot{y}) - \sigma_{b_{1_\phi}}(K_{1_\phi}(y - y_d)),\end{aligned} \tag{5.13d}$$

where $\bar{u} = u - g$. The development of the control law and the stability analysis are presented in detail in [13].

In Fig. 5.8 we introduce the vehicle behavior when the quadrotor prototype autonomously and in real time follows a square with 2 m sides using

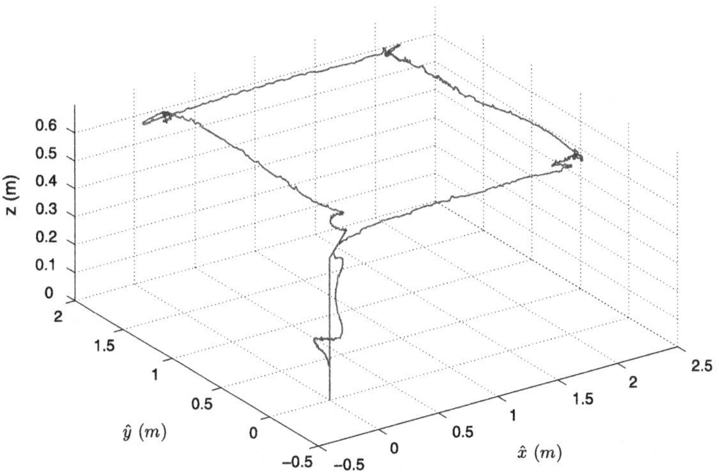

Figure 5.8 $\hat{x}$, $\hat{y}$, and z responses when the flying vehicle realizes a square trajectory autonomously.

the previous control law. The trajectory coordinates for this mission are $(x_0, y_0, z_0) = (0, 0, 0.6)$, $(x_1, y_1, z_1) = (2, 0, 0.6)$, $(x_2, y_2, z_2) = (2, 2, 0.6)$, and $(x_3, y_3, z_3) = (0, 2, 0.6)$, all in meters. Observe that even if the flying vehicle is not equipped with a position sensor, the EKF is capable of estimating the $\hat{x}$ and $\hat{y}$ position with an excellent precision using essentially the translational velocities in this plane and the attitude of the vehicle.

In order to prove the behavior in more complex trajectories and validate the absence of accumulative errors in the estimation, some user-free paths were realized. To overwhelm the mission, four pillar-like obstacles were placed as Fig. 5.9 shows. Remember that the vehicle is not equipped with other sensors to avoid obstacles. Then bad estimation and worse tuning of the control parameters will imply a crash of the quadrotor into the columns. The objective of the mission was for the helicopter to follow a trajectory given by a set of desired coordinates. In this mission the aerial vehicle was to cross the four pillars, round one of them, go to the center of the pillars and land, all in autonomous mode. The desired coordinates were: $(x_0, y_0, z_0) = (0, 0, 0)$, $(x_1, y_1, z_1) = (0, 0, 0.6)$, $(x_2, y_2, z_2) = (2, 0, 0.6)$, $(x_3, y_3, z_3) = (2, 4, 0.6)$, $(x_4, y_4, z_4) = (4, 4, 0.6)$, $(x_5, y_5, z_5) = (4, 2, 0.6)$, $(x_6, y_6, z_6) = (2, 2, 0)$, and $(x_f, y_f, z_f) = (2, 2, 0)$, all in meters.

A 3D view of the $\hat{x}$, $\hat{y}$, and z response is introduced in Fig. 5.10. Notice in this figure that the quadrotor realizes the desired mission very well. The

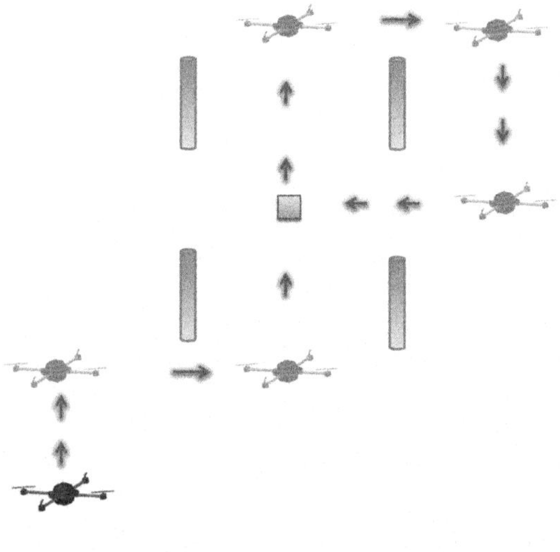

Figure 5.9 The desired mission with obstacles.

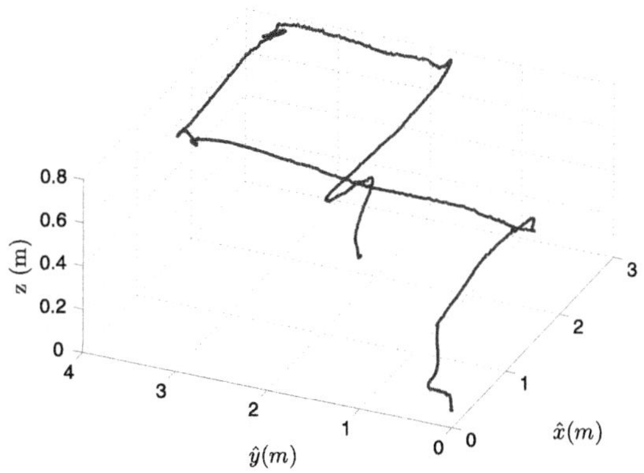

Figure 5.10 $\hat{x}, \hat{y}$, and z responses when the vehicle realizes the trajectory autonomously.

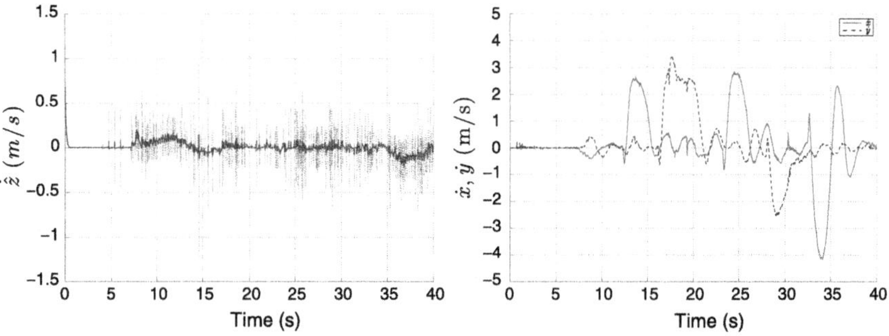

Figure 5.11 Translational velocity responses: (Left) $\dot{\hat{z}}$ response when using the EKF (solid line) and a classical Euler derivation (dashed line); (Right) $\dot{x}$ and $\dot{y}$ responses obtained from the optical flow algorithm.

$\dot{\hat{z}}$ estimation is shown in Fig. 5.11. For comparative analysis, two performances of this estimated state using different approaches are shown in this figure where the dashed line represents the behavior when using the classical Euler derivation while the solid line illustrates the performance when the EKF is used and given a better behavior. In a closed-loop system the EKF estimation was the only employed.

Observe from Fig. 5.9 that the aerial vehicle first moves in the positive longitudinal axis, then in the positive lateral axis, after that in the positive longitudinal axis, later in the negative lateral axis, and finally in the negative longitudinal axis. These displacements imply movements in the pitch and roll angles and in the angular rate of the vehicle, see Fig. 5.12.

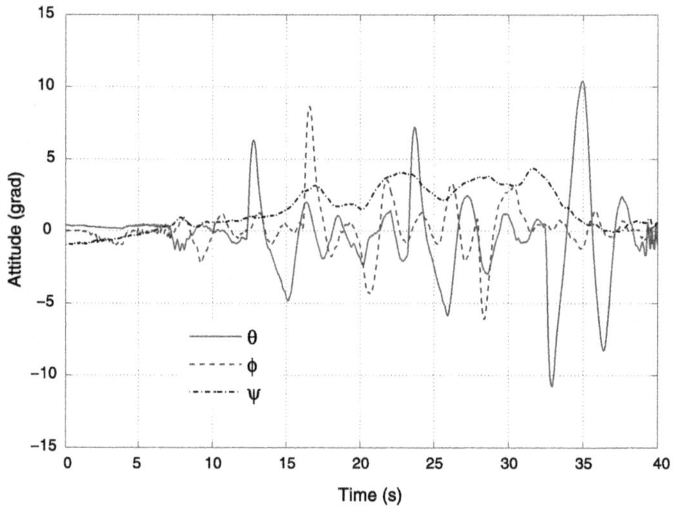

Figure 5.12 Attitude behavior.

5.4 OPTITRACK MEASUREMENTS VS EKF ESTIMATION

Finally, the proposed EKF data fusion estimation was compared with the measurements coming from the OptiTrack system. It consists of an array of infrared cameras, two synchronized camera hubs, and the OptiTrack Tracking Tools processing software. The OptiTrack system provides, with good precision, the 6 DOF position and orientation of the quadrotor in real time. The mission for this comparative flight test was: (i) take-off and hover at $x_1 = y_1 = 0$ and $z_1 = 0.8$ m, (ii) move to $(x_2 = 0,\ y_2 = 2,\ z_2 = 0.8)$ m, (iii) fly back to (x_1, y_1, z_1) m, and finally, (iv) land, see Figs. 5.13–5.14. Notice in this figure that both responses (EKF and OptiTrack) are very similar. These measurements indicate good accuracy of the EKF algorithm when estimating the missing states.

Figs. 5.15 and 5.16 introduce our main result, the estimation of the horizontal position. Observe in these figures that the $\hat{x}$ and $\hat{y}$ estimates (solid lines) are very close to the measurements x and y (dashed lines) coming from the OptiTrack system. This fact helps us better illustrate the good precision when estimating these missing states using the data fusion with the EKF. The small errors that we can see in Fig. 5.16 could come from the optical flow. When the vehicle moves (in this case in the y axis), it changes its roll angle, and this variation will result in a velocity estimate of the optical flow in the opposite direction. Despite these errors, the estimation

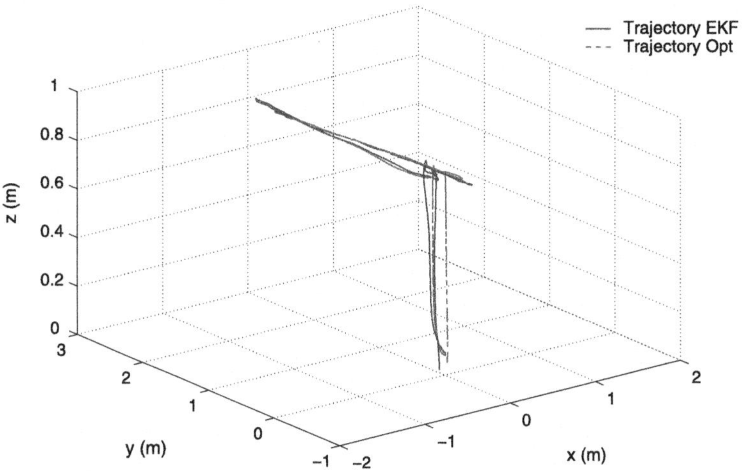

Figure 5.13 System response when the aerial vehicle follows desired coordinates. The solid line represents the estimated states using the EKF, while the dashed line depicts the measured states using the OptiTrack system.

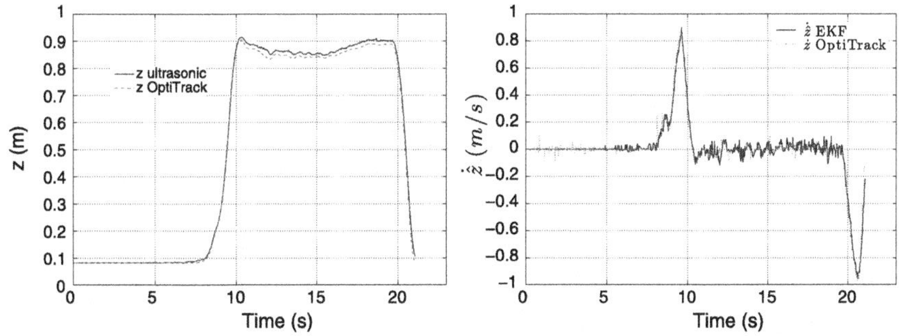

Figure 5.14 z and $\dot{z}$ response. The solid lines represent the estimated states using the EKF (rigth) and the ultrasonic sensor (left), while the dashed lines show the measured states using the OptiTrack system.

is fairly good, but better results could be obtained if the rotational effects were compensated in the optical flow algorithm.

Also in Figs. 5.15 and 5.16 the translational velocities of the aerial vehicle are presented. The dashed red lines introduce the $\dot{x}$ and $\dot{y}$ responses measured by the OptiTrack system. The solid lines represent the estimates of these states when using the optical flow algorithm. Remember that the mission was to follow a straight line along the y axis, and thus there is no movement in the x axis.

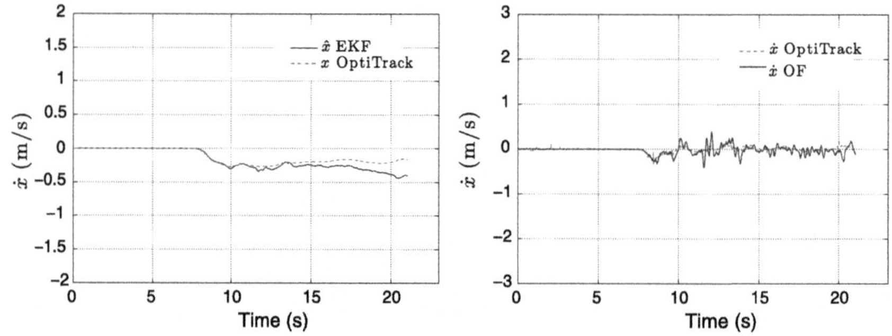

Figure 5.15 x and $\dot{x}$ responses coming from the EKF and the OptiTrack system.

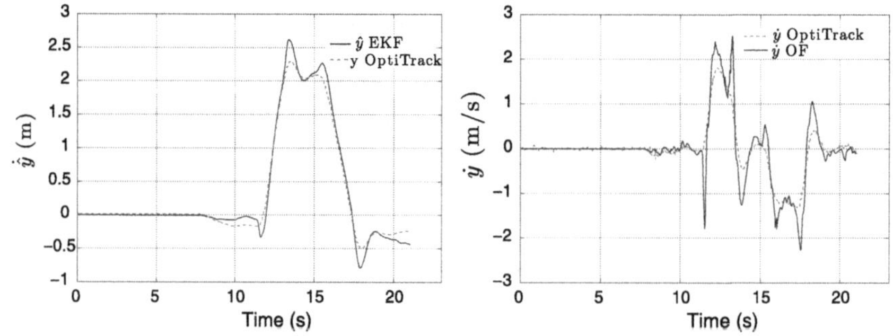

Figure 5.16 y and $\dot{y}$ responses coming from the EKF and the OptiTrack system.

5.5 ROTATIONAL OPTICAL FLOW COMPENSATION

To compensate the rotational component of (5.3), the algorithm developed in [15] was implemented numerically. Based on the KF, the algorithm fuses the angular measurements of an IMU and the optical flow. Consider the following KF equations:

$$\begin{aligned} \hat{\boldsymbol{x}}_{k+1} &= \boldsymbol{A}_k \hat{\boldsymbol{x}}_k + \boldsymbol{\lambda}_{\xi k}, \qquad & \mathbb{E}[\boldsymbol{\lambda}_{\xi k} \boldsymbol{\lambda}_{\xi k}^T] = \boldsymbol{Q}_k, \\ \boldsymbol{y}_k &= \boldsymbol{H}_k \hat{\boldsymbol{x}}_k + \boldsymbol{\lambda}_{vk}, \qquad & \mathbb{E}[\boldsymbol{\lambda}_{vk} \boldsymbol{\lambda}_{vk}^T] = \boldsymbol{R}_k, \end{aligned} \tag{5.14}$$

with

$$\boldsymbol{A}_k = \begin{bmatrix} I & 0 \\ 0 & I \end{bmatrix}, \quad \hat{\boldsymbol{x}}_k = \begin{bmatrix} \hat{\boldsymbol{V}}_{OF} \\ \hat{\boldsymbol{W}} \end{bmatrix}, \quad \boldsymbol{H}_k = \begin{bmatrix} I & \boldsymbol{K}_R^T \\ 0 & I \end{bmatrix}, \quad \boldsymbol{y}_k = \begin{bmatrix} \boldsymbol{OF} \\ \boldsymbol{\Omega}_{imu} \end{bmatrix},$$

where $\hat{\boldsymbol{V}}_{OF}$ is the vector of translational components of (5.3), $\hat{\boldsymbol{W}}$ is the vector of rotational components, $\boldsymbol{K}_R$ is the constant scale factors depending of the intrinsic parameters of the camera [15], $\boldsymbol{OF}$ is the optical flow, and

$\boldsymbol{\Omega}_{imu}$ are the angular velocities measured by the IMU. Using Eqs. (3.13) and (3.14), the following KF equations can be derived:

$$\begin{aligned}
\hat{\boldsymbol{x}}_k^- &= \begin{bmatrix} I & 0 \\ 0 & I \end{bmatrix} \begin{bmatrix} \hat{\boldsymbol{V}}_{OF} \\ \hat{\boldsymbol{W}} \end{bmatrix}_{k-1}, \\
\hat{\boldsymbol{x}}_k^+ &= \begin{bmatrix} \hat{\boldsymbol{V}}_{OF} \\ \hat{\boldsymbol{W}} \end{bmatrix}^- + \boldsymbol{K}_k \begin{bmatrix} \boldsymbol{OF} - \hat{\boldsymbol{V}}_{OF} - \boldsymbol{K}_R^T \hat{\boldsymbol{W}} \\ \boldsymbol{\Omega}_{imu} - \hat{\boldsymbol{W}} \end{bmatrix}, \\
\boldsymbol{P}_k^- &= \begin{bmatrix} I & 0 \\ 0 & I \end{bmatrix}_{k-1} + \boldsymbol{P}_{k-1} + \begin{bmatrix} I & 0 \\ 0 & I \end{bmatrix}_{k-1}^T + \boldsymbol{Q}_k, \\
\boldsymbol{S}_k &= \boldsymbol{H}_k \boldsymbol{P}_k^- \boldsymbol{H}_k^T + \boldsymbol{R}_k, \\
\boldsymbol{K}_k &= \boldsymbol{P}_k^- \boldsymbol{H}_k^T \boldsymbol{S}_k^{-1}, \\
\boldsymbol{P}_k^+ &= (\boldsymbol{I} - \boldsymbol{K}_k \boldsymbol{H}_k) \boldsymbol{P}_k^-.
\end{aligned}$$

The KF was implemented in a test-bed, based on the Computer-on-Module Gumstix Overo Fire, an IMU 3DMGX3-25 and a downwards-pointing Chameleon Mono Point Grey camera. The results are depicted in Figs. 5.17–5.18. Fig. 5.17 shows the OF computation $\dot{x}_{im}$ and $\dot{y}_{im}$ and the translational component of the OF calculated using the KF approach. When the test-bed quickly changes its orientation, as it can observed in Fig. 5.18, the rotational component has an important contribution to the total OF computation. Therefore, if there is no rotational compensation, the translational velocity of the UAV is erroneously calculated.

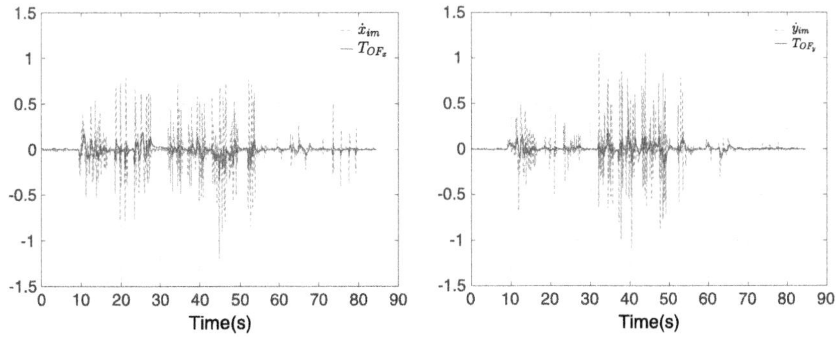

Figure 5.17 Optical Flow computation of ($\dot{x}_{im}$, $\dot{y}_{im}$) and their translational components T_{OF_x} and T_{OF_y}.

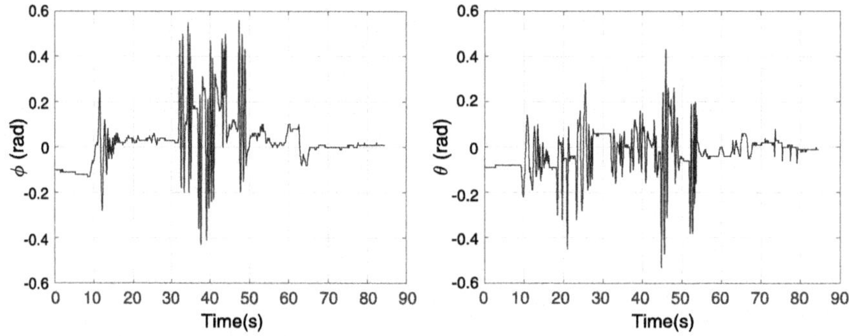

Figure 5.18 ϕ and θ angle responses.

5.6 DISCUSSION

An alternative localization system for GPS-denied environments was developed in this chapter. To estimate the horizontal position (x, y) measurements from an IMU, an ultrasonic rangefinder and a camera were fused using the EKF. The camera was used to compute the optical flow and estimate the horizontal velocities of the vehicle. The estimation algorithm was validated in open-loop, closed-loop, and in real-time flights, and the results were compared to the OptiTrack system. The results have shown the effectiveness and efficiency of the proposed scheme.

Notice that the user-free path mission can be realized only if the position of the vehicle is well known or estimated (our case) with good precision. If it is not the case, the vehicle could crash into the pillars. In order to carry out such missions, some teams use indoor measurement systems to locate the vehicle, but the main drawback of these solutions is that they are limited to their workspaces (VICON, OptiTrack, etc.).

REFERENCES

1. Y. Song, B. Xian, Y. Zhang, X. Jiang, X. Zhang, Towards autonomous control of quadrotor unmanned aerial vehicles in a GPS-denied urban area via laser ranger finder, Optik, International Journal for Light and Electron Optics 126 (2015) 3877–38882.
2. F. Wang, J.-Q. Cui, B.-M. Chen, T.H. Lee, A comprehensive UAV indoor navigation system based on vision optical flow and laser FastSLAM, Acta Automatica Sinica 39 (2013) 1889–1899.
3. Y.M. Mustafah, A.W. Azman, F. Akbar, Indoor UAV positioning using stereo vision sensor, in: International Symposium on Robotics and Intelligent Sensors 2012 (IRIS 2012), Procedia Engineering 41 (2012) 575–579.

4. C. Troiani, A. Martinelli, C. Laugier, D. Scaramuzza, Low computational-complexity algorithms for vision-aided inertial navigation of micro aerial vehicles, Robotics and Autonomous Systems 69 (2015) 80–97, selected papers from 6th European Conference on Mobile Robots.
5. S. Fux, Development of a planar low cost Inertial Measurement Unit for UAVs and MAVs, Master's thesis, Swiss Federal Institute of Technology, Zurich, 2008.
6. B. Lucas, T. Kanade, An iterative image registration technique with an application to stereo vision, in: Proceedings of International Joint Conference on Artificial Intelligence, vol. 2, Vancouver, British Columbia, 1981, pp. 674–679.
7. A. Eresen, N. İmamoğlu, M.O. Efe, Autonomous quadrotor flight with vision-based obstacle avoidance in virtual environment, Expert Systems with Applications 39 (2012) 894–905.
8. F. Raudies, Optic flow, Scholarpedia 8 (7) (2013) 30724.
9. K.-I. Kanatani, Transformation of optical flow by camera rotation, IEEE Transactions on Pattern Analysis and Machine Intelligence 10 (1988) 131–143.
10. T. Kailath, Lectures on Wiener and Kalman Filtering, Springer Verlag, 1981.
11. H. Goldstein, Classical Mechanics, Addison Wesley Series in Physics, Addison–Wesley, USA, 1980.
12. M. Gerdts, Optimal Control of ODEs and DAEs, Walter de Gruyter, 2012.
13. G. Sanahuja, P. Castillo, A. Sanchez, Stabilization of n integrators in cascade with bounded input with experimental application to a VTOL laboratory system, International Journal of Robust and Nonlinear Control 20 (2010) 1129–1139.
14. A. Sanchez, P. Garcia, P. Castillo Garcia, R. Lozano, Simple real-time stabilization of vertical take-off and landing aircraft with bounded signals, Journal of Guidance, Control, and Dynamics 31 (2008) 1166–1176.
15. E. Rondon, Navigation d'un vehicule aerien par flux optique, PhD thesis, Université de Tehcnologié de Compiègne, France, 2010.

PART 3

Navigation Schemes & Control Strategies

The increasing complexity of missions that UAVs must perform has resulted in the development of more complex control techniques. Over the years linear and nonlinear Proportional-Derivative (PD) controllers have been widely used to control the attitude and position of an UAV. Nevertheless, when the system is constantly disturbed or there are uncertainties in the model, controllers are not capable to achieve the desired position or their performance is degraded considerably. In this part of the book we focus on the development of different control techniques to improve the performance of the closed-loop system and the navigation mission.

The first three chapters cover the development of the control schemes. Five approaches are addressed in these chapters: two are based on the saturation function (Chaps. 6 and 8), one is based on the sliding modes (Chap. 7), one uses backstepping (Chap. 6), and the last includes a feedback in combination with the Uncertainty Disturbance Estimator (Chap. 8).

Finally, the last three chapters focus on how to improve a mission. For example, on the one hand, some inspection and surveillance tasks require that the aerial vehicle passes through specific points with constraints on the velocity or time. Thus the trajectory generation and path tracking problem needs to be solved here. In Chap. 9 this problem is addressed and a solution is proposed. On the other hand, autonomous navigation in an unstructured environment also needs algorithms for obstacle avoidance. In Chap. 10 a control scheme is proposed to avoid fixed obstacles using the Artificial Potential Field (APF) and the limit cycle approaches. Notions for beginners in this field are introduced here. Finally, in Chap. 11 a tool to improve manual flight when the vehicle is out of reach is discussed. A haptic teleoperation algorithm is proposed for collision-free navigation. In this scheme a solution is also presented for pose estimation in indoor environments using visual information.

CHAPTER 6

Nonlinear Control Algorithms with Integral Action*

Working with UAVs (Unmanned Aerial Vehicles) is not an easy task. Sometimes when researchers or students propose controllers, the most popular suggestion is to use linear controllers like feedback or PD (Proportional-Derivative) controllers. In some cases, to correctly apply these controllers, it is necessary to modify the nonlinear mathematical equations using some assumptions, to develop a simplified nonlinear or even, in the best case from a control design point of view, a linear model. The latter for some researches is inconceivable; nevertheless, it gives an easier way to understand the system redline in several cases. Linearizing the quadcopter model is not a problem because, when the vehicle is hovering using small angle movements to perform a trajectory, its dynamics are strongly simplified and resemble a linear system. In general, two integrators or four integrators in cascade are the most used linear models.

Even if the linear mathematical equations are used, this does not signify that the controller has to be linear. There exist several control algorithms conceived from linear models but directly applied to nonlinear systems. And when such a controller is validated in real time, it captures the attention of the UAV community.

In addition, it is common in practical applications, such as the stabilization at a given position, that the closed-loop system performance is degraded and the states do not converge to the desired values when model uncertainties or unknown constant perturbations are present. The main motivation for this chapter is to improve the behavior of the popular control schemes by adding an integral part.

* The results in this chapter were developed in collaboration with A.G. Alatorre and S. Mondié from CINVESTAV-IPN, Mexico, and with J. Colmenares and N. Marchand from GIPSA Lab, France.

Indoor Navigation Strategies for Aerial Autonomous Systems.
DOI: http://dx.doi.org/10.1016/B978-0-12-805189-4.00009-3

6.1 FROM PD TO PID CONTROLLERS

The simplest mathematical representation of the attitude of a VTOL vehicle (quadcopter included) is given by two integrators in cascade. From (5.7) we can deduce that[1]

$$\ddot{\eta} = u_\eta + w_\eta,$$

where η defines an Euler angle (yaw, pitch, or roll), u_η represents the control input, and w_η stands for an unknown perturbation.

Stabilizing the previous system is an easy task: in general, the disturbance is neglected and then a PD controller is enough. Then we propose taking

$$u_\eta = -k_p e(t) - k_d \frac{de(t)}{dt} \tag{6.1}$$

with k_p defining the proportional gain, k_d the derivative gain, and $e(t) = \eta - \eta_d$ being the error between the state η and its desired value η_d. Notice in Fig. 6.1 that with $w_n = 0$ the control law stabilizes the system (solid line). Stability analysis of the closed-loop system is obvious and not necessary to develop here. For simulation purposes the initial conditions were $\eta(0) = 10$ and $\eta_d = 0$. Nevertheless, if $w_\eta \neq 0$, a steady state error could be observed, see Fig. 6.1-dotted line (for simulation, $w_\eta = 4$). The previous problem can be solved by adding an integral term to the controller, see Fig. 6.3. This control algorithm is known as a PID controller (proportional-integral-derivative).

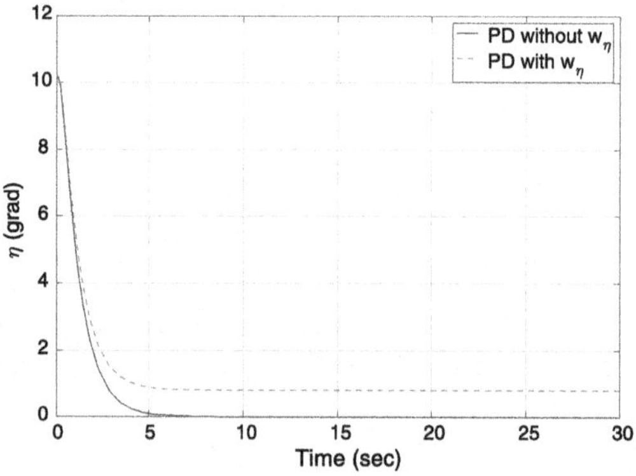

Figure 6.1 η response when applying a PD controller with and without perturbation w_η in the system.

[1] With $I_\eta = 1$.

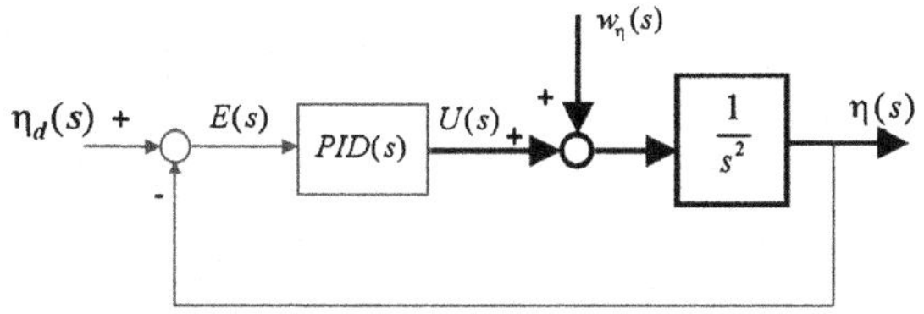

Figure 6.2 PID control structure.

The PID controller is by far the most common control algorithm used in industrial processes [1]. The most popular expression for this algorithm is

$$u(t) = -k_p e(t) - k_i \int e(t)dt - k_d \frac{de}{dt}, \tag{6.2}$$

where k_i defines the integral gain. Its diagram representation is illustrated in Fig. 6.2.

From Fig. 6.2 notice that the error expression is

$$e_\infty = \lim_{s \longrightarrow 0} e(s) = \frac{s^2}{s^2 + PDI(s)} \eta_d(s) + \frac{1}{s^2 + PDI(s)} w_\eta(s),$$

and it can be concluded that, as the control law has an integral term, $e_\infty = 0$.

There exist several expressions derived from (6.2), for example, the PID with set-point weighting algorithm is becoming popular because it can be deduced as a regulator with 2 DOF (Degrees Of Freedom), namely

$$u(t) = K_p(b\eta_d - \eta) + K_i \int e(t)dt + K_p \frac{d(c\eta_d - \eta)}{dt}, \tag{6.3}$$

where η_d represents the set-point and η is the measurement variable. The response to the set-point changes will depend on the magnitudes of b and c while its stability and robustness with respect to perturbations and the measurement noise will be the same as that for the algorithm without set-point weighting (6.2).

Notice that, when controlling the quadcopter attitude, it is possible to measure the angular rate $\dot{\eta}$. In this way, from (6.3) and with $c = 0$ and $b = 1$, we get

$$u(t) = k_p e(t) - k_d \dot{\eta}(t) + k_i \int e(t)dt. \tag{6.4}$$

In Fig. 6.3, it can be observed how the attitude performance is restored when using a PID controller. Following these ideas, the goal of this chapter is to introduce two nonlinear controllers having integral terms to stabilize the quadcopter system or only some of its dynamics.

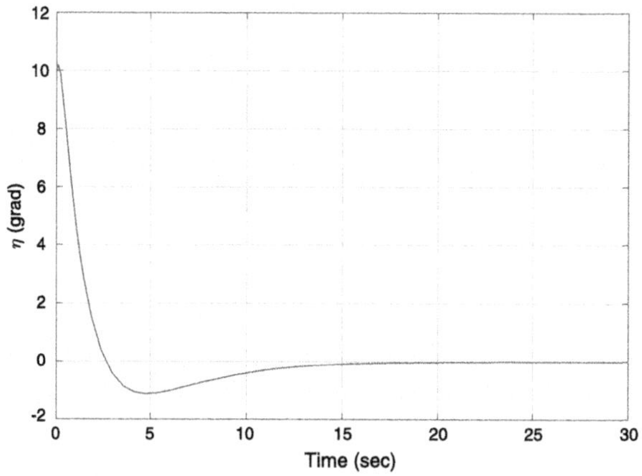

Figure 6.3 η response when applying a PID controller with constant perturbation $w_\eta = 4$ in the system.

6.2 SATURATED CONTROLLERS WITH INTEGRAL COMPONENT

The purpose of this section is to present a summary of the control algorithms based on saturation functions with integral action to stabilize a quadcopter represented by a chain of integrators in cascade[2]. A detailed version of this result can be found in [2].

The algorithms stabilize a chain of integrators of n states described by

$$\begin{aligned} \dot{x}_1 &= x_2, \\ \dot{x}_2 &= x_3, \\ \vdots &= \vdots \\ \dot{x}_n &= u(t), \end{aligned}$$

or, in the classical form, by

$$\dot{\mathbf{x}}(t) = A\mathbf{x}(t) + Bu(t), \tag{6.5}$$

[2] Notice that in the previous section the attitude of a VTOL vehicle could be represented by two integrators in cascade.

with

$$A = \begin{pmatrix} 0 & 1 & 0 & \dots & 0 \\ 0 & 0 & 1 & \dots & 0 \\ \vdots & \vdots & \vdots & \ddots & \vdots \\ 0 & 0 & 0 & \dots & 0 \end{pmatrix}, \quad B^T = \begin{pmatrix} 0 & 0 & 0 & \dots & 1 \end{pmatrix},$$

where $\mathbf{x}(t) \in \mathbb{R}^n$ defines the states of the system, $u(t) \in \mathbb{R}^1$ is called the input, $A \in \mathbb{R}^{n \times n}$ represents the state matrix, and $B \in \mathbb{R}^n$ denotes the input matrix. An essential element of these control laws is the saturation function that can be defined as follows.

Definition 6.1. Given a positive constant b, a function $\sigma_b(s) : \mathbb{R} \to \mathbb{R}$ is said to be a linear saturation for s if it is continuous and nondecreasing function satisfying:

- $s\sigma_b(s) > 0$ for all $s \neq 0$,
- $\sigma_b(s) = s$ for $|s| \leq b$,
- $\sigma_b(s) = \frac{bs}{|s|}$ for $|s| > b$.

Several controllers have been proposed using saturation functions to stabilize system (6.5); see, for example, [3–5]. Nevertheless, these controllers did not include an integral term. In the following, three main configurations of nonlinear controllers based on saturation functions with an integral part are presented. For the stability analysis the following change of coordinates is necessary:

$$\begin{aligned} y_1 &= \int_0^t x_1, \\ y_2 &= x_1, \\ \vdots &= \vdots \\ y_{n'} &= x_n. \end{aligned}$$

Then (6.5) yields

$$\dot{\mathbf{y}}(t) = A'\mathbf{y}(t) + B'u(t), \tag{6.6}$$

where $n' = n + 1$, $A' \in \mathbb{R}^{n' \times n'}$, and $B' \in \mathbb{R}^{n'}$. Indeed, the stability of system (6.6) implies the stability of system (6.5).

6.2.1 Nested Saturation with Integral Part Controller – NSIP

The following controller asymptotically stabilizes system (6.6):

$$u_{NSIP} = -\sigma_{b_{n'}}(z_{n'} + \sigma_{b_{n'-1}}(z_{n'-1} + \cdots + \sigma_{b_1}(z_1))), \tag{6.7}$$

where

$$\begin{aligned} z_i &= z_{i+1} + \sum_{j=1}^{n'-i} \begin{pmatrix} n'-i-1 \\ j-1 \end{pmatrix} y_{n'-j} \quad \forall i \in [1, n'-1], \\ z_{n'} &= y_{n'}. \end{aligned}$$

Notice that $\begin{pmatrix} * \\ * \end{pmatrix}$ is the binomial coefficient.

Stability Analysis

Rewriting (6.7) as

$$u_{NSIP} = -\sigma_{b_{n'}}(z_{n'} + \xi_{n'}),$$

we can deduce that $|\xi_{n'}| \leq b_{\xi'_n}$ for $b_{\xi'_n}$ which will be defined later to assure convergence. Take $V_{n'} = \frac{1}{2}z_{n'}^2$, then $\dot{V}_{n'} = -z_n\sigma_{b_{n'}}(z_{n'} + \xi_{n'})$. If $|z_{n'}| > b_{\xi_{n'}}$ then $\mathrm{sgn}(\sigma_{b_{n'}}(z_{n'} + \xi_{n'})) = \mathrm{sgn}(z_{n'})$ and so for $b_{\sigma_{b_{n'}}} \geq 2b_{\xi_{n'}}$ we obtain $u_{NSIP} = -z_{n'} - \xi_{n'}$, $\forall t > T_1$. This proves that (6.7) stabilizes system (6.6) for $i = n'$. The next step is to prove that the same is also true for $i-1$. We propose $V_{i-1} = \frac{1}{2}z_{i-1}^2$ and hence $\dot{V}_{i-1} = -z_{i-1}\dot{z}_{i-1}$. To compute $\dot{z}_{i-1}$ remember that[3]

$$z_{i-1} = z_l + \sum_{l=i-1}^{n'-1}\sum_{j=1}^{n'-l} \begin{pmatrix} n'-l-1 \\ j-1 \end{pmatrix} y_{n'-j},$$

then

$$\dot{z}_{i-1} = u + \sum_{l=i-1}^{n'-1}\sum_{j=1}^{n'-l} \begin{pmatrix} n'-l-1 \\ k'-1 \end{pmatrix} y_{n'-j+1} = -\xi_i$$

with $\xi_i = \sigma_{b_{i-1}}(z_{i-1} + \xi_{i-1})$, and therefore $\dot{V}_{i-1} = -z_{i-1}\sigma_{i-1}(z_{i-1} + \xi_{i-1})$. As $|\xi_{i-1}| \leq b_{\xi_{i-1}}$, if $|z_{i-1}| > b_{\xi_{i-1}}$, $\mathrm{sgn}(\sigma_{b_{i-1}}(z_{i-1} + \xi_{i-1})) = \mathrm{sgn}(z_{i-1})$ which

[3] sgn denotes the *sign* (or *signum*) function.

implies that $\dot{V}_{i-1} < 0$ and $\exists\ T_{n'-i+2} > T_{n'-i+1}$ such that $\forall t > T_{n'-i+2}$, $|z_{i-1}| < b_{\xi_{i-1}}$, $\forall i \in [2, n']$. Therefore, if $b_i \geq 2b_{i-1}$, it follows that

$$u = -z_{n'} - \xi_{n'} \qquad \text{with} \qquad \xi_l = -z_{l-1} - \xi_{l-1}, \quad \forall l \in \left[i-1, n'\right].$$

So we proved that the closed loop (6.6)–(6.7) is stable for $i-1$, then, using the recurrence theorem, (6.6)–(6.7) holds $\forall i \in [1, n']$.

From the previous analysis we have that $\dot{z}_1 = \sigma_{b_1}(z_1)$ and the last positive-definite function is $V_1 = \frac{1}{2}z_1^2$; therefore, $\dot{V}_1 = -z_1\sigma_{b_1}(z_1)$, which implies that $\dot{V}_1 < 0$ and z_1 goes to zero. Hence, from the definition of z_1, we have that $\xi_2 \to 0$. If we assume that $z_i \to 0$ as $t \to \infty$ which implies that $\xi_{i+1} \to 0$, then for $i+1$ it follows that $V_{i+1} = \frac{1}{2}z_{i+1}^2$ and then $\dot{V}_{i+1} = -z_{i+1}\sigma_{b_{i+1}}(z_{i+1} + \xi_{i+1})$. This means that $z_{i+1} \to 0$ as $t \to \infty$. Finally, as z_{i+1} tends to zero, using the recurrence theorem, we obtain $z_k \to 0$ as $t \to \infty$ $\forall k \in [1, n']$. Observe that as $z_{n'} = y_{n'}$, $y_{n'} \to 0$.

From the above, we notice that $z_l \to 0$ for $l \in [i, n']$. This implies that y_l also goes to zero. Now we have that

$$z_{i-1} \quad = \quad \sum_{j=0}^{n'-i} \begin{pmatrix} n' - i + 1 \\ j \end{pmatrix} y_{n'-j} + y_{i-1}. \tag{6.8}$$

Analyzing the previous equation, we can deduce that $\sum(\cdot)y_{n'-j}$ contains z_i variables, $i \in [1, n']$. It was proved that these variables converge to zero, so the above implies that y_{i-1} also goes to zero, and we can conclude that every y_k, $\forall k \in [1, n']$ goes to zero.

6.2.2 Separated Saturation with Integral Part Controller – SSIP

The following controller stabilizes system (6.6) globally asymptotically:

$$u = -\sigma_{b_{n'}}(z_{n'}) - \sigma_{b_{n'-1}}(z_{n'-1}) - \cdots - \sigma_{b_1}(z_1), \tag{6.9}$$

where

$$z_i \quad = \quad \sum_{j=0}^{n'-i} \begin{pmatrix} n' - i \\ j \end{pmatrix} y_{n'-j}, \quad \forall i \in \left[1, n'\right]. \tag{6.10}$$

Stability analysis is similar to that proposed for the NSIP algorithm, the main difference is the saturation constraints required to perform the Lyapunov analysis and defined as

$$b_i \geq b_{i-1} + b_{i-2} + \cdots + b_1. \tag{6.11}$$

6.2.3 Saturated State Feedback with Integral Part Controller – SSFIP

The following controller can be seen as a PID algorithm with bounded states. The algorithm has the form

$$u_{SSFIP} = -\sum_{j=1}^{n} \sigma_j(k_j x_j) + \sigma_{\int b_1}\left(\int_0^t x_1\right), \tag{6.12}$$

where x_i denotes the state i and $k_j > 0$ is the constant gain controller. This control law stabilizes system (6.5) globally asymptotically. For the stability analysis, we rewrite (6.12) in the following form:

$$u_{SSFIP} = -\sum_{i=1}^{n'} \sigma_{b_i}(k_i y_i). \tag{6.13}$$

Rewrite (6.13) for $i = n'$ as

$$u_{SSFIP} = -\sigma_{b_{n'}}(k_{n'} y_{n'}) - \xi_{n'} \tag{6.14}$$

with $\xi_n' = \sum_{j=1}^{n'-1} \sigma_{b_j}(k_j y_j)$ and define

$$\begin{aligned} z_{n'} &= y_{n'}, \\ z_i &= k_{i+1} y_i + z_{i+1}, \quad \forall i \in \left[1, n'-1\right]. \end{aligned}$$

Propose $V_{n'} = \frac{1}{2} z_{n'}^2$, and then $\dot{V}_{n'} = -y_{n'}(\sigma_{b_{n'}}(k_{n'} y_{n'}) + \xi_{n'})$. Assuming that $b_{n'} > b_{\xi_{n'}}$, we get $|k_{n'} y_{n'}| > b_{\xi_{n'}}$ which implies that $\dot{V}_{n'} < 0$. Thus, there exist a time T_1 such that $\forall t > T_1$, $|y_{n'}| \leq \frac{b_{\xi_{n'}}}{k_{n'}}$ and, as a consequence, $u_{SSFIP} = -k_{n'} y_{n'} - \xi_{n'}$.

It is assumed that for a given l, (6.13) is true and $|y_l|$ is bounded $\forall l \in [i, n']$ and $i \neq 1$. Now, we prove that (6.13) is also true for $i - 1$. Take $V_{i-1} = \frac{1}{2} z_{i-1}^2$, then $\dot{V}_{i-1} = z_{i-1} \dot{z}_{i-1}$. We have that $z_{i-1} - z_{n'} = \sum_{j=i-1}^{n'-1} k_{j+1} y_j$, and then $z_{i-1} = \sum_{j=i-1}^{n'-1} k_{j+1} y_j + y_{n'}$. Therefore $\dot{z}_{i-1} = -\xi_i$.

The above implies

$$\dot{V}_{i-1} = -\left(k_i y_{i-1} + \sum_{j=i}^{n'-1} k_{j+1} y_j + y_{n'}\right)\left(\sigma_{b_{i-1}}(k_{i-1} y_{i-1}) + \xi_{i-1}\right). \tag{6.15}$$

Assuming $b_{i-1} > b_{\xi_{i-1}}$ gives $|y_{i-1}| > \frac{b_{\xi_{i-1}}}{k_{i-1}}$, hence $\mathrm{sgn}(\sigma_{b_{i-1}}(k_{i-1}y_{i-1}) + \xi_{i-1}) = \mathrm{sgn}(y_{i-1})$. Notice that $\left(\sum_{j=i}^{n'-1} k_{j+1}y_j + y_{n'}\right) = z_i$ and z_i is bounded. Thus, if $|k_i y_{i-1}| > b_{z_i}$, then $\mathrm{sgn}(k_i y_{i-1} + z_i) = \mathrm{sgn}(y_{i-1})$. Therefore $\dot{V}_{i-1} < 0$ and $\exists\ T_{n'-i+2} > T_{n'-i+1}$ such that $\forall t > T_{n'-i+2}$,

$$\begin{aligned} |y_{i-1}| &< \frac{b_{\xi_{i-1}}}{k_{i-1}}, \quad \forall i \in [2, n'], \\ |y_1| &< \frac{b_1}{k_1}. \end{aligned}$$

Since it was proved that (6.13) is true for $i-1$, using the recurrence theorem, the same is also true $\forall i \in [1, n']$. This means that for $i = 1$ there exists a time $T_{n'}$ such that

$$u = -\sum_{i=1}^{n'} k_i y_i, \quad \forall t > T_{n'}. \tag{6.16}$$

The previous equations hold if the following conditions are satisfied: $\frac{b_{\xi_i}}{k_i} > \frac{b_{z_{i+1}}}{k_{i+1}}$, $\forall i \in [2, n'-1]$; $\frac{b_1}{k_1} > \frac{b_{z_2}}{k_2}$, $b_i > b_{\xi_i}$, $\forall i \in [2, n']$; $b_{x_{1_i}} = \sum_{j=2}^{i} b_{j-1}$, $b_{z_i} = \sum_{j=i}^{n'-1} \frac{k_{j+1}b_{\xi_j}}{k_j} + \frac{b_{\xi_{n'}}}{k_{n'}}$.

Observe that for $t > T_{n'}$ the closed-loop system is $\dot{\mathbf{y}}(t) = \bar{A}\mathbf{y}(t)$ and, as $\bar{A}$ is Hurwitz, stability is proved.

The previous stability analysis demonstrated the convergence of the states to zero, $y_i \to 0$, implying $x_i \to 0$. Nevertheless, we can consider system (6.6) as an error system such that $y_1 = \int_0^t e_1, y_2 = e_1, \ldots, y_{n'} = e_n$, with $e_1 = x_1 - x_{1_d}, \ldots, e_n = x_n - x_{n_d}$; and then this implies that $e_i \to 0$ and $x_1 \to x_{1_d}, \ldots, x_n \to x_{n_d}$. The previous holds for x_{i_d} constant or bounded $|x_{i_d}| \leq b_i$.

More details about the previous controllers can be found in [2].

6.2.4 Validation in a Quadcopter Vehicle

The control algorithms were validated in simulations using the simplified quadcopter model described in Sect. 5.1.5 of Chap. 5. To develop the algorithms, the following hypotheses were used:

- $m, I_j = 1$
- quasi-stationary movements
- no external and unknown disturbances present

Notice from the equations in the simplified model that the altitude and the attitude can be represented by two integrators in cascade as

$$\ddot{z} = u_z, \qquad \ddot{\psi} = u_\psi, \tag{6.17}$$

where z is the altitude of the aerial vehicle, ψ the yaw angle, u_ψ defines the control input for the yaw angle, and $u_z = u\cos(\theta)\cos(\phi) - g$, where θ, ϕ denote the pitch and roll angles, g expresses the gravitational acceleration, and u represents the main control input.

Due to chapter-length limitation, we only use the SSFIP algorithm; similar results can be obtained using the NSIP and SSIP algorithms. Therefore the control scheme to stabilize the previous system takes the form

$$u_\gamma := -\sigma_{\dot{\gamma}}\left(k_{3_\gamma}\dot{\gamma}\right) - \sigma_\gamma\left(k_{2_\gamma}\gamma\right) - \sigma_{e_\gamma}\left(k_{1_\gamma}\int_0^t e_\gamma dt\right), \qquad \gamma : z, \psi. \tag{6.18}$$

Once the altitude and the yaw movement are stabilized, the next step is to stabilize the longitudinal and lateral dynamics. They are typically coupled and both have high impact on the complete vehicle dynamics. Thus their classical simple representation is denoted by four integrators in cascade in the following form

$$\ddot{s} \approx \tan(\rho),$$
$$\ddot{\rho} \approx u_\rho,$$

where s defines the state x or y, and ρ represents the angle θ or ϕ, respectively. Notice that the controller needs to be developed for a linear system. Then linearizing the above, it follows that $\ddot{s} \approx \rho$. Using the proposed methodology, the controller SSFIP for the previous system is

$$u_\rho := -\sigma_{\dot{\rho}}(k_{5_\rho}\dot{\rho}) - \sigma_\rho(k_{4_\rho}\rho) - \sigma_{\dot{s}}(k_{3_\rho}\dot{s}) - \sigma_s(k_{2_\rho}s) - \sigma_{e_s}\left(k_{1_\rho}\int_0^t e_s dt\right)$$

with $e_s = s - s_{ref}$ and s_{ref} being the desired position. Remark that for the longitudinal model $\ddot{x} \approx -\tan(\theta)$, and consequently for the longitudinal case, the last three terms of the above expression are positive.

6.2.4.1 Simulation Results

The simulation objective is to accurately follow a desired trajectory in the presence of uncertainties (white noise) in the states x, y, and z. The following graphs introduce good behavior when applying the SSFIP control to the quadcopter system, see Figs. 6.4–6.7. Notice the good performance of the closed-loop system, the boundedness of the states, and the smallness of control responses. Observe also that only the θ and ϕ angles are affected by the uncertainties; this fact is normal because the noise in the x or y axis produces uncertainties in the longitudinal or lateral dynamics. The desired altitude is $z_d = 20$ m, and in the horizontal plane the desired trajectory is to follow a square trajectory with 6 m on each side. The initial conditions were zero for all the states except for the yaw angle which was $\psi(0) = 10$ grad.

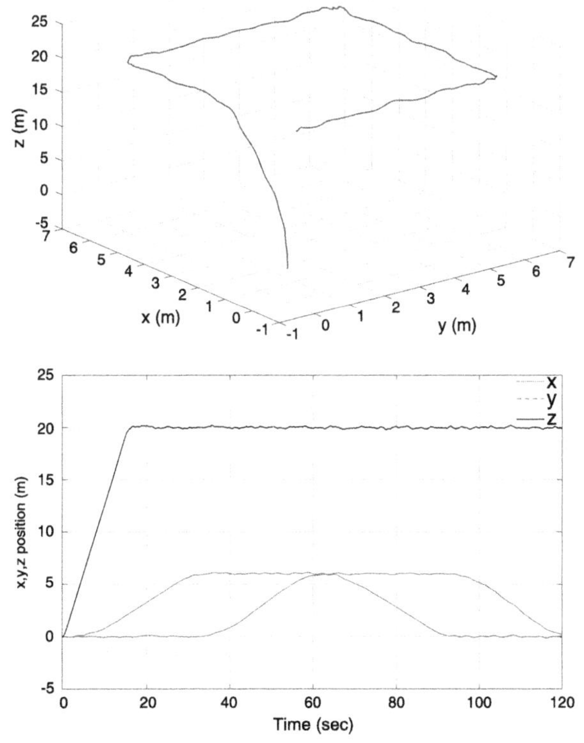

Figure 6.4 3D trajectory when using the saturated control algorithm with integral part.

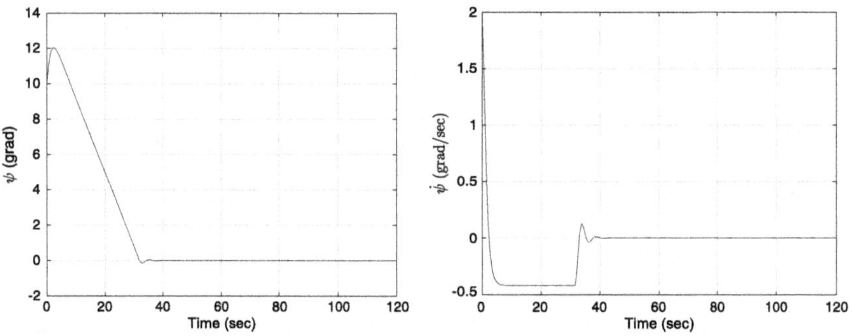

Figure 6.5 ψ and $\dot{\psi}$ responses, the initial condition for yaw was 10 grad.

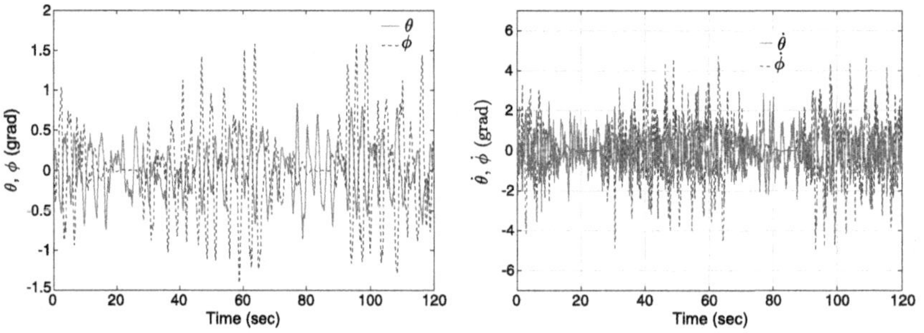

Figure 6.6 Attitude and angular rates performances.

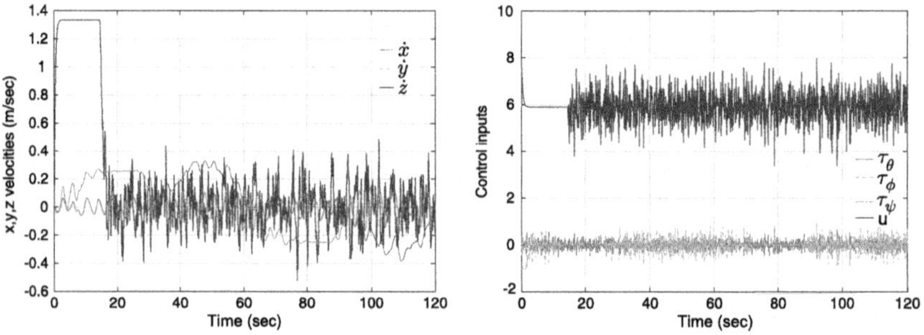

Figure 6.7 Velocity responses and control inputs behavior.

6.3 INTEGRAL AND ADAPTIVE BACKSTEPPING CONTROL – IAB

The next controller is developed using the quadcopter model defined in Sect. 2.3 of Chap. 2. Introducing uncertainties in the model, Eqs. (2.17)–(2.18) become

$$m\ddot{\xi} = R\hat{F} + f_g + \delta_u, \tag{6.19}$$

$$\mathrm{I}\dot{\Omega} = -\Omega \times \mathrm{I}\Omega + \hat{\tau} + \delta_\tau, \tag{6.20}$$

where ξ defines the position of the quadrotor, $\hat{F}$ and $\hat{\tau}$ express the thrust and torque vectors generated by the rotation of helices, f_g denotes the gravity force. R is the rotation matrix, Ω introduces the angular velocity, and I represents the inertia matrix of the quadrotor. The terms δ_u and δ_τ define the non-modeled dynamics and the perturbations of the aerial vehicle (drag effects included).[4]

Adding an integral term in the classical backstepping approach increases the robustness of the closed-loop system. In addition, in this algorithm an adaptive sense is introduced, i.e., the adaptive part will help to estimate some unknown parameters and to counteract them.

6.3.1 Quadcopter IAB Algorithm

To stabilize the quadcopter, let us define the position error as follows:

$$e_\xi = \xi - \xi_{ref} \quad \Longrightarrow \quad \dot{e}_\xi = \dot{\xi} - \dot{\xi}_{ref} = \nu - \dot{\xi}_{ref}. \tag{6.21}$$

Now propose a positive definite function to design a virtual velocity ν^ν that will ensure the convergence to the desired position[5]

$$V_{Lr} = \frac{1}{2}\left(\mathbb{K}_{Ir}\chi_1, \chi_1\right) + \frac{1}{2}\left(e_\xi, e_\xi\right) \tag{6.22}$$

where $(\cdot, \cdot)$ represents the inner vector product, $\mathbb{K}_{Ir}$ is a positive constant matrix, and $\chi_1 = \int_0^t e_\xi \, d\tau$. Differentiating V_{Lr} results in $\dot{V}_{Lr} = (\mathbb{K}_{Ir}\chi_1, e_\xi) + (e_\xi, \dot{e}_\xi)$, and taking the virtual velocity as $\nu^\nu = \dot{\xi}_{ref} - \mathbb{K}_{Ir}\chi_1 - \mathbb{K}_r e_\xi$, with $\mathbb{K}_r$ a positive constant matrix, gives

$$\dot{V}_{Lr}\big|_{\nu=\nu^\nu} = -(\mathbb{K}_r e_\xi, e_\xi) \leq 0 \quad \forall\, t \geq 0.$$

[4] In δ_i, the external and unknown disturbances, w_i, are also included.

[5] To avoid confusion, in this chapter we use ν to denote the linear velocity of the vehicle expressed in the inertial frame and V for a Lyapunov or positive function.

Define the velocity error as

$$e_v = v - v^v \quad \Longrightarrow \quad \dot{e}_v = \dot{v} - \dot{v}^v = \frac{1}{m}(\mathbf{u} + \delta_u) - \dot{v}^v$$

with $\mathbf{u} = R\hat{F} + f_g$. Now a positive definite function is proposed to ensure convergence of the position error to zero

$$V_{Lv} = V_{Lr} + \frac{1}{2}(e_v, e_v) \quad \Longrightarrow \quad \dot{V}_{Lv} = \dot{V}_{Lr} + (e_v, \dot{e}_v).$$

From the velocity error definition, it follows that $v = v^v + e_v$, and then the above can be expressed as $\dot{V}_{Lv} = -(\mathbb{K}_r e_\xi, e_\xi) + (e_\xi, e_v) + (e_v, \dot{e}_v)$, by choosing

$$\mathbf{u} = -\hat{\delta}_u + m\left(\dot{v}^v - e_\xi - \mathbb{K}_v e_v\right)$$

with $\mathbb{K}_v$ denoting a positive constant matrix and $\hat{\delta}_u$ the estimated value of δ_u. Rewriting $\dot{V}_{Lv}$, it follows that

$$\dot{V}_{Lv} = -(\mathbb{K}_r e_\xi, e_\xi) - (\mathbb{K}_v e_v, e_v) + \frac{1}{m}(e_v, e_{\delta_u})$$

with $e_{\delta_u} = \delta_u - \hat{\delta}_u$. Suppose that δ_u is constant, thus $\dot{e}_{\delta_u} = -\dot{\hat{\delta}}_u$. Consider the following candidate Lyapunov function $V_{Lu} = V_{Lv} + \frac{1}{2}({\Gamma_1}^{-1} e_{\delta_u}, e_{\delta_u})$, with Γ_1 being a positive diagonal constant matrix. Then

$$\dot{V}_{Lu} = -(\mathbb{K}_r e_\xi, e_\xi) - (\mathbb{K}_v e_v, e_v) + \frac{1}{m}(e_v, e_{\delta_u}) + ({\Gamma_1}^{-1} e_{\delta_u}, -\dot{\hat{\delta}}_u).$$

Now propose $\dot{\hat{\delta}}_u = \frac{1}{m}\Gamma_1 e_v$ as the desired dynamics for $\hat{\delta}_u$. Thus

$$\dot{V}_{Lu} = -(\mathbb{K}_r e_\xi, e_\xi) - (\mathbb{K}_v e_v, e_v) \leq 0 \quad \forall\, t \geq 0.$$

There exists a relationship between $\mathbf{u}$, the main force $\hat{F}$, and the quadcopter attitude η so that it is possible to find $(\eta, \hat{F})$ from $\mathbf{u} = [u_x, u_y, u_z]^T$. Here the z–y–x Euler convention is used. Consequently, the rotation matrix, R, has the following form:

$$\mathbb{R} = \begin{pmatrix} c_\psi c_\theta & -s_\psi c_\phi + c_\psi s_\theta s_\phi & s_\psi s_\phi + c_\psi s_\theta c_\phi \\ s_\psi c_\theta & c_\psi c_\phi + s_\psi s_\theta s_\phi & -c_\psi s_\phi + s_\psi s_\theta c_\phi \\ -s_\theta & c_\theta s_\phi & c_\theta c_\phi \end{pmatrix}, \tag{6.23}$$

where the c and s stand for the cosine and sine functions, respectively. From (6.19) and (6.23), the following relations for $(\eta_{ref}, \hat{F}_{ref})$ can be obtained:

$$\theta_{ref} = \arctan\left(\frac{u_y s_\psi + u_x c_\psi}{u_z + mg}\right), \tag{6.24}$$

$$\phi_{ref} = \arctan\left(c_{\theta_{ref}} \cdot \frac{u_x s_\psi - u_y c_\psi}{u_z + mg}\right), \tag{6.25}$$

$$f_{ref} = \frac{u_z + mg}{c_{\theta_{ref}} \cdot c_{\phi_{ref}}}. \tag{6.26}$$

ψ_{ref} is assigned arbitrarily, and the main thrust is taken as $\hat{F}_{ref} = [0, 0, f_{ref}]^T$. Let us define the error of Euler angles as

$$e_\eta = \eta - \eta_{ref} \quad \Longrightarrow \quad \dot{e_\eta} = \dot{\eta} - \dot{\eta}_{ref} = W_\eta^{-1}\Omega - \dot{\eta}_{ref}.$$

Propose the next positive definite function $V_{L\eta} = \frac{1}{2}\left(\mathbb{K}_{I\eta}\chi_2, \chi_2\right) + \frac{1}{2}(e_\eta, e_\eta)$, where $\mathbb{K}_{I\eta}$ is a positive constant matrix and $\chi_2 = \int_0^t e_\eta d\tau$. Thus, $\dot{V}_{L\eta} = (\mathbb{K}_{I\eta}\chi_2, e_\eta) + (e_\eta, \dot{e}_\eta)$. Define the virtual angular velocity as $\Omega^v = W_\eta(\dot{\eta}_{ref} - \mathbb{K}_{I\eta}\chi_2 - \mathbb{K}_\eta e_\eta)$, with $\mathbb{K}_\eta$ being a positive constant matrix. This implies

$$V_{L\eta}|_{\Omega=\Omega^v} = -(\mathbb{K}_\eta e_\eta, e_\eta) \leq 0 \quad \forall\, t \geq 0.$$

Define the angular velocity error as $e_\Omega = \Omega - \Omega^v \quad \Longrightarrow \quad \dot{e}_\Omega = \dot{\Omega} - \dot{\Omega}^v$. Remember that $\Omega = \Omega^v + e_\Omega$, then $\dot{\Omega} = \mathrm{I}^{-1}(\hat{\tau} - \Omega \times \mathrm{I}\Omega + \delta_\tau)$. Propose the following positive definite function, $V_{L\Omega} = V_{L\eta} + \frac{1}{2}(e_\Omega, e_\Omega)$, then get

$$\dot{V}_{L\Omega} \quad = \quad \dot{V}_{L\eta} + (e_\Omega, \dot{e}_\Omega) = -(\mathbb{K}_\eta e_\eta, e_\eta) + (e_\eta, W_\eta^{-1}e_\Omega) + (e_\Omega, \dot{e}_\Omega).$$

Choose the control input as

$$\hat{\tau} = -\hat{\delta}_\tau + \Omega \times \mathrm{I}\Omega + \mathrm{I}\left(\dot{\Omega}^v - (W_\eta^{-1})^T e_\eta - \mathbb{K}_\Omega e_\Omega\right)$$

with $\mathbb{K}_\Omega$ being a positive constant matrix and $\hat{\delta}_\tau$ the estimated value of δ_τ. Then

$$\dot{V}_{L\Omega} = -(\mathbb{K}_\eta e_\eta, e_\eta) - (\mathbb{K}_\Omega e_\Omega, e_\Omega) + (e_\Omega, \mathrm{I}^{-1}e_{\delta_\tau})$$

with $e_{\delta_\tau} = \delta_\tau - \hat{\delta}_\tau$. Consider δ_τ constant, then $\dot{e}_{\delta_\tau} = -\dot{\hat{\delta}}_\tau$. Propose the following candidate Lyapunov function, $V_{L\tau} = V_{L\Omega} + \frac{1}{2}({\Gamma_2}^{-1}e_{\delta_\tau}, e_{\delta_\tau})$, where ${\Gamma_2}^{-1}$ is a positive diagonal constant matrix. It implies that

$$\dot{V}_{L\tau} = -(\mathbb{K}_\eta e_\eta, e_\eta) - (\mathbb{K}_\Omega e_\Omega, e_\Omega) + (\mathrm{I}^{-1}e_\Omega, e_{\delta_\tau}) + ({\Gamma_2}^{-1}e_{\delta_\tau}, -\dot{\hat{\delta}}_\tau).$$

Define $\dot{\hat{\delta}}_\tau = \Gamma_2 \mathrm{I}^{-1} e_\Omega$ as the desired dynamics for $\hat{\delta}_\tau$. Thus

$$\dot{V}_{L\tau} = -(\mathbb{K}_\eta e_\eta, e_\eta) - (\mathbb{K}_\Omega e_\Omega, e_\Omega) \leq 0 \quad \forall\, t \geq 0.$$

6.3.2 Simulation with Perturbations

Several simulations were carried out to validate the IAB control. The simulation parameters are written in Table 6.1, these parameters take into account the characteristics of the Quadcopter Flexbot platform shown in Table 6.2. In Fig. 6.8, the performance of the closed-loop system response can be observed. In this simulation no perturbations were considered. The

Table 6.1 Simulation parameters

Parameter	Value	Parameter	Value
I	$\mathrm{diag}(8, 8, 6) \cdot 10^{-4}$	g	9.81
ξ_{ref}	$[\cos(0.1t),\ \sin(0.1t),\ 1]$	$\mathbb{K}_r$	$\mathrm{diag}(1, 1, 1) \cdot 0.35$
$\mathbb{K}_v$	$\mathrm{diag}(1, 1, 1) \cdot 16$	$\mathbb{K}_\eta$	$\mathrm{diag}(1, 1, 1) \cdot 2$
$\mathbb{K}_\Omega$	$\mathrm{diag}(1, 1, 1) \cdot 16$	Γ_1	$\mathrm{diag}(1, 1, 1) \cdot 57 \cdot 10^{-4}$
Γ_2	$\mathrm{diag}(32, 32, 24) \cdot 10^{-6}$	$\mathbb{K}_{Ir}$	$\mathrm{diag}(1, 1, 1) \cdot 0.05$
$\mathbb{K}_{I\eta}$	$\mathrm{diag}(1, 1, 1) \cdot 0.1$	Initial conditions	Zero

Table 6.2 Quadcopter flexbot parameters

Parameter	Value
Mass	0.057 kg
Payload	0.023 kg
Max diameter	0.12 m
Helix length	0.05 m

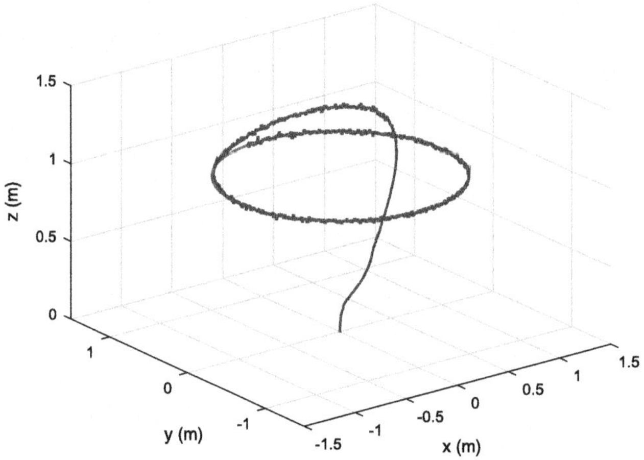

Figure 6.8 Quadrotor position without perturbations.

desired mission was to realize a circle trajectory of radius 1 m and altitude of 1 m. To keep the chapter of acceptable length, only the quadcopter's position responses are presented.

The next simulation takes into account two perturbations when the UAV is following a circular trajectory. One of them considers an extra torque which could arise from a displacement of its battery from the gravity center of the vehicle. The other considers the vehicle moving in the pres-

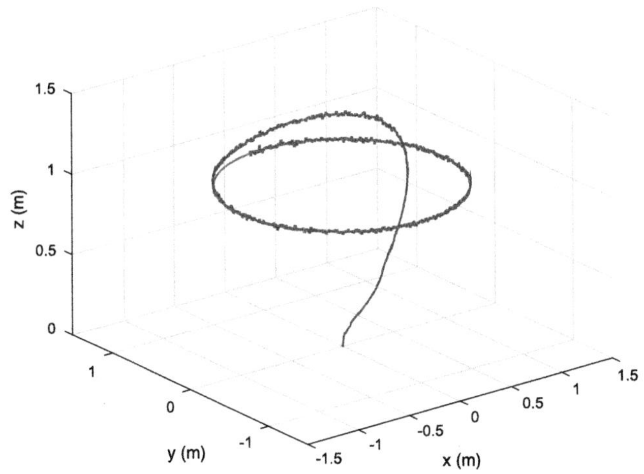

Figure 6.9 Position response of the quadrotor under perturbations.

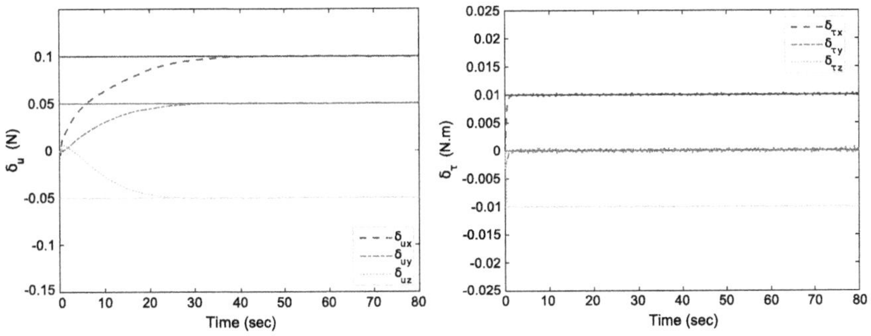

Figure 6.10 Estimated force and torque perturbations.

ence of a crosswind which produces an external force acting on the quadcopter. The force perturbation $\delta_u = [0.1, 0.05, -0.5]^T$ represents about 25% of the vehicle weight, and the torque perturbation $\delta_\tau = [0.01, 0, -0.01]^T$ is about the 25% of the total torque that quadcopter can manage. The systems response can be appreciated in Figs. 6.9 and 6.10. Notice that even if there are perturbations in the system, the controller stabilizes the desired position well. The estimates of perturbations are presented in Fig. 6.10. As before, notice that the adaptive part estimates the unknown perturbations well.

6.3.3 Simulation with Inaccurate Mass

Other simulations were realized to validate the performance of the controller when the vehicle mass is changing (it could be that the vehicle is transporting an object). Then the next simulation introduces the behavior of the vehicle which is tracking a circular trajectory and after a certain time it drops the object it was carrying. In simulation, the object represents 25% of the initial mass and is dropped at 40 seconds. The presence of perturbations of the same magnitudes as before is considered as well, see Fig. 6.11. Notice here that the controller is capable of stabilizing the ve-

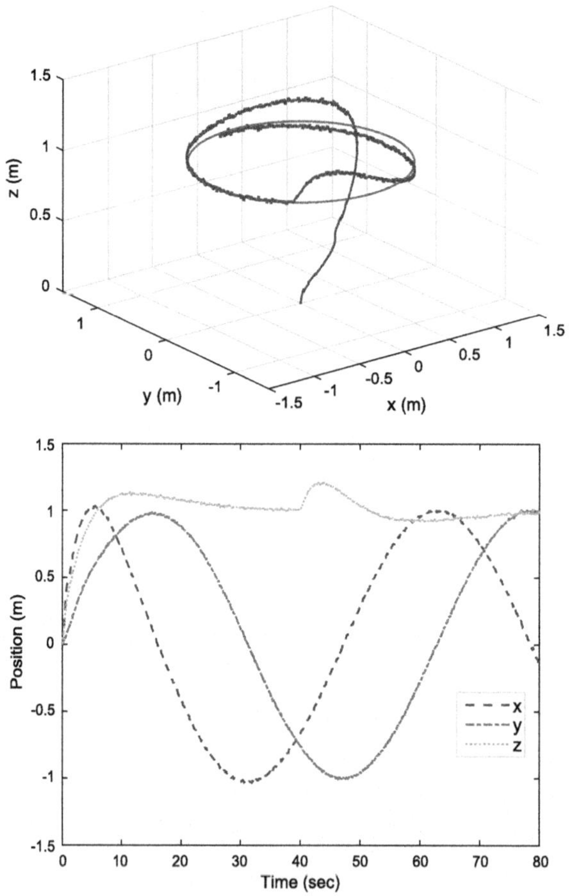

Figure 6.11 Position response of the quadrotor with perturbations and an inaccurate mass value.

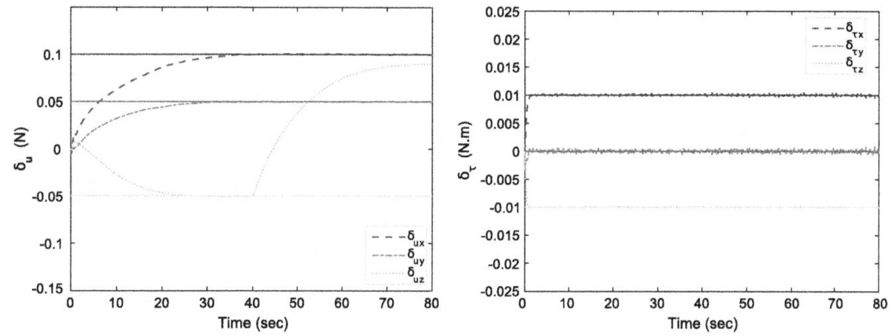

Figure 6.12 Estimates of force and torque perturbations.

hicle and tracking the trajectory even in the presence of perturbations and with a mass loss. In Fig. 6.12 the estimates of perturbations are illustrated. Observe that the effect of weight loss is included in the force estimates.

6.3.4 Experimental Results

The integral adaptive backstepping algorithm proposed previously was validated in flight tests. Three experiments were carried out to observe the performance of the closed-loop system. The first one was to follow a circular trajectory, the second experiment consisted in stabilizing the aerial vehicle at hover, and the last flight test was to keep the vehicle hovering in the presence of external and unknown wind. The vehicle used in the experiments was a quadcopter with the parameters shown in Table 6.3. A radio antenna was added to the quadrotor in order to get access to the torques estimated by the adaptive algorithm in the platform. This antenna weighs 15 g and was placed in one arm of the vehicle in order to generate the torque perturbation.

Table 6.3 Quadcopter parameters

Parameter	Value
Mass	0.257 kg
Payload	0.070 kg
Max diameter	0.20 m
Helix length	0.125 m
$\mathbb{I}$	$\mathrm{diag}(8, 8, 6) \cdot 10^{-2}$

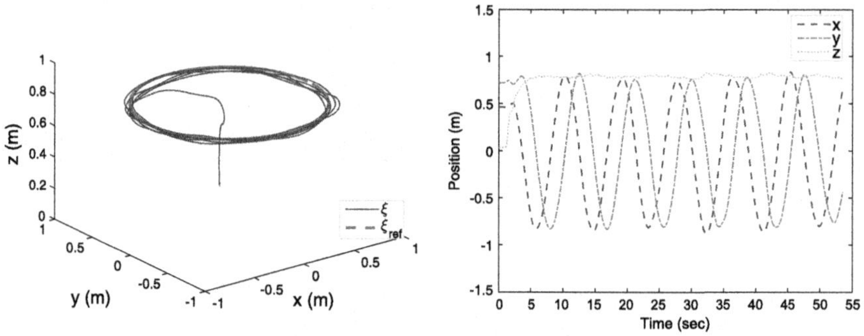

Figure 6.13 Position response of the quadrotor.

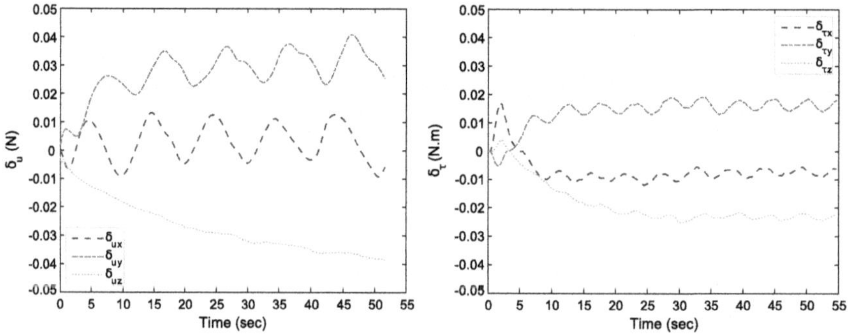

Figure 6.14 Estimation of force and torque perturbations.

Circular Trajectory

The reference path is a circle of 0.8 m radius whose center is at $(x, y, z) = (0, 0, 0.8)$ m, the result can be seen in Fig. 6.13.

Fig. 6.14 shows the estimates of the force and torque perturbations, respectively. These estimates take into account the dynamics not modeled, as well the additional torque generated by the antenna.

Stabilization at Hover

In this experiment the objective was to maintain the quadcopter at hover, the desired position was $\xi_{ref} = (0, 0, 1)$ m. Figs. 6.15 and 6.16 describe the behavior of the vehicle during the test.

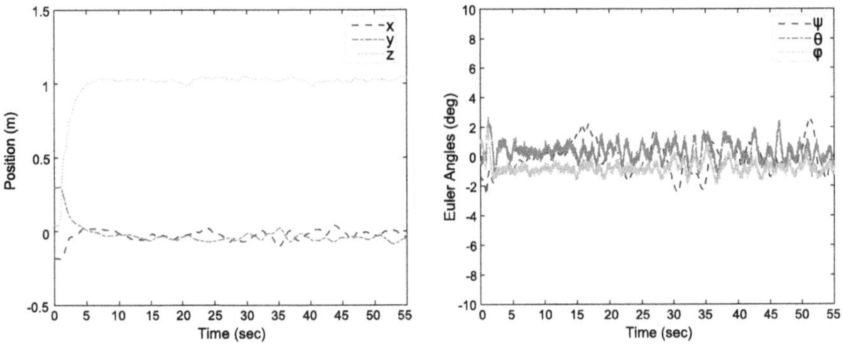

Figure 6.15 Position responses of the quadrotor.

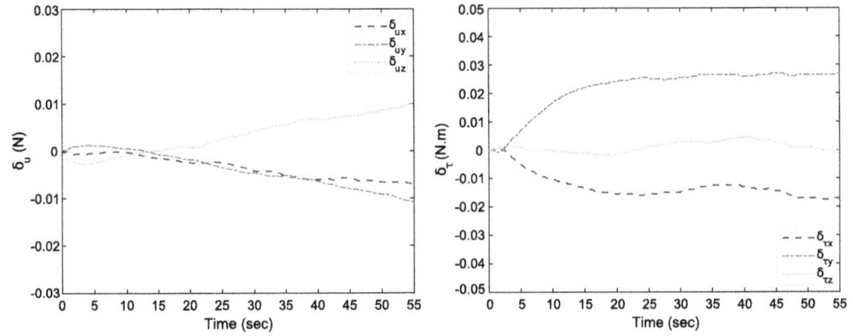

Figure 6.16 Estimation of force and torque perturbations.

Hover with External Perturbations

In this section the last experiment is repeated; however, in this case a disturbance is introduced by a wind gust generated with a fan for 17 s. Figs. 6.17–6.18 show the behavior of the quadrotor. We remark that the estimates and Euler angles oscillate more because they try to compensate the perturbations introduced in this experiment.

6.4 DISCUSSION

Two control algorithms containing an integral term were proposed to stabilize or navigate an aerial vehicle. Saturation functions were used to conceive three nonlinear control laws and applied to stabilize a quadcopter vehicle. Simulation results indicated good performance of the closed-loop system

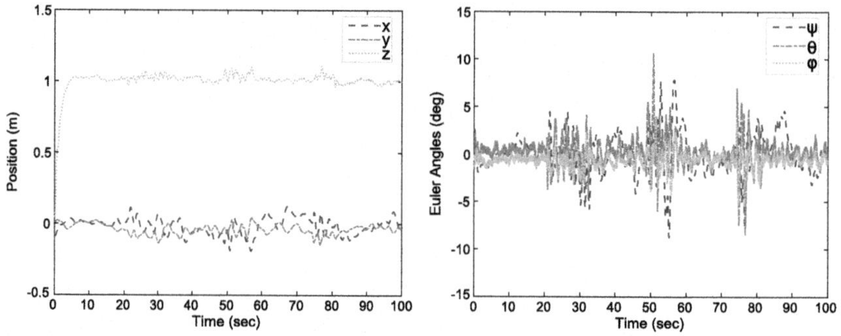

Figure 6.17 Position responses of the quadrotor.

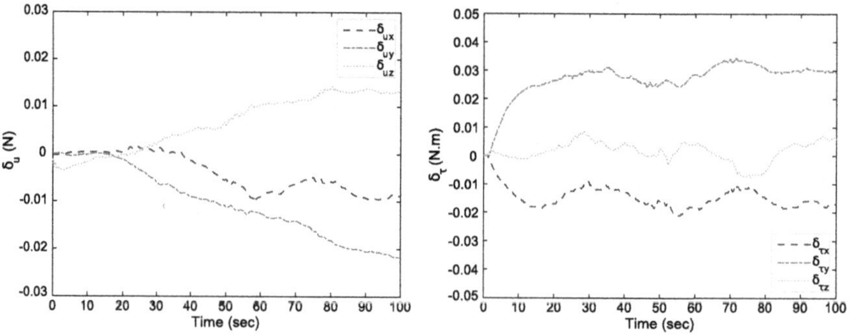

Figure 6.18 Force and torque perturbation responses.

when following a desired trajectory. The control inputs and states were bounded without degrading the vehicle performance.

The second controller is based on the backstepping approach to propose an integral and adaptive nonlinear control law. This controller was also applied to the quadcopter vehicle and validated in simulation and real-time flight tests. Three cases were studied and analyzed to demonstrate good behavior of the closed-loop system. Simulations and experimental results demonstrated this fact.

REFERENCES

1. K.J. Åström, T. Hägglund, Advanced PID Control, ISA – The Instrumentation, Systems, and Automation Society, Research Triangle Park, NC 27709, 2006.
2. A. Alatorre, P. Castillo, S. Mondié, Saturations-based nonlinear controllers with integral term: validation in real-time, TCON, International Journal of Control 89 (5) (2016) 879–891.

3. A.R. Teel, Global stabilization and restricted tracking for multiple integrators with bounded controls, Systems & Control Letters 18 (1992) 165–171.
4. G. Sanahuja, P. Castillo, A. Sanchez, Stabilization of n integrators in cascade with bounded input with experimental application to a VTOL laboratory system, International Journal of Robust and Nonlinear Control 20 (2010) 1129–1139.
5. J. Guerrero, P. Castillo, S. Salazar, R. Lozano, Mini rotorcraft flight formation control using bounded inputs, Journal of Intelligent & Robotic Systems 65 (2011) 175–186.

CHAPTER 7

Sliding Mode Control*

Controlling the attitude dynamics of a drone is an essential task when working with aerial vehicles. This dynamics has a nonlinear nature included in the Coriolis matrix, see Chap. 2. It was demonstrated in previous chapters that generally this dynamics is used to conceive control algorithms in its linear or simplified nonlinear mode. Nevertheless, it is also possible to represent these nonlinear equations as linear and perturbed equations, implying the design of controllers in an easy way. Nonlinear controllers are becoming popular when working with UAVs because they can be robust with respect to unknown perturbations such as wind present in the environment. Several researchers have proposed countless algorithms to stabilize the attitude of the aerial vehicle. The sliding mode approach becomes an essential tool due to its robustness and quick dynamics to converge the states. This methodology has been extensively studied in many works, see [1–5].

In this chapter the sliding mode and singular optimal control methodologies are used to design nonlinear controllers to stabilize the nonlinear attitude of a quadcopter vehicle. Two results are introduced here: first, a new form to represent the nonlinear equations for the orientation of a VTOL[1] vehicle in the presence of unknown disturbances as a linear MIMO and perturbed system is presented, and second, a control law is proposed and validated in simulations and in real time to stabilize the attitude of a quadcopter.

7.1 FROM THE NONLINEAR ATTITUDE REPRESENTATION TO LINEAR MIMO EXPRESSION

The simplest form to represent the orientation of a quadcopter or VTOL vehicle is to consider two integrators in cascade with external perturbations

* This chapter was developed in collaboration with Efrain Ibarra from the Laboratoire Heudiasyc at the Université de Technologie de Compiegne in France and with M. Jimenez from the Universidad Autonoma de Nuevo Leon in Mexico.

[1] Vertical Take-Off and Landing.

Indoor Navigation Strategies for Aerial Autonomous Systems.
DOI: http://dx.doi.org/10.1016/B978-0-12-805189-4.00010-X

as follows:

$$\ddot{\eta} = u_\eta + w_\eta, \tag{7.1}$$

or

$$\dot{\mathbf{x}} = \bar{A}\mathbf{x} + \bar{B}_1 u_\eta + \bar{B}_2 w_\eta \quad \text{with} \quad \mathbf{x} = [\eta_1 \;\; \eta_2], \tag{7.2}$$

where η represents the attitude vector with η being ϕ, θ, or ψ, i.e., the Euler angles, roll, pitch, and yaw, respectively, u_η defines the control input, and w_η the unknown and external perturbation. Even if this representation is experimentally valid for small angles, it does not represent the Coriolis and aerodynamic effects that the aerial vehicle experiences and can generate undesirable dynamics in flight when the vehicle moves quickly.

Studying the complete nonlinear attitude equations to design the controller can be an arduous task; nevertheless, some authors prefer to consider the strongest terms in the orientation to represent their model. These equations are described in the following and, considering our experience with quadcopters, they closely represent the attitude of a quadrotor vehicle:

$$\begin{aligned}
\ddot{\phi} &= \dot{\theta}\dot{\psi}\left(\frac{I_y - I_z}{I_x}\right) - \frac{I_r}{I_x}\dot{\theta}\Omega + \frac{l}{I_x}u_\phi + w_\phi, \\
\ddot{\theta} &= \dot{\phi}\dot{\psi}\left(\frac{I_z - I_x}{I_y}\right) - \frac{I_r}{I_y}\dot{\phi}\Omega + \frac{l}{I_y}u_\theta + w_\theta, \\
\ddot{\psi} &= \dot{\theta}\dot{\phi}\left(\frac{I_x - I_y}{I_z}\right) + \frac{l}{I_z}u_\psi + w_\psi,
\end{aligned} \tag{7.3}$$

where the distance from each motor to the gravity center of the vehicle is denoted by l. The inertia of the vehicle in each axis is defined by I_x, I_y, and I_z while the inertia of the motor is represented by I_r, and the speed of the rotor is defined by Ω.

Notice that system (7.3) is quite different from (7.1) even if the unknown disturbances or uncertainties, w_η, are also considered in the model. We study the system with bounded perturbations because it is obvious that the physical characteristics of the vehicle (power motors, etc.) are not unlimited, and, as a consequence, the perturbations need to be bounded, i.e., $|w_\eta| \leqq L_\eta$, where L_η is a constant that defines the amplitude of each perturbation.

Theorem 7.1. *System (7.3) is equivalent to system*

$$\dot{\boldsymbol{x}} = A\boldsymbol{x} + B(\boldsymbol{u} + \bar{\boldsymbol{w}}) \tag{7.4}$$

with $\boldsymbol{x} = (\phi_1 \quad \theta_1 \quad \psi_1 \quad \phi_2 \quad \theta_2 \quad \psi_2)^T$, $\boldsymbol{u} = (u_\phi \quad u_\theta \quad u_\psi)^T$,

$$\bar{\boldsymbol{w}} = \begin{pmatrix} \dfrac{\theta_2\left(\psi_2\left(\dfrac{I_y - I_z}{I_x}\right) - \dfrac{I_r}{I_x}\Omega\right) - \phi_2 + w_\phi}{\dfrac{l}{I_x}} \\ \dfrac{\phi_2\left(\psi_2\left(\dfrac{I_z - I_x}{I_y}\right) - \dfrac{I_r}{I_y}\Omega\right) - \theta_2 + w_\theta}{\dfrac{l}{I_y}} \\ \dfrac{\theta_2\phi_2\left(\dfrac{I_x - I_y}{I_z}\right) - \psi_2 + w_\psi}{\dfrac{l}{I_z}} \end{pmatrix}, \quad A = \begin{pmatrix} 0_{3\times 3} & \mathbb{I}_{3\times 3} \\ 0_{3\times 3} & \mathbb{I}_{3\times 3} \end{pmatrix}, \quad B = \begin{pmatrix} 0_{3\times 3} \\ B_2 \end{pmatrix}, \quad B_2 = \begin{pmatrix} \dfrac{l}{I_x} & 0 & 0 \\ 0 & \dfrac{l}{I_y} & 0 \\ 0 & 0 & \dfrac{l}{I_z} \end{pmatrix}.$$

Proof. Consider $\gamma_1 = \left(\dfrac{I_y - I_z}{I_x}\right)$, $\gamma_2 = \left(\dfrac{I_z - I_x}{I_y}\right)$, $\gamma_3 = \left(\dfrac{I_x - I_y}{I_z}\right)$, $\beta_1 = \dfrac{I_r}{I_x}\Omega$, $\beta_2 = \dfrac{I_r}{I_y}\Omega$, $b_1 = \dfrac{l}{I_x}$, $b_2 = \dfrac{l}{I_y}$, and $b_3 = \dfrac{l}{I_z}$. Define $\phi_1 = \phi$, $\theta_1 = \theta$, $\psi_1 = \psi$, $\dot{\phi}_1 = \phi_2$, $\dot{\theta}_1 = \theta_2$, and $\dot{\psi}_1 = \psi_2$.

Then, rewriting (7.3) it follows that

$$\begin{aligned} \dot{\phi}_1 &= \phi_2, & \dot{\phi}_2 &= \theta_2\left(\psi_2\gamma_1 - \beta_1\right) + b_1 u_\phi + w_\phi, \\ \dot{\theta}_1 &= \theta_2, & \dot{\theta}_2 &= \phi_2\left(\psi_2\gamma_2 - \beta_2\right) + b_2 u_\theta + w_\theta, \\ \dot{\psi}_1 &= \psi_2, & \dot{\psi}_2 &= \theta_2\phi_2\gamma_3 + b_3 u_\psi + w_\psi. \end{aligned} \tag{7.5}$$

To simplify the analysis, define $f_1 = \theta_2\left(\psi_2\gamma_1 - \beta_1\right)$, $f_2 = \phi_2\left(\psi_2\gamma_2 - \beta_2\right)$, and $f_3 = \theta_2\phi_2\gamma_3$. Taking the three right equations of (7.5), we obtain

$$\begin{aligned} \dot{\phi}_2 &= \phi_2 - \phi_2 + f_1 + b_1 u_\phi + w_\phi, \\ \dot{\theta}_2 &= \theta_2 - \theta_2 + f_2 + b_2 u_\theta + w_\theta, \\ \dot{\psi}_2 &= \psi_2 - \psi_2 + f_3 + b_3 u_\psi + w_\psi. \end{aligned}$$

Define $\Phi_1 = f_1 - \phi_2$, $\Phi_2 = f_2 - \theta_2$, and $\Phi_3 = f_3 - \psi_2$. Then

$$\begin{aligned} \dot{\phi}_2 &= \phi_2 + b_1(u_\phi + \bar{w}_\phi), \\ \dot{\theta}_2 &= \theta_2 + b_2(u_\theta + \bar{w}_\theta), \\ \dot{\psi}_2 &= \psi_2 + b_3(u_\psi + \bar{w}_\psi) \end{aligned}$$

with $\bar{w}_i = (\Phi_i + w_\eta)/b_i$. Finally, define $\boldsymbol{\eta}_i = (\phi_i \quad \theta_i \quad \psi_i)^T$, $i = 1, 2$. Then we can write

$$\begin{aligned} \dot{\boldsymbol{\eta}}_1 &= \boldsymbol{\eta}_2, \\ \dot{\boldsymbol{\eta}}_2 &= \boldsymbol{\eta}_2 + B_2(\boldsymbol{u} + \bar{\boldsymbol{w}}), \end{aligned}$$

which is equivalent to system (7.4). □

7.2 NONLINEAR OPTIMAL CONTROLLER WITH INTEGRAL SLIDING MODE DESIGN

The goal is to stabilize the quadcopter attitude using an optimal control $\boldsymbol{u}$ that is robust with respect to perturbations and parameter variations. For this it is necessary to minimize the following singular quadratic cost:

$$J(\boldsymbol{x}(t)) = \frac{1}{2} \int_{t_1}^{\infty} \left[\boldsymbol{x}(t)^T Q \boldsymbol{x}(t)\right] dt, \tag{7.6}$$

with $Q = Q^T > 0$. The minimization of (7.6) is subject to

$$\dot{\boldsymbol{\eta}}_1 = \boldsymbol{\eta}_2. \tag{7.7}$$

Developing (7.6), it follows that

$$J = \frac{1}{2} \int_{t_1}^{\infty} \left(\boldsymbol{\eta}_1^T Q_{11} \boldsymbol{\eta}_1 + 2\boldsymbol{\eta}_1^T Q_{12} \boldsymbol{\eta}_2 + \boldsymbol{\eta}_2^T Q_{22} \boldsymbol{\eta}_2\right) dt. \tag{7.8}$$

To eliminate the cross terms, the Utkin variable $\boldsymbol{v} = \boldsymbol{\eta}_2 + Q_{22}^{-1} Q_{12}^T \boldsymbol{\eta}_1$ is used. Then

$$J = \frac{1}{2} \int_{t_1}^{\infty} (\boldsymbol{\eta}_1^T Q_1 \boldsymbol{\eta}_1 + \boldsymbol{v}^T Q_{22} \boldsymbol{v}) dt \tag{7.9}$$

with $Q_1 = Q_{11} - Q_{12} Q_{22}^{-1} Q_{12}^T$. Rewriting (7.7) with the Utkin variable yields

$$\dot{\boldsymbol{\eta}}_1 = A_1 \boldsymbol{\eta}_1 + \boldsymbol{v}, \tag{7.10}$$

where $A_1 = -Q_{22}^{-1} Q_{12}^T$. Then (7.9) is not singular with respect to the variable $\boldsymbol{v}$, so that $\boldsymbol{v}$ is taken as an optimal virtual control variable and is given by

$$\boldsymbol{v} = -Q_{22}^{-1} P \boldsymbol{\eta}_1, \tag{7.11}$$

where $P \in \mathbb{R}^{3\times3}$ is the solution to the Riccati equation $PA_1 + A_1^T P - P Q_{22}^{-1} P + Q_1 = 0$. Introducing the Utkin variable into (7.11), we have

$$\eta_2 + Q_{22}^{-1}\left(Q_{12}^T + P\right)\boldsymbol{\eta}_1 = \mathbf{0}. \tag{7.12}$$

Notice that (7.11) and (7.12) are only true if (7.9) is minimized for all $t_1 \geq 0$.

Observe that (7.12) is an optimal vector that can be used for designing the vector $\boldsymbol{S} \in \mathbb{R}^3$ given by

$$\boldsymbol{S} = \boldsymbol{\eta}_2 + Q_{22}^{-1}\left(Q_{12}^T + P\right)\boldsymbol{\eta}_1, \tag{7.13}$$

where (7.13) represents the sliding surface vector. Taking the derivative of the previous equation with respect to time gives

$$\dot{S} = \left[\mathbb{I}_{3\times3} + Q_{22}^{-1}\left(Q_{12}^T + P\right)\right]\boldsymbol{\eta}_2 + B_2\left(\boldsymbol{u} + \bar{\boldsymbol{w}}\right). \tag{7.14}$$

To remove the linear parts, take $\boldsymbol{u}$ as

$$\boldsymbol{u} = B_2^{-1}\left\{\bar{\boldsymbol{u}} - \left[\mathbb{I}_{3\times3} + Q_{22}^{-1}\left(Q_{12}^T + P\right)\right]\boldsymbol{\eta}_2\right\}. \tag{7.15}$$

Thus

$$\dot{S} = \bar{\boldsymbol{u}} + B_2\,\boldsymbol{w}, \tag{7.16}$$

where $\bar{\boldsymbol{u}}$ is the new controller to assure the convergence of the system. Developing the above, it follows that

$$\begin{pmatrix} \dot{S}_1 \\ \dot{S}_2 \\ \dot{S}_3 \end{pmatrix} = \begin{pmatrix} \bar{u}_1 + f_1 - \phi_2 + w_\phi \\ \bar{u}_2 + f_2 - \theta_2 + w_\theta \\ \bar{u}_3 + f_3 - \psi_2 + w_\psi \end{pmatrix}.$$

To remove the linearities $\boldsymbol{\eta}_2$, we propose $\bar{\boldsymbol{u}}$ as

$$\bar{\boldsymbol{u}} = \begin{pmatrix} \bar{u}_1 \\ \bar{u}_2 \\ \bar{u}_3 \end{pmatrix} = \begin{pmatrix} \bar{v}_1 + \phi_2 \\ \bar{v}_2 + \theta_2 \\ \bar{v}_3 + \psi_2 \end{pmatrix}.$$

Then, rewrite $\dot{\boldsymbol{S}}$ as

$$\dot{\boldsymbol{S}} = \begin{pmatrix} \bar{v}_1 + f_1 + w_\phi \\ \bar{v}_2 + f_2 + w_\theta \\ \bar{v}_3 + f_3 + w_\psi \end{pmatrix},$$

where each $\dot{S}_i \in \dot{\boldsymbol{S}}$ is represented as

$$\dot{S}_i = \bar{v}_i + f_i\left(\boldsymbol{\eta}_1, \boldsymbol{\eta}_2, t\right) + w_\eta. \tag{7.17}$$

7.2.1 Convergence of the Sliding Surfaces

In the conventional sliding mode control the robustness property is not guaranteed from the first time instant because the robustness is only guaranteed when the sliding surface reaches zero.

With the integral sliding mode we will be able to compensate nonlinear terms and bounded uncertainties, also the robustness will be guaranteed from the initial time instance. In the following, we introduce variable $\bar{v}_i$ to stabilize the sliding surfaces.

Then, we propose

$$\bar{v}_i = \bar{v}_{i1} + \bar{v}_{i2}. \tag{7.18}$$

- Part $\bar{v}_{i1}$ will be responsible of compensating the nonlinear terms f_i and the bounded disturbance w_η from the beginning.
- Component $\bar{v}_{i2}$ will make sure that each sliding surface S_i reaches the optimal surface $S_i = 0$ at a defined finite time t_1 taking in consideration that the perturbations f_i and w_η have been compensated from the initial time instance $t = 0$.

Design of $\bar{v}_{i1}$

For $\bar{v}_{i1}$ we propose a new auxiliary surface σ_i with $i = 1, 2, 3$ given by[2]

$$\begin{cases} \sigma_i = S_i - Z_i, \\ \dot{Z}_i = \bar{v}_{i2}. \end{cases} \tag{7.19}$$

It should be mentioned that $\bar{v}_{i1}$ is designed as a conventional sliding mode control. This means that for the stability analysis a candidate Lyapunov function could be used. Therefore

$$V(\sigma_i) = \frac{1}{2}\sigma_i^2 > 0. \tag{7.20}$$

The asymptotic stability of (7.19) at the equilibrium point 0 can be proved if the following conditions are satisfied:

$$\text{(a)} \lim_{|\sigma_i| \to \infty} V = \infty, \tag{7.21}$$

$$\text{(b)} \ \dot{V} < 0 \text{ for } \sigma_i \neq 0. \tag{7.22}$$

[2] In this chapter variable σ is used only as an auxiliary sliding surface and not as a saturation function.

Condition (a) is obviously satisfied by V in (7.20). Nevertheless, the finite-time convergence (global finite-time stability) could be achieved if condition (b) is modified as

$$\dot{V} \leq -\alpha_i V^{1/2}, \quad \alpha_i > 0. \tag{7.23}$$

Indeed, separating variables and integrating inequality (7.23) over the time interval $0 \leq \tau \leq t$, we obtain

$$V^{1/2}(\sigma_i(t)) \leq -\frac{1}{2}\alpha_i t + V^{1/2}(\sigma_i(0)), \tag{7.24}$$

and, considering that $V(\sigma_i(t))$ reaches zero in finite time t_r, get

$$t_r \leq \frac{2V^{1/2}(\sigma_i(0))}{\alpha_i}. \tag{7.25}$$

Therefore control $\bar{v}_{i1}$ that satisfies (7.23) will drive σ_i to zero in finite time t_r and will keep it at zero $\forall t \geq t_r$.

Now notice that (7.23) can be written as

$$\sigma_i\dot{\sigma}_i \leq -\bar{\alpha}_i |\sigma_i|, \qquad \bar{\alpha}_i = \frac{\alpha_i}{\sqrt{2}}, \qquad \bar{\alpha}_i > 0. \tag{7.26}$$

Consequently, from the above and with (7.17), (7.18) and (7.19) we have

$$\begin{aligned}\sigma_i * \left(\dot{S}_i - \dot{Z}_i\right) &= \sigma_i * \left(\bar{v}_{i1} + \bar{v}_{i2} + f_i\left(\boldsymbol{\eta}_1, \boldsymbol{\eta}_2, t\right) + w_\eta - \bar{v}_{i2}\right)\\ &= \sigma_i * \left(\bar{v}_{i1} + f_i\left(\boldsymbol{\eta}_1, \boldsymbol{\eta}_2, t\right) + w_\eta\right),\end{aligned}$$

and, selecting $\bar{v}_{i1} = -\rho_i \operatorname{sgn}(\sigma_i)$, condition (7.26) is fulfilled if and only if

$$\rho_i = \bar{\alpha}_i + \left|f_i\left(\boldsymbol{\eta}_1, \boldsymbol{\eta}_2, t\right)\right| + L_\eta. \tag{7.27}$$

Notice that (7.27) represents the necessary gains for ensuring the finite time stability in a bounded finite time t_r, which means

$$t_r \leq \frac{2V^{1/2}(\sigma_i(0))}{\alpha_i} = \frac{|\sigma_i(0)|}{\bar{\alpha}}. \tag{7.28}$$

The above implies that $\sigma_i = \dot{\sigma}_i = 0$ for all $t \geq t_r$, then the condition $\dot{\sigma}_i = 0$ produces

$$\dot{\sigma}_i = \underbrace{-\rho_i \operatorname{sgn}(\sigma_i)}_{\bar{v}_{i1}} + f_i\left(\boldsymbol{\eta}_1, \boldsymbol{\eta}_2, t\right) + w_\eta = 0 \quad \forall t \geq t_r,$$

meaning that $-\rho_1 \operatorname{sgn}(\sigma_i)$ will compensate the perturbative terms $f_i(\boldsymbol{\eta}_1, \boldsymbol{\eta}_2, t) + w_\eta$ only during the reaching phase.

In the following, to eliminate the reaching phase, we observe in (7.28) that proposing $\sigma_i(0) = 0$ will imply $t_r = 0$ and as a result $\sigma_i = \dot{\sigma}_i = 0$ for all $t \geq 0$.

Based on the above considerations, we present the next result

$$\bar{v}_{i1} = -\rho_i \operatorname{sgn}(\sigma_i) = -\left(f_i\left(\boldsymbol{\eta}_1, \boldsymbol{\eta}_2, t\right) + w_\eta\right), \quad \forall t \geq 0$$

if and only if $\sigma_i(0) = 0$.

Therefore control $\bar{v}_{i1} = -\rho_i \operatorname{sgn}(\sigma_i)$ compensates $f_i(\boldsymbol{\eta}_1, \boldsymbol{\eta}_2, t) + w_\eta$ for all $t \geq 0$ if and only if $\sigma_i(0) = 0$.

Now considering that control $\bar{v}_{i1}$ accomplishes $\sigma_i(t) = 0$ for all $t \geq 0$ and due to (7.19), it follows that $S_i(t) = Z_i(t)$ for all $t \geq 0$. Therefore (7.19) can be rewritten as

$$\begin{cases} S_i = Z_i, \\ \dot{Z}_i = \bar{v}_{i2}, \end{cases} \quad \text{with } Z_i(0) = S_i(0). \tag{7.29}$$

Then considering (7.29) we have

$$\dot{S}_i = \bar{v}_{i2}.$$

The next step is to design $\bar{v}_{i2}$ such that S_i converges to zero in finite time.

Design of $\bar{v}_{i2}$

In order to achieve global finite-time stability at the optimal sliding surfaces $S_i = 0$, we introduce $\bar{v}_{i2} = -k_i |S_i|^{1/2} \operatorname{sgn}(S_i)$; then $\dot{S}_i$ is written as

$$\dot{S}_i = -k_i |S_i|^{1/2} \operatorname{sgn}(S_i). \tag{7.30}$$

The following Lyapunov function is proposed to prove that (7.30) converges to zero in finite time:

$$V(S_i) = |S_i| > 0. \tag{7.31}$$

The previous equations must satisfy conditions (7.22). Observe that, by definition (7.31), condition (a) is achieved; to fulfill condition (b), we use (7.23). Notice that when introducing (7.31) into (7.23) an equivalent modified condition is obtained, which is given by

$$\frac{S_i \dot{S}_i}{|S_i|} \leq -\alpha_i |S_i|^{1/2}. \tag{7.32}$$

Introducing (7.30) into (7.32), the previous inequality becomes

$$-k_i |S_i|^{1/2} \leq -\alpha_i |S_i|^{1/2}. \tag{7.33}$$

Therefore to satisfy (7.33) each gain k_i must be equal to α_i, meaning that $k_i > 0$, and this implies that

$$\dot{V} = -\alpha_i V^{1/2} \quad \text{if} \quad k_i = \alpha_i > 0.$$

Observe that the finite-time convergence time t_{r_i} will not be bounded, it will be exactly equal to

$$t_{r_i} = \frac{2V^{1/2}(S_i(0))}{k_i} = \frac{2|S_i(0)|^{1/2}}{k_i}; \tag{7.34}$$

consequently, $S_i = 0$ in a finite time t_{r_i}. Observe that (7.34) represents the finite-time convergence of each S_i to the optimal sliding surface $S_i = 0$. With the purpose that each S_i has the same finite time convergence, we fix $t_{r_i} = t_1 = \text{cte}$,[3] then we will be able to design the gains k_i in order to have $S_1(t_1) = S_2(t_1) = \cdots = S_n(t_1) = 0$.

Fixing $t_{r_i} = t_1 = \text{cte}$, the necessary gains for getting $S_i(t_1) = 0$ can be obtained with $k_i = \dfrac{2|S_i(0)|^{1/2}}{t_1}$.

Summarizing the methodology, it follows that when considering (7.17) and the fact that $\bar{v}_i = \bar{v}_{i1} + \bar{v}_{i2}$, we have

$$\dot{S}_i = \underbrace{-\rho_i \operatorname{sgn}(S_i - Z_i)}_{\bar{v}_{i1}} + \underbrace{\left(-k_i |S_i|^{1/2} \operatorname{sgn}(S_i)\right)}_{\bar{v}_{i2}} + f_i + w_\eta$$

with gains

$$\rho_i = \bar{\alpha}_i + \left|f_i(\boldsymbol{\eta}_1, \boldsymbol{\eta}_2, t)\right| + L_\eta \quad \text{and} \quad k_i = \frac{2|S_i(0)|^{1/2}}{t_1}. \tag{7.35}$$

- Component $\bar{v}_{i1}$ will compensate the perturbative terms $f_i(\boldsymbol{\eta}_1, \boldsymbol{\eta}_2, t) + w_\eta$ for all $t \geq 0$ if and only if $Z_i(0) = S_i(0)$.
- Considering that the perturbative terms have been compensated from the initial time, control $\bar{v}_{i2}$ will ensure that every S_i will converge to the optimal sliding surface $S_i = 0$ in a fixed reaching time t_1.

Therefore, component $\bar{u}_i$ from (7.15) is given by

$$\begin{aligned} \bar{u}_i &= -\rho_i \operatorname{sgn}(S_i - Z_i) - k_i |S_i|^{1/2} \operatorname{sgn}(S_i) + \eta_{2_i}, \\ Z_i &= -k_i \textstyle\int |S_i|^{1/2} \operatorname{sgn}(S_i)\, dt \quad \text{with } Z_i(0) = S_i(0), \end{aligned} \tag{7.36}$$

and gains $t_1 = \text{cte}$ and $\bar{\alpha}_i = \dfrac{\alpha_i}{\sqrt{2}}$ in (7.35).

[3] cte = constant.

7.3 NUMERICAL VALIDATION

Simulations are realized to validate the proposed controllers. From section 7.1 notice that (7.3) can be expressed in regular form as

$$\begin{aligned} \dot{\eta}_1 &= \eta_2, \\ \dot{\eta}_2 &= \eta_2 + B_2(\boldsymbol{u} + \bar{\boldsymbol{w}}), \end{aligned}$$

where $\boldsymbol{\eta_1} = (\phi_1, \theta_1, \psi_1)^T$ and $\boldsymbol{\eta_2} = (\phi_2, \theta_2, \psi_2)^T$.

Following the previous control procedure, some matrices are necessary to compute $\boldsymbol{u}$. These matrices are proposed as follows:

$$Q = Q^T = \begin{pmatrix} 17 & 8 & 7 & -6 & -4 & 2 \\ 8 & 26 & 6 & 4 & 4 & 13 \\ 7 & 6 & 11 & 9 & 1 & 2 \\ -6 & 4 & 9 & 23 & 9 & 4 \\ -4 & 4 & 1 & 9 & 9 & 3 \\ 2 & 13 & 2 & 4 & 3 & 8 \end{pmatrix} > 0,$$

thus

$$Q_{11} = \begin{pmatrix} 17 & 8 & 7 \\ 8 & 26 & 6 \\ 7 & 6 & 11 \end{pmatrix}, \quad Q_{12} = Q_{12}^T = \begin{pmatrix} -6 & -4 & 2 \\ 4 & 4 & 13 \\ 9 & 1 & 2 \end{pmatrix},$$

$$Q_{22} = \begin{pmatrix} 23 & 9 & 4 \\ 9 & 9 & 3 \\ 4 & 3 & 8 \end{pmatrix}.$$

Therefore, from (7.9) and (7.10),

$$Q_1 = \begin{pmatrix} 13.1924 & 3.9244 & 8.0378 \\ 3.9244 & 4.5773 & 3.7113 \\ 8.0378 & 3.7113 & 6.1443 \end{pmatrix},$$

$$A_1 = \begin{pmatrix} 0.1787 & 0.1203 & -0.5601 \\ 0.4330 & -0.0034 & 0.5017 \\ -0.5017 & -1.6838 & -0.1581 \end{pmatrix}.$$

Solving the Riccati equation, it follows that

$$P = \begin{pmatrix} 25.3132 & 9.9395 & -0.5181 \\ 9.9395 & 8.9679 & -1.5217 \\ -0.5181 & -1.5217 & 4.4597 \end{pmatrix},$$

and finally $\boldsymbol{S}$ can be written as

$$\boldsymbol{S} = \boldsymbol{\eta_2} + \underbrace{Q_{22}^{-1}\left(Q_{12}^{T} + P\right)}_{M}\boldsymbol{\eta_1}, \tag{7.37}$$

where

$$M = \begin{pmatrix} M_{11} & M_{12} & M_{13} \\ M_{21} & M_{22} & M_{23} \\ M_{31} & M_{32} & M_{33} \end{pmatrix} = \begin{pmatrix} 0.9702 & -0.0036 & 0.5814 \\ -0.2404 & 1.1036 & -0.9275 \\ -0.2097 & 1.0228 & 0.8646 \end{pmatrix}.$$

Then, the sliding surfaces are given by

$$\begin{aligned} S_1 &= \phi_2 + M_{11}\phi_1 + M_{12}\theta_1 + M_{13}\psi_1, \\ S_2 &= \theta_2 + M_{21}\phi_1 + M_{22}\theta_1 + M_{23}\psi_1, \\ S_3 &= \psi_2 + M_{31}\phi_1 + M_{32}\theta_1 + M_{33}\psi_1. \end{aligned}$$

Rewriting $\boldsymbol{u} \in \mathbb{R}^3$, it follows that

$$\begin{aligned} u_\phi &= (I_x/l)\left(\bar{u}_1 - M_{13}\psi_2 - M_{12}\theta_2 - (M_{11}+1)\phi_2\right), \\ u_\theta &= (I_y/l)\left(\bar{u}_2 - M_{21}\phi_2 - M_{23}\psi_2 - (M_{22}+1)\theta_2\right), \\ u_\psi &= (I_z/l)\left(\bar{u}_3 - M_{31}\phi_2 - M_{32}\theta_2 - (M_{33}+1)\,\psi_2\right), \end{aligned}$$

with

$$\begin{aligned} \bar{u}_1 &= -\rho_1 \operatorname{sgn}(S_1 - Z_1) - k_1 |S_1|^{1/2} \operatorname{sgn}(S_1) + \phi_2, \\ Z_1 &= -k_1 \int |S_1|^{1/2} \operatorname{sgn}(S_1)\, dt \quad \text{with } Z_1(0) = S_1(0), \\ \bar{u}_2 &= -\rho_2 \operatorname{sgn}(S_2 - Z_2) - k_2 |S_2|^{1/2} \operatorname{sgn}(S_2) + \theta_2, \\ Z_2 &= -k_2 \int |S_2|^{1/2} \operatorname{sgn}(S_2)\, dt \quad \text{with } Z_2(0) = S_2(0), \\ \bar{u}_3 &= -\rho_3 \operatorname{sgn}(S_3 - Z_3) - k_3 |S_3|^{1/2} \operatorname{sgn}(S_3) + \psi_2, \\ Z_3 &= -k_3 \int |S_3|^{1/2} \operatorname{sgn}(S_3)\, dt \quad \text{with } Z_3(0) = S_3(0), \end{aligned}$$

and gains given by

$$\begin{aligned} \rho_1 &= \left(\frac{\alpha_1}{\sqrt{2}} + |\theta_2(\psi_2\gamma_1 - \beta_1)| + L_1\right), \\ \rho_2 &= \left(\frac{\alpha_2}{\sqrt{2}} + |\phi_2(\psi_2\gamma_2 - \beta_2)| + L_2\right), \\ \rho_3 &= \left(\frac{\alpha_3}{\sqrt{2}} + |\theta_2\phi_2\gamma_3| + L_3\right). \end{aligned}$$

$$\left.\begin{aligned} k_1 &= \frac{2\,|S_1(0)|^{1/2}}{t_1} \\ k_2 &= \frac{2\,|S_2(0)|^{1/2}}{t_1} \\ k_3 &= \frac{2\,|S_3(0)|^{1/2}}{t_1} \end{aligned}\right| , \qquad t_1 = \text{cte} \tag{7.38}$$

The initial conditions were set as $\phi_1(0) = -5$, $\theta_1(0) = 2$, $\psi_1(0) = -3$, all in grad, $\phi_2(0) = 6$, $\theta_2(0) = 4$, $\psi_2(0) = 6$ in grad/s. These conditions imply $S_1(0) = -0.6026$, $S_2(0) = 10.1917$, and $S_3(0) = 6.5004$. We can also define $t_1 = 1$ s as a finite-time convergence instance for the sliding surfaces S_1, S_2, and S_3. Thus, from (7.38) we obtained $k_1 = 1.5525$, $k_2 = 6.3849$, and $k_3 = 5.0992$. Furthermore, for simulation purposes we assumed $\alpha_1 = 0.12$, $\alpha_2 = 0.14$, and $\alpha_3 = 0.2$. The bounded disturbances were selected as follows:

$$\begin{aligned} w_\phi &= 2\sin(t)\,\text{sgn}(S_1), & |w_\phi| &\leqq L_\phi = 2, \\ w_\theta &= -1.5\cos(2t)\,\text{sgn}(S_2), & |w_\theta| &\leqq L_\theta = 1.5, \\ w_\psi &= -0.5\exp(\cos(t))\,\text{sgn}(S_3), & |w_\psi| &\leqq L_\psi = 0.5\exp(1). \end{aligned}$$

The above disturbances were considered multiplied by $\text{sgn}(S_i)$ for the following reasons:

1. To observe that the effect chattering produced by $\text{sgn}(S_i)$ is not relevant for the compensation of such disturbances. We are only interested in the knowledge of the maximum amplitude that each perturbation can reach, and not in the high frequency that it can produce.
2. To graphically validate the theory. It is expected to observe an evident chattering effect in w_ϕ, w_θ, and w_ψ when $S_i = 0$ at the desired finite time $t_1 = 1$.

The following graphs were obtained when applying the proposed control scheme. From Fig. 7.1 observe that the auxiliary sliding surfaces σ_1, σ_2, and σ_3 are zero from the initial time instance, meaning that the robustness is guaranteed with respect to bounded uncertainties all the time. Besides, S_1, S_2, and S_3 converge to zero in a desired finite time $t_1 = 1$; this means that the vector of sliding surface $\boldsymbol{S} = \boldsymbol{\eta_2} + Q_{22}^{-1}(Q_{12}^T + P)\boldsymbol{\eta_1}$ converges to the optimal vector of sliding surfaces $\boldsymbol{S} = \boldsymbol{0}$ in finite time $t_1 = 1$, and with this fact every solution $\boldsymbol{\eta_1} = (\phi, \theta, \psi)^T$ and $\boldsymbol{\eta_2} = (\dot\phi, \dot\theta, \dot\psi)^T$ that belongs to $\boldsymbol{S} = \boldsymbol{0}$ will be called an optimal sliding mode because it will be able to min-

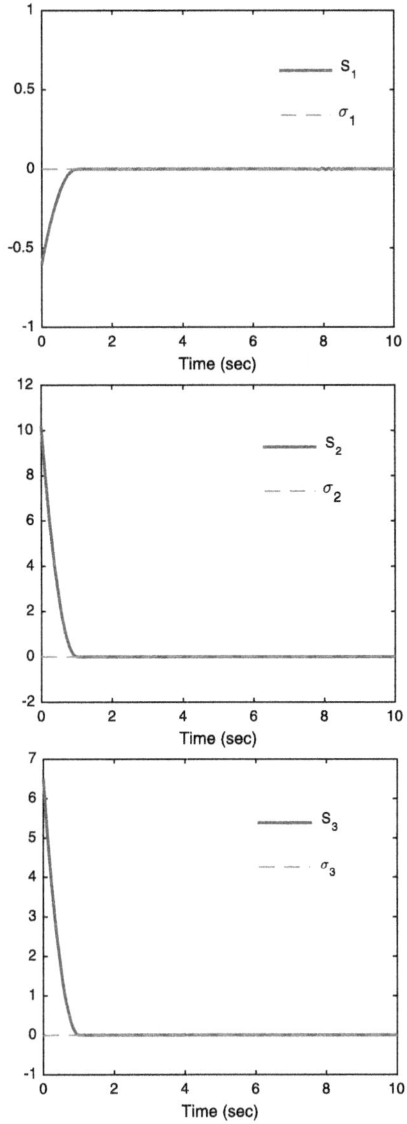

Figure 7.1 Convergence of auxiliary surfaces σ_i and sliding surfaces S_i.

imize the cost function (7.9) for all $t \geq t_1$ and in this manner solve the LQR problem.[4]

[4] This means that the state variables which minimize the quadratic cost function will be asymptotically stable.

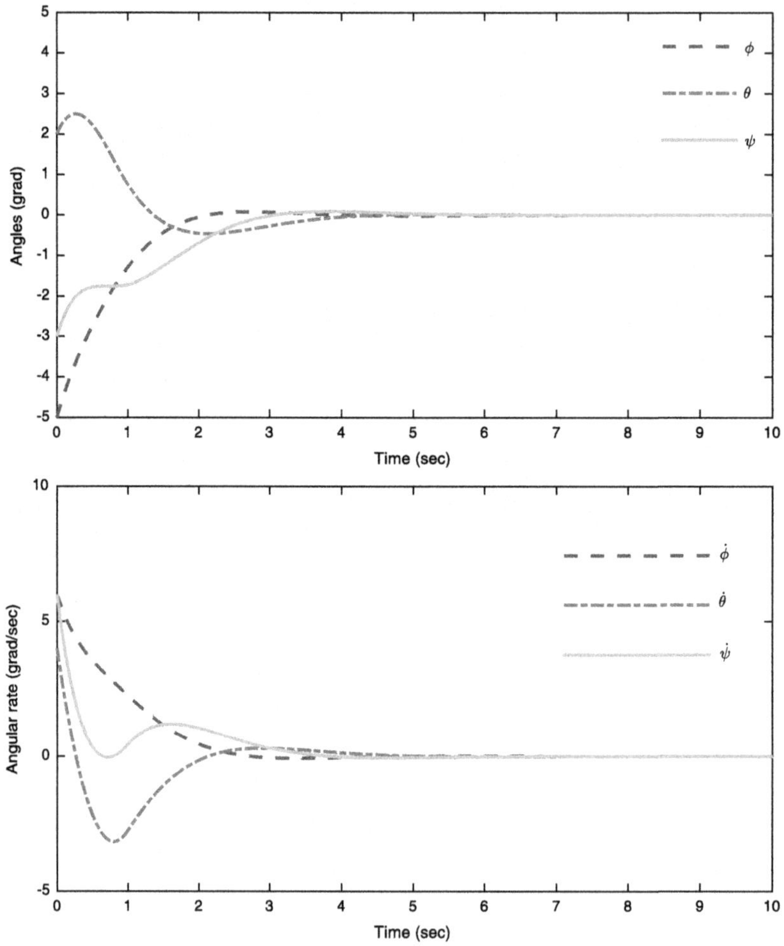

Figure 7.2 Stabilization of the dynamics of ϕ, θ, ψ and $\dot{\phi}, \dot{\theta}, \dot{\psi}$.

In Fig. 7.2 an asymptotic stabilization is clearly visible for the dynamics of ϕ, θ, ψ and $\dot{\phi}, \dot{\theta}, \dot{\psi}$ due to the convergence of the vector $\boldsymbol{S}$ to the optimal sliding vector $\boldsymbol{S} = \boldsymbol{0}$ in finite time $t_1 = 1$.

In Fig. 7.3 it can be observed that the bounded uncertainties w_ϕ, w_θ, and w_ψ present an evident chattering effect by considering the factor $\operatorname{sgn}(S_i)$, and also that this chattering effect appears at time $t_1 = 1$. However, although these uncertainties present a high frequency, the control signal responses u_ϕ, u_θ, and u_ψ shown in Fig. 7.4 compensate such perturbations from the initial time instance $t = 0$.

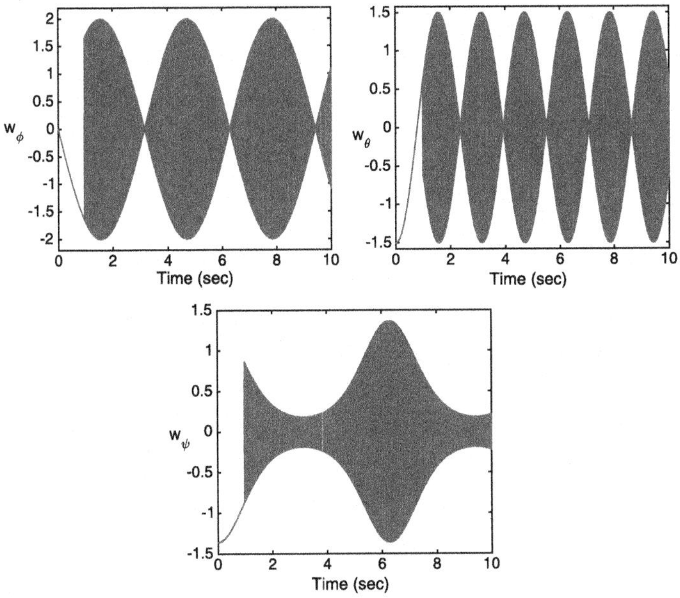

Figure 7.3 Bounded uncertainties w_ϕ, w_θ, and w_ψ.

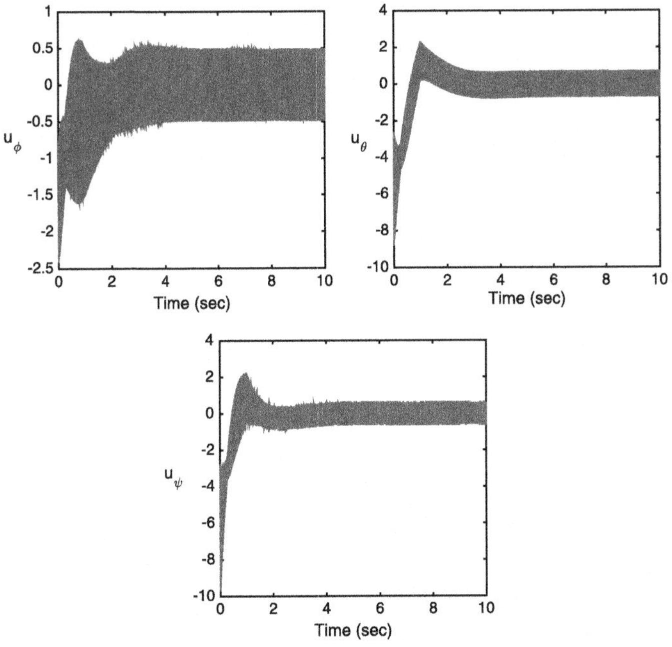

Figure 7.4 Control signals u_ϕ, u_θ, and u_ψ.

In Fig. 7.4 it is shown that the control signal responses u_ϕ, u_θ, and u_ψ have a chattering effect from the beginning, meaning that such controllers are compensating the proposed bounded uncertainties w_ϕ, w_θ, and w_ψ for all $t \geq 0$.

7.3.1 Emulation Results

One of the problems existing today is due to the discontinuity produced by the sgn function. This discontinuity produces a high frequency, normally called the chattering effect, which in practical applications is not convenient because it could produce unwanted vibrations that could damage the instruments in real-time implementations. Nevertheless, there exists extensive literature on techniques that help diminish the chattering effect.

Notice the chattering effect produced by the sgn function in Fig. 7.4. If we want to improve the performance of the control inputs, several tricks could be used. One technique used frequently is to approximate the sgn function; see, for example, [6, Chap. 1]. This approximation can be applied as

$$\operatorname{sgn}(\sigma) \approx \frac{\sigma}{|\sigma| + \epsilon}. \tag{7.39}$$

For our simulations purposes we consider $\epsilon = 0.0007$.

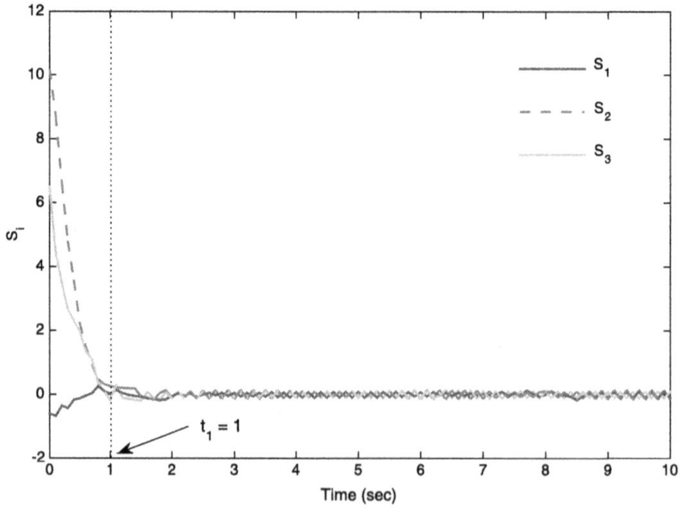

Figure 7.5 Convergence of the sliding surfaces S_1, S_2, and S_3.

The idea when validating the proposed algorithm is to have similar results in simulations and/or experiments. For this we consider that all states ϕ, θ, ψ, $\dot{\phi}$, $\dot{\theta}$, and $\dot{\psi}$ are affected by some kind of white noise (that it is true when using inertial sensors). Under these assumptions the following graphs are obtained.

In Fig. 7.5 we observe that the sliding surfaces S_1, S_2, and S_3 converge to zero in finite time t_1, although the nonlinear system is being affected by white noise. In Fig. 7.6 asymptotic convergence is observed for the dynamics of ϕ, θ, ψ and $\dot{\phi}, \dot{\theta}, \dot{\psi}$ in spite of having white noise in the internal dynamics of the system. Due to this, the controls shown in Fig. 7.7 achieve the convergence of the vector $\boldsymbol{S}$ to the optimal sliding vector $\boldsymbol{S} = \mathbf{0}$ in

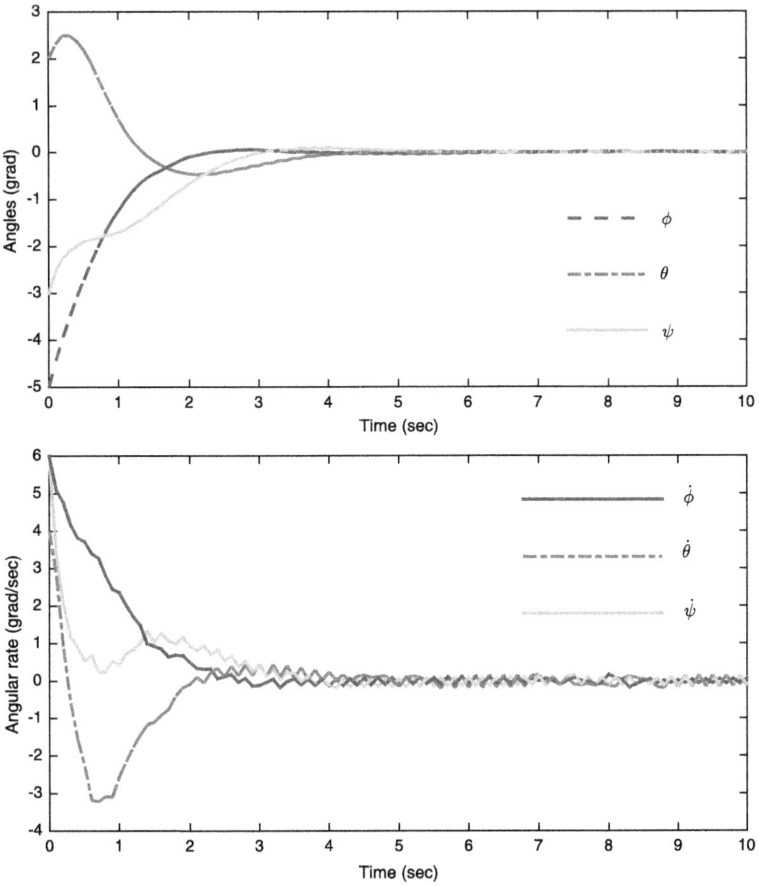

Figure 7.6 Dynamic stabilization of ϕ, θ, ψ and $\dot{\phi}, \dot{\theta}, \dot{\psi}$ with sensor noise.

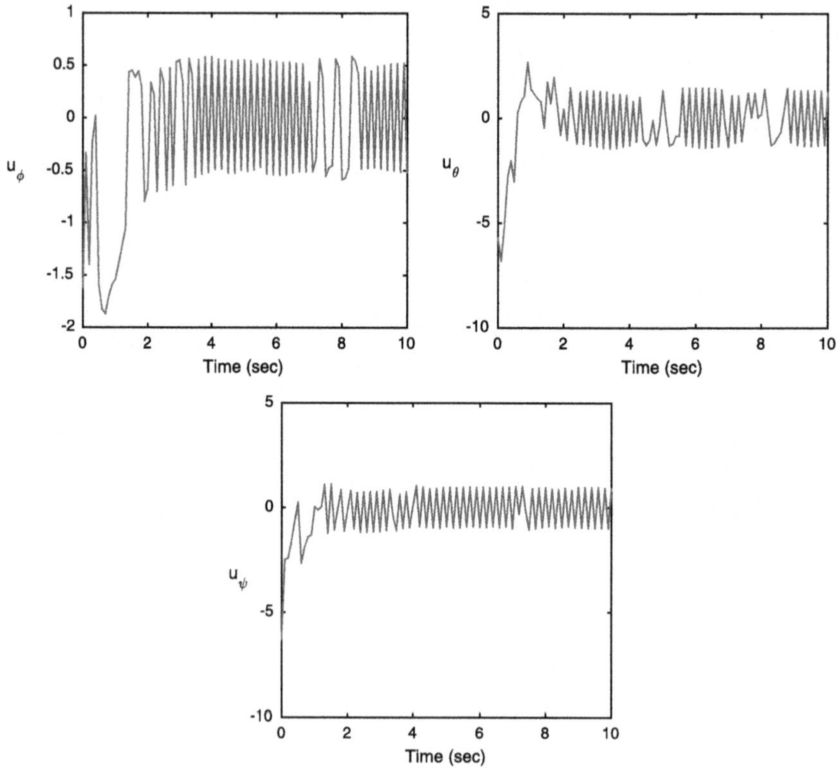

Figure 7.7 Control signals u_ϕ, u_θ, and u_ψ.

finite time. In Fig. 7.7 a representation of the controllers u_ϕ, u_θ, and u_ψ is shown; they drive $\boldsymbol{S}$ to zero in finite time in the presence of white noise and bounded uncertainties.

7.4 REAL-TIME VALIDATION

The previous controller was validated on our platform (see Sect. 4.1.5 of Chap. 4) to analyze the attitude performance of a quadcopter. Manual and aggressive references were given to test the behavior of the controller. The following main graphs illustrate the results.

Applying the controller as obtained in Sect. 7.3 in real time is very difficult, and the system is very sensible to small changes. This effect appears because even if the sliding surfaces go to zero, the cross terms of other variables are present in these surfaces, e.g., recall that surface S_1 is for assuring the roll angle convergence and notice that pitch and yaw terms are also included. In addition, if the terms M_{ij} of these variables (pitch and yaw) are bigger then they will strongly influence the controller performance. Therefore, in this part we will include a theoretical–practical tune up that could be used for sliding controllers.

Observe that $\boldsymbol{S} = \boldsymbol{v_q} + M\boldsymbol{q}$. Then we can choose M not only to assure convergence of the sliding surfaces but also to facilitate the implementation in real time. Therefore the goal will be to weigh the main diagonal of M to assure good convergence of each state. Remember that M is given by $M = Q_{22}^{-1}\left(Q_{12}^T + P\right)$ and Q is given by

$$Q = Q^T = \begin{pmatrix} Q_{11} & Q_{12} \\ Q_{12}^T & Q_{22} \end{pmatrix} > 0, \quad Q_{11}, Q_{12}, Q_{22} \in \mathbb{R}^{3\times 3}.$$

For a practical tune up we can consider $Q_{12} = \boldsymbol{0_{3\times 3}}$ and Q_{22} as

$$Q_{22} = \begin{pmatrix} r & 0 & 0 \\ 0 & s & 0 \\ 0 & 0 & g \end{pmatrix}$$

with r, s, and g being positive numbers to guarantee $Q_{22} = Q_{22}^T > 0$. Hence M becomes

$$M = \begin{pmatrix} P_{11}/r & 0 & 0 \\ 0 & P_{22}/s & 0 \\ 0 & 0 & P_{33/g} \end{pmatrix} = \begin{pmatrix} M_{11} & 0 & 0 \\ 0 & M_{22} & 0 \\ 0 & 0 & M_{33} \end{pmatrix},$$

and this implies that each S_i is given by

$$\begin{aligned} S_1 &= \phi_2 + M_{11}\phi_1, \\ S_2 &= \theta_2 + M_{22}\theta_1, \\ S_3 &= \psi_2 + M_{33}\psi_1. \end{aligned}$$

Then $\boldsymbol{u} \in \mathbb{R}^3$ for practical validation is given as

$$\begin{aligned} u_\phi &= (I_x/l)\left(\bar{u}_1 - (M_{11}+1)\,\phi_2\right), \\ u_\theta &= (I_y/l)\left(\bar{u}_2 - (M_{22}+1)\theta_2\right), \\ u_\psi &= (I_z/l)\left(\bar{u}_3 - (M_{33}+1)\,\psi_2\right), \end{aligned}$$

with

$$\begin{aligned} \bar{u}_1 &= -\rho_1 \operatorname{sgn}(S_1 - Z_1) - k_1 |S_1|^{1/2} \operatorname{sgn}(S_1) + \phi_2, \\ Z_1 &= -k_1 \int |S_1|^{1/2} \operatorname{sgn}(S_1)\, dt \quad \text{with } Z_1(0) = S_1(0), \\ \bar{u}_2 &= -\rho_2 \operatorname{sgn}(S_2 - Z_2) - k_2 |S_2|^{1/2} \operatorname{sgn}(S_2) + \theta_2, \end{aligned}$$

$$
\begin{aligned}
Z_2 &= -k_2 \int |S_2|^{1/2} \operatorname{sgn}(S_2)\, dt \quad \text{with } Z_2(0) = S_2(0), \\
\bar{u}_3 &= -\rho_3 \operatorname{sgn}(S_3 - Z_3) - k_3 |S_3|^{1/2} \operatorname{sgn}(S_3) + \psi_2, \\
Z_3 &= -k_3 \int |S_3|^{1/2} \operatorname{sgn}(S_3)\, dt \quad \text{with } Z_3(0) = S_3(0).
\end{aligned}
$$

The new gains are given by

$$
\begin{aligned}
\rho_1 &= \left(\frac{\alpha_1}{\sqrt{2}} + |\theta_2 (\psi_2 \gamma_1 - \beta_1)| + L_1 \right), \\
\rho_2 &= \left(\frac{\alpha_2}{\sqrt{2}} + |\phi_2 (\psi_2 \gamma_2 - \beta_2)| + L_2 \right), \\
\rho_3 &= \left(\frac{\alpha_3}{\sqrt{2}} + |\theta_2 \phi_2 \gamma_3| + L_3 \right),
\end{aligned}
$$

and other parameters are

$$
\left.
\begin{aligned}
k_1 &= \frac{2|S_1(0)|^{1/2}}{t_1} \\
k_2 &= \frac{2|S_2(0)|^{1/2}}{t_1} \\
k_3 &= \frac{2|S_3(0)|^{1/2}}{t_1}
\end{aligned}
\right| \quad t_1 = \text{cte},
$$

$$
\begin{aligned}
\beta_1 &= \beta_2 (\gamma_1 + 1), & \gamma_3 &= -\gamma_1 \left(\frac{1}{\gamma_2 (\gamma_1 + 1) + 1} \right), \\
b_2 &= \frac{b_1}{(\gamma_1 + 1)}, & b_3 &= \frac{b_1}{[\gamma_2 (\gamma_1 + 1) + 1]}, \\
b_2^{-1} &= \frac{(\gamma_1 + 1)}{b_1}, & b_3^{-1} &= \frac{[\gamma_2 (\gamma_1 + 1) + 1]}{b_1}.
\end{aligned}
$$

Analyzing the previous gains, we can observe that the gains for heuristic tune up are β_2, γ_1, γ_2, and b_1. Thus, the gains can be proposed as

$$
\begin{aligned}
\rho_1 &= (|\theta_2 (\psi_2 \gamma_1 - \beta_1)| + L_x), \\
\rho_2 &= (|\phi_2 (\psi_2 \gamma_2 - \beta_2)| + L_y), \\
\rho_3 &= (|\theta_2 \phi_2 \gamma_3| + L_z),
\end{aligned}
$$

and if we want $\cancelto{0}{\psi_2 \gamma_1 - \beta_1}$ and $\cancelto{0}{\psi_2 \gamma_2 - \beta_2}$ then for the tune up we can use the following expressions:

$$
\gamma_3 = -\frac{\gamma_1}{\gamma_1 + 1}, \quad \gamma_1, b_1 = \text{any number}, \quad b_2 = b_3 = \frac{b_1}{(\gamma_1 + 1)},
$$

and then the gains become $\rho_1 = (L_x)$, $\rho_2 = (L_y)$, and $\rho_3 = (|\theta_2 \phi_2 \gamma_3| + L_z)$.

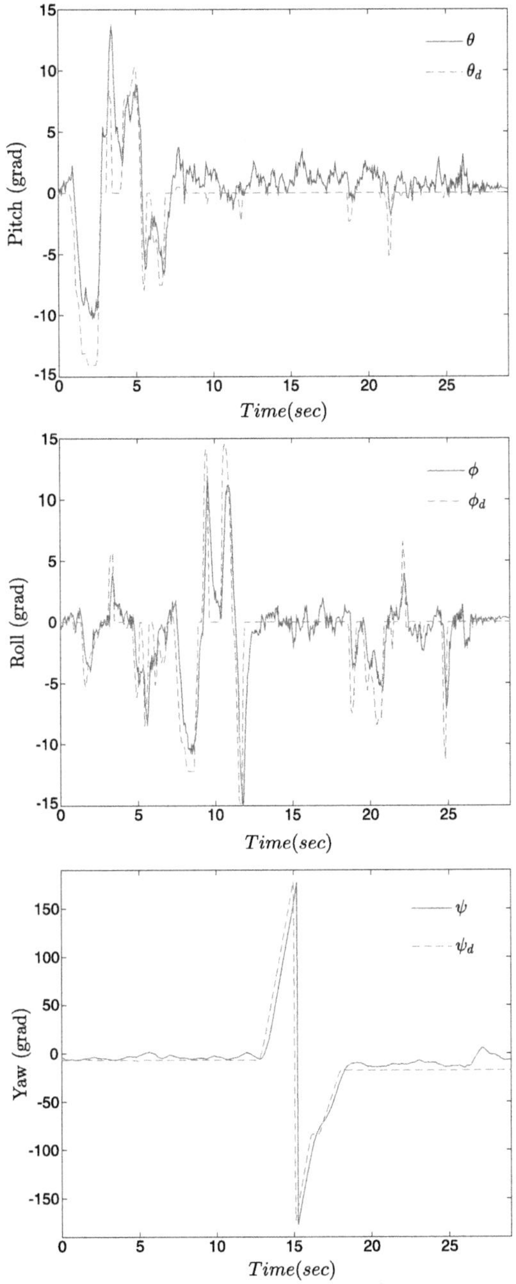

Figure 7.8 Attitude response when applying the controller in real time.

Fig. 7.8 introduces the behavior of the controllers when they are applied in a quadcopter prototype. Observe in this figure that the controller performs well in practice and that the closed-loop system is guaranteed even in the presence of aggressive maneuvers. Also notice that the procedure to tune the gains in the controller works pretty well and can be applied to others sliding controllers. Desired references were given manually by the pilot. Notice that several changes were produced to observe the performance of the controller.

7.5 DISCUSSION

Sliding mode control is becoming a popular tool when working with UAVs since robustness and quick convergence properties make such controllers very interesting to apply in autonomous vehicles. On the negative side, the main problem is the chattering effect produced in the control responses. This effect could damage the physical parts of the system and could require lots of energy for good efficiency. New controllers as proposed in this chapter try to reduce these drawbacks and improve the performance of such algorithms. Nevertheless, many issues still need to be solved and remain an open research topic.

The controller presented in this chapter was designed to be robust with respect to unknown and bounded perturbations and to guarantee convergence in finite time. This fact is not typical in controllers. From emulation results we could observe good performance of the algorithms. The next step for this controller will be to reduce the chattering effect in its design. In addition, a methodology to theoretically tune sliding algorithms was also presented in this chapter. It is very useful when applying the controller in real time. The graphs obtained when implementing the proposed controller demonstrated good performance in closed-loop system.

REFERENCES

1. T. Ledgerwood, M.E., Controllability and nonlinear control of rotational inverted pendulum, in: Advances in Robust and Nonlinear Control Systems, ASME Journal on Dynamic Systems and Control 43 (1992) 81–88.
2. A. Lukyanov, Optimal nonlinear block-control method, in: Proceedings of the 2nd European Control Conference, Groningen, Netherlands, 1993, pp. 1853–1855.
3. A. Lukyanov, S.J. Dodds, Sliding mode block control of uncertain nonlinear plants, in: Proceedings of the IFAC World Congress, San Francisco, CA, USA, 1996, pp. 241–246.
4. V. Utkin, Sliding Modes in Control and Optimization, Springer-Verlag, 1992.

5. V. Utkin, J. Guldner, J. Shi, Sliding Mode Control in Electromechanical Systems, Taylor and Francis, 1999.
6. Y. Shtessel, C. Edwards, L. Fridman, A. Levant, Sliding Mode Control and Observation, Birkhäuser, 2014.

CHAPTER 8

Robust Simple Controllers*

In any realistic problem, uncertainties (modeling errors or unknown dynamics) and external disturbances exist and they could compromise the resulting classical control design and therefore the stability of the closed-loop system. Robustness is a property that guarantees that the essential functions of the designed system are maintained under adverse conditions in which the model no longer accurately reflects reality [1]. In other words, robust control refers to the control of a nominal plant accompanied by unknown dynamics (uncertainties) and/or subject to unknown and external disturbances. These uncertainties and disturbances are taken into account during the design process.

In the specific case of the quadrotor aerial vehicle, in Sect. 2.3 of Chap. 2 we showed that the quadrotor undergoes a blade flapping effect and drag dynamics. Moreover, in several real (outdoor) missions the quadrotor could be exposed to wind. Hence the first proposed approach takes into account such perturbations. The control algorithm is based on the saturation function and is used to stabilize a chain of integrators in cascade.

In Chap. 4 we emphasized that one common problem when applying control strategies in real time is the external wind. When this unknown perturbation is significant, the performance of the closed-loop system is compromised. Based on the Uncertainty and Disturbance Estimator (UDE), a robust control scheme is proposed to reinforce the closed loop when using linear controllers.

The main advantage of both control strategies is that they are linear controllers with only a few tuning parameters and therefore their implementation in quadrotor vehicles is easy. This fact makes them an interesting choice for controlling quadrotors in real applications and for dealing with model uncertainties and persistent disturbances with outstanding performance [2].

* The results in this chapter were developed in collaboration with O. Santos from the Universidad Autónoma del Estado de Hidalgo, México and R. Sanz from the Universitat Politécnica de Valéncia, Spain.

Indoor Navigation Strategies for Aerial Autonomous Systems.
DOI: http://dx.doi.org/10.1016/B978-0-12-805189-4.00011-1

8.1 NONLINEAR ROBUST ALGORITHMS BASED ON SATURATION FUNCTIONS

Remember that the saturation function is defined as follows: Given a positive constant b, a function $\sigma_b : \mathbb{R} \to \mathbb{R}$ is said to be a linear saturation for s if it is a continuous and nondecreasing function satisfying

$$s\sigma_b(s) > 0 \text{ for all } s \neq 0; \quad \sigma_b(s) = s \text{ when } |s| \leq b; \quad |\sigma_b(s)| \leq b \text{ for all } s \in \mathbb{R}.$$

Theorem 8.1. *Consider the system*

$$\dot{r}_1 = r_2, \tag{8.1}$$

$$\dot{r}_2 = u_r + g_w, \tag{8.2}$$

where g_w is an unknown and bounded perturbation such that $|g_w| < \gamma$, with γ being a positive constant. Then the following nonlinear controller

$$u_r = -\sigma_{b_2}(K_2 r_2) - \sigma_{b_1}(K_1 r_1), \qquad K_{1,2} > 0 \tag{8.3}$$

stabilizes (8.1) *and* (8.2).

Proof. Define $V_2 = \frac{1}{2} r_2^2$. Then

$$\dot{V}_2 = \dot{r}_2 r_2 = -r_2\left(\sigma_{b_2}(K_2 r_2) + \sigma_{b_1}(K_1 r_1) - g_w\right).$$

Choosing $b_2 > | b_1 - \gamma |$ means that $|K_2 r_2| > |\sigma_{b_1}(K_1 r_1) - g_w|$. Then

$$\operatorname{sgn}\left(\sigma_{b_2}(K_2 r_2) + \sigma_{b_1}(K_1 r_1) - g_w\right) = \operatorname{sgn}(r_2),$$

implying $\dot{V}_2 \leq 0$. The latter signifies that there exists a time $t > T_1$ such that $| r_2(t) | \leq \frac{|b_1 - \gamma|}{K_2}$ and $u_r = -K_2 r_2 - \sigma_{b_1}(K_1 r_1)$.

Define $v_1 = K_2 r_1 + r_2$. Then $\dot{v}_1 = -\sigma_{b_1}(K_1 r_1) + g_w$. Consider $V_1 = \frac{1}{2} v_1^2$, and then

$$\begin{aligned} \dot{V}_1 = v_1 \dot{v}_1 &= -(K_2 r_1 + r_2)\left(\sigma_{b_1}(K_1 r_1) - g_w\right) \\ &\leq -\left(K_2 r_1 + \frac{| b_1 - \gamma |}{K_2}\right)\left(\sigma_{b_1}(K_1 r_1) + \gamma\right). \end{aligned}$$

Observe that if $| K_1 r_1 | > \gamma$ then $\operatorname{sgn}(\sigma_{b_1}(K_1 r_1) + \gamma) = \operatorname{sgn}(r_1)$. On the other hand, if $| K_2 r_1 | > |r_2|$ then $\operatorname{sgn}(K_2 r_1 + r_2) = \operatorname{sgn}(r_1)$, implying $\dot{V}_1 \leq 0$. The latter yields $b_1 > \frac{\gamma(K_1 - K_2^2)}{K_1}$. Hence v_1 is bounded and, from its definition, r_1 is also bounded, so there exists a time $t > T_2 > T_1$ large enough such that $| r_1(t) | < \frac{b_1}{K_1}$ and $u_r = -K_2 r_2 - K_1 r_1$.

Rewriting the closed-loop system for $t > T_2$, we have

$$\dot{r}(t) = \mathbf{A}r(t) + g_w(t) \tag{8.4}$$

with $\boldsymbol{r} = [r_1, r_2]^T$, $A = \begin{bmatrix} 0 & 1 \\ -K_1 & -K_2 \end{bmatrix}$. Notice that this system could be seen as a perturbation of the nominal system $\dot{\boldsymbol{r}}(t) = \boldsymbol{A}\boldsymbol{r}(t)$.

Notice that g_w is an unknown and bounded perturbation, thus, from the stability theory and the stability of perturbed systems [3], it follows that if $g_w = 0$ then (8.4) yields $\dot{\boldsymbol{r}} = \mathbf{A}\boldsymbol{r}$. Hence if every eigenvalue, λ, of the matrix $\mathbf{A}$ has a strictly negative real part, i.e., $Re(\lambda) < 0$, the system is stable. Nevertheless, if g_w is a vanishing perturbation, and taking into account that any vanishing perturbation term that is locally Lipschitz in $\boldsymbol{r}$ uniformly in t in a bounded neighborhood of the origin satisfies the linear growth condition over that neighborhood, the complete system is also stable. On the other hand, if g_w is a non-vanishing perturbation, our solution is ultimately bounded by a bound that depends on the size of the perturbation. □

Theorem 8.2. *Consider the nonlinear system*

$$\dot{r}_1 = \kappa r_2 + r_2 g_w, \tag{8.5}$$

$$\dot{r}_2 = r_3, \tag{8.6}$$

$$\dot{r}_3 = u_r + g_w \epsilon, \tag{8.7}$$

where g_w is an unknown and bounded perturbation such that $|g_w| < \gamma$, with $\gamma, \kappa > 0$ being constants, and $0 < \epsilon < 1$. Then the following nonlinear controller

$$u_r = -\sigma_{b_3}(K_3 r_3) - \sigma_{b_2}(K_2 r_2) - \sigma_{b_1}(K_1 r_1), \qquad K_{1,2,3} > 0 \tag{8.8}$$

stabilizes (8.5)–(8.7).

Proof. Rewrite (8.8) as

$$u_r = -\sigma_{b_3}(K_3 r_3) - \zeta_3, \tag{8.9}$$

where $| \zeta_3 | \leq b_{\zeta_3}$ is a function that will be defined later and $b_{\zeta_i} > 0$ is constant. Define $\mathcal{V}_3 = \frac{1}{2} r_3^2$. Then

$$\dot{\mathcal{V}}_3 = r_3 \dot{r}_3 = -r_3 \left(\sigma_{b_3}(K_3 r_3) + \zeta_3 - g_w \epsilon \right).$$

Choosing $b_3 > | b_{\zeta_3} - \epsilon\gamma |$ implies that $|K_3 r_3| > |\zeta_3 - g_w \epsilon|$, meaning that $\dot{\mathcal{V}}_3 \leq 0$. So it follows that there exists a large enough time $\bar{T}_1$ such that

$\mid r_3(t) \mid \leq \frac{|b_{\zeta_3} - \epsilon\gamma|}{K_3}$ and $u_r = -K_3 r_3 - \zeta_3$. Denote $\beta_2 = K_3 r_2 + r_3$, thus $\dot{\beta}_2 = -\zeta_3 + \epsilon g_w$. Introduce

$$\zeta_3 = \sigma_{b_2}(K_2 r_2) + \zeta_2$$

with $\mid \zeta_2 \mid \leq b_{\zeta_2}$ to be defined later. From above observe that $b_{\zeta_3} = b_2 + b_{\zeta_2}$. Let $\mathcal{V}_2 = \frac{1}{2}\beta_2^2$. Then

$$\dot{\mathcal{V}}_2 = -\left(K_3 r_2 + r_3\right)\left(\sigma_{b_2}(K_2 r_2) + \zeta_2 - \epsilon g_w\right).$$

Notice that $\mid K_2 r_2 \mid > \mid b_{\zeta_2} - \epsilon g_w \mid$ implies that $\text{sgn}(\sigma_{b_2}(K_2 r_2) + \zeta_2 - \epsilon g_w) = \text{sgn}(r_2)$. Besides, the inequality $\mid K_3 r_2 \mid > |r_3|$, i.e., $\mid r_2 \mid > \frac{|b_{\zeta_3} - \epsilon\gamma|}{K_3^2}$, signifies that $\text{sgn}\left(K_3 r_2 + r_3\right) = \text{sgn}(r_2)$. Define $b_2 > \mid b_{\zeta_2} - \epsilon\gamma \mid$, then the above inequality holds and implies that $\dot{\mathcal{V}}_2 \leq 0$. This indicates that β_2 is bounded and, from its definition, r_2 is also bounded. Therefore there exists a time $t > \bar{T}_2 > \bar{T}_1$ such that $\mid r_2(t) \mid \leq \frac{|b_{\zeta_2} - \epsilon\gamma|}{K_2}$ and $u_r = -K_3 r_3 - K_2 r_2 - \zeta_2$. Define $\beta_1 = \frac{K_2}{\kappa} r_1 + K_3 r_2 + r_3$, consequently $\dot{\beta}_1 = -\zeta_2 + \frac{K_2 r_2 g_w}{\kappa} + \epsilon g_w$. Propose

$$\zeta_2 = \sigma_{b_1}(K_1 r_1),$$

thus $b_{\zeta_2} = b_1$. Let us now consider $\mathcal{V}_1 = \frac{1}{2}\beta_1^2$. Therefore

$$\dot{\mathcal{V}}_1 = -\left(\frac{K_2}{\kappa} r_1 + K_3 r_2 + r_3\right)\left(\sigma_{b_1}(K_1 r_1) - \frac{K_2 r_2 g_w}{\kappa} - \epsilon g_w\right).$$

Notice that $\left| -\frac{K_2 r_2 g_w}{\kappa} - \epsilon g_w \right| \leq \frac{\gamma|b_1 - \epsilon\gamma|}{\kappa} + \epsilon\gamma$ and observe that if $\mid K_1 r_1 \mid > \frac{\gamma|b_1 - \epsilon\gamma|}{\kappa} + \epsilon\gamma$, then $\text{sgn}\left(\sigma_{b_1}(K_1 r_1) - \frac{K_2 r_2 g_w}{\kappa} - \epsilon g_w\right) = \text{sgn}(r_1)$. On the other hand, remember that after a time $t > \bar{T}_2$, r_2 and r_3 are bounded and small, therefore defining $b_1 > \frac{\gamma|b_1 - \epsilon\gamma|}{g} + \epsilon\gamma$ implies also that $\left|\frac{K_2}{\kappa} r_1\right| > |K_3 r_2 + r_3|$, and as a consequence $\text{sgn}\left(\frac{K_2}{\kappa} r_1 + K_3 r_2 + r_3\right) = \text{sgn}(r_1)$, signifying $\dot{\mathcal{V}}_1 \leq 0$. Then there exists a time $t > \bar{T}_3 > \bar{T}_2$ such that $\mid r_1(t) \mid < \frac{b_1}{K_1}$ and $u_r = -K_3 r_3 - K_2 r_2 - K_1 r_1$. Thus for $t > \bar{T}_3$ it follows that the closed loop could be written as a perturbed linear system of (8.4). Likewise, from the stability theory [3] and the stability of perturbed systems, the system is stable or ultimately bounded. □

8.1.1 Robustness Under Nonlinear Uncertainties

Notice that the proposed nonlinear control laws are robust, by construction, with respect to bounded and unknown disturbances. Nevertheless, some assumptions (as small angles, neglected some parameters, etc.) could produce nonlinear uncertainties. Now, we will prove the closed-loop robustness under nonlinear uncertainties.

Proposition 8.1. *Consider the following disturbed system:*

$$\dot{r}_1 = r_2, \tag{8.10}$$
$$\dot{r}_2 = u_r + g_w + f(r_2), \tag{8.11}$$

where $f(r_2)$ satisfies

$$\left|f(r_2)\right| \leq \alpha_1 |r_2| \qquad \forall \alpha_1 > 0, \tag{8.12}$$

and consider the control law given in (8.3). *Then there exists a sufficiently small constant such that* (8.10)–(8.11) *is robustly stable.*

Proof. Let us assume that the nominal system (8.10)–(8.11) is stable in closed loop with the control law (8.3). Take $V_2 = \frac{1}{2}r_2^2$. Then using (8.11), (8.3) and (8.12), it follows that

$$\begin{aligned}\dot{V}_2 &= \dot{r}_2 r_2 = -r_2\left(\sigma_{b_2}(K_2 r_2) + \sigma_{b_1}(K_1 r_1) - g_w - f(r_2)\right)\\ &\leq -r_2\left(\sigma_{b_2}(K_2 r_2) + b_1 - g_w\right) + \alpha_1 |r_2|\,|r_2|\\ &\leq -r_2\left(\sigma_{b_2}(K_2 r_2) + b_1 - g_w\right) + \alpha_1 r_2^2.\end{aligned}$$

A sufficient condition to obtain a stable system under nonlinear uncertainties is

$$-r_2\left(\sigma_{b_2}(K_2 r_2) + b_1 - g_w\right) + \alpha_1 r_2^2 \leq 0. \tag{8.13}$$

The above implies that

$$r_2 \leq \frac{\sigma_{b_2}(K_2 r_2) + b_1 - g_w}{\alpha_1}. \tag{8.14}$$

Remember that r_2 is bounded, i.e., $|r_2| \leq \frac{|b_1 - \gamma|}{K_2}$. Define $v_1 = K_2 r_1 + r_2$. Then $\dot{v}_1 = -\sigma_{b_1}(K_1 r_1) + g_w + f(r_2)$. Consider $V_1 = \frac{1}{2}v_1^2$, and get

$$\begin{aligned}\dot{V}_1 = v_1\dot{v}_1 &= -(K_2 r_1 + r_2)\left(\sigma_{b_1}(K_1 r_1) - g_w - f(r_2)\right)\\ &= -(K_2 r_1 + r_2)\left(\sigma_{b_1}(K_1 r_1) - g_w\right) + (K_2 r_1 + r_2) f(r_2)\\ &\leq -\left(K_2 r_1 + \frac{|\,b_1 - \gamma\,|}{K_2}\right)\left(\sigma_{b_1}(K_1 r_1) + g_w\right)\\ &\quad + \alpha_1 (K_2 |r_1| + |r_2|)\,|r_2|\\ &\leq -\left(K_2 r_1 + \frac{|\,b_1 - \gamma\,|}{K_2}\right)\left(\sigma_{b_1}(K_1 r_1) + g_w\right)\\ &\quad + \alpha_1\left(K_2 |r_1| + \frac{|\,b_1 - \gamma\,|}{K_2}\right)\frac{|\,b_1 - \gamma\,|}{K_2}.\end{aligned}$$

Observe that b_1 was chosen such that $\left(K_2 r_1 + \frac{|b_1-\gamma|}{K_2}\right)(\sigma_{b_1}(K_1 r_1) + g_w) > 0$. Thus a sufficient condition in order to conclude that $\dot{V}_1 \leq 0$ is

$$-\left(K_2 r_1 + \frac{|\,b_1-\gamma\,|}{K_2}\right)\left(\sigma_{b_1}(K_1 r_1) + g_w\right) + \alpha_1\left(K_2\,|r_1| + \frac{|\,b_1-\gamma\,|}{K_2}\right)\frac{|\,b_1-\gamma\,|}{K_2} \leq 0. \tag{8.15}$$

As r_2 is a bounded function, $r_1 = \int_0^t r_2 dt$ is also a bounded function. This implies that $\left(K_2 r_1 + \frac{|b_1-\gamma|}{K_2}\right)(\sigma_{b_1}(K_1 r_1) + \gamma) < \delta$ with $\delta > 0$, thus

$$|r_1| \leq \frac{1}{\alpha_1}\left(\frac{\delta}{|\,b_1-\gamma\,|} - \frac{|\,b_1-\gamma\,|}{K_2^2}\right), \tag{8.16}$$

and the robust stability is proved. □

Proposition 8.2. *Consider the nonlinear system*

$$\dot{r}_1 = \kappa r_2 + r_2 g_w, \tag{8.17}$$
$$\dot{r}_2 = r_3, \tag{8.18}$$
$$\dot{r}_3 = u_r + g_w \epsilon + f(r_3), \tag{8.19}$$

where $f(r_3)$ satisfies the following property:

$$\left|f(r_3)\right| \leq \alpha_2\,|r_3|, \qquad \alpha_2 > 0. \tag{8.20}$$

Then (8.8) *stabilizes* (8.17)–(8.19) *for a sufficiently small α_2.*

Proof. Introducing (8.9) into (8.19), we have

$$\dot{r}_1 = \kappa r_2 + r_2 g_w, \tag{8.21}$$
$$\dot{r}_2 = r_3, \tag{8.22}$$
$$\dot{r}_3 = -\sigma_{b_3}(K_3 r_3) - \zeta_3 + g_w \epsilon + f(r_3). \tag{8.23}$$

Consider $V_3 = \frac{1}{2} r_3^2$, then

$$\begin{aligned}\dot{V}_3 &= r_3 \dot{r}_3 = -r_3\left(\sigma_{b_3}(K_3 r_3) + \zeta_3 - \epsilon g_w - f(r_3)\right)\\ &\leq -r_3\left(\sigma_{b_3}(K_3 r_3) + \zeta_3 - \epsilon g_w\right) + \alpha_2\,|r_3(t)|^2.\end{aligned}$$

Notice that $|r_3(t)|^2 = r_3(t)^2$, implying that

$$\dot{V}_3 \leq -r_3\left(\sigma_{b_3}(K_3 r_3) + \zeta_3 - \epsilon g_w\right) + \alpha_2 r_3(t)^2. \tag{8.24}$$

Remember that b_3 was chosen such that $|K_3 r_3| > |\zeta_3 - \epsilon g_w|$. Hence a sufficient condition to conclude robust stability is

$$-r_3\left(\sigma_{b_3}(K_3 r_3) + \zeta_3 - \epsilon g_w\right) + \alpha_2 r_3(t)^2 \leq 0, \tag{8.25}$$

or

$$r_3 \leq \frac{\sigma_{b_3}(K_3 r_3) + \zeta_3 - \epsilon g_w}{\alpha_2}. \tag{8.26}$$

Notice that r_3 is bounded. Following a similar procedure as in Proposition 8.1, we can obtain that r_2 and r_1 are also bounded, and then robust stability is proved. □

8.1.2 Validation in a Quadrotor Aerial Vehicle

The proposed controllers have been obtained from linear and perturbed systems, yet, it is not possible to apply the previous approach to the nonlinear quadcopter equations (2.17) and (2.18). Nevertheless, such methodology could be applied to this vehicle when linearizing its nonlinear equations.

Usually the quadrotor operates in an attitude range within $\pm 30°$, and therefore the equations of motion can be approximately decoupled about each attitude axis. The dominant dynamics are associated with the longitudinal/lateral dynamics of the vehicle. Indeed, around hover the motion of this aerial vehicle is largely decoupled in each axis, and then, due to its configuration, the Coriolis matrix becomes very small and can be neglected for control design. In addition, the symmetry of the quadrotors means that the significant attitude dynamics could be described by a single equation ($\ddot{\eta} \approx \tau_\eta$). Similarly, at hover it can be assumed that the rotor thrust is proportional to the square of the propellers' rotation speed and that the rotor and body planes are aligned. Consequently, the flapping angles are small, and then the longitudinal and lateral rotor thrust could be neglected.

The previous reasoning implies that, when considering quasi-stationary maneuvers, small flapping angles, and the same thrust being applied to each motor, the nonlinear quadcopter dynamics can be studied as n integrators in cascade (linear system). And from the physical characteristics of the vehicle, its dynamics could be analyzed separately for control design.

Nevertheless, we remark from proposed theorems that the control scheme can impose bounds on each state. This signifies that the bounds of the quadcopter controllers could be chosen small enough to guarantee that the advanced ratio is small and the motion is constrained in pitch and roll angles (small angles). This means that the control algorithms can impose linear behavior to the system, and then the previous hypothesis needed to apply the proposed theorems can be satisfied. Simulation tests and experiments have supported this fact.

Let us consider that the aerial vehicle moves with small angles and is kept by its controller so that $\cos(\theta) \approx 1$ and $\sin(\theta) \approx \theta$, thus the altitude ($z$) and yaw dynamics ($\psi$) could be rewritten as

$$\ddot{z} \approx \bar{u} + f_{d_z}, \tag{8.27}$$
$$\ddot{\psi} \approx \tau_\psi + \tau_{d_\psi}, \tag{8.28}$$

where $\bar{u} = u - g$. On the one hand, remember from Chap. 2 that f_{d_k} and τ_{d_j} denote the unknown disturbances due to the wind, flapping and drag dynamics. On the other hand, observe from (6.19) and Fig. 2.2 that f_{d_k} and τ_{d_j} are produced essentially by the wind (without wind these parameters mainly vanish at hover), and then they can be expressed for control design or simulations purposes as proportional to the wind V_w, i.e., $f_{d_k}, \tau_{d_j} \approx \epsilon_{\{k,j\}} V_w$, where $0 < \epsilon_{\{k,j\}} < 1$. Considering that the wind is bounded, this implies that $|\epsilon_{\{k,j\}} V_w| \leq \epsilon_{\{k,j\}} \gamma_w$, where $\gamma_w > 0$ is constant.

Eqs. (8.27) and (8.28) correspond to two integrators in cascade like (8.1), therefore using Theorem 8.1, we get

$$\bar{u} = -\sigma_{b_{2_z}}\left(K_{2_z}\dot{z}\right) - \sigma_{b_{1_z}}\left(K_{1_z}(z - z_d)\right), \tag{8.29}$$
$$\tau_\psi = -\sigma_{b_{2_\psi}}\left(K_{2_\psi}\dot{\psi}\right) - \sigma_{b_{1_\psi}}\left(K_{1_\psi}(\psi - \psi_d)\right). \tag{8.30}$$

Introducing the above into (8.27) and (8.28), respectively, implies that $\{\dot{z}, \dot{\psi}\} \to 0$, $\psi \to \psi_d$, and $z \to z_d$, with z_d being the desired altitude and ψ_d representing the desired yaw angle.

Following this methodology, the longitudinal and lateral subsystems can be rewritten as

$$\ddot{x} \approx -g\theta - \theta f_{d_x}, \tag{8.31}$$
$$\ddot{\theta} \approx \tau_\theta + \tau_{d_\theta}, \tag{8.32}$$

$$\ddot{y} \approx g\phi + \phi f_{d_y}, \tag{8.33}$$
$$\ddot{\phi} \approx \tau_\phi + \tau_{d_\phi}. \tag{8.34}$$

A popular outdoor application is to fly quadcopter vehicles in semi-autonomous mode, i.e., when the pilot gives translational movements to the vehicle. Nevertheless, autonomous flights become really needed. One challenge to solve when realizing autonomous flight is to locate the vehicle. GPS is the most common solution even if the data are noised or corrupted in several cases. It is well know that if in some applications the GPS does not work properly (GPS-denied environments), then the position is imprecise or not available. One alternative to know the quadrotor position is to use vision systems as the VICON or the OptiTrak (for indoor validation tests). However, such vision systems are not always an affordable option.

Most quadcopter platforms are equipped with an IMU (inertial measurement unit), an ultrasound sensor (for the altitude), and a camera (often used to estimate the translational velocity of the vehicle). Following this trend, we will develop the horizontal controller only to stabilize the attitude and the translational velocity of the quadcopter.[1] This controller could be easily implemented in the test-beds.

Thus, defining $x_1 = \dot{x} - V_{x_r}$, $x_2 = \theta$, and $x_3 = \dot{\theta}$, with V_{x_r} being the desired constant translational velocity in the x axis, (8.31) yields

$$\begin{aligned} \dot{x}_1 &\approx -gx_2 - x_2\epsilon_x V_w, \\ \dot{x}_2 &= x_3, \\ \dot{x}_3 &= \tau_\theta + \epsilon_\theta V_w. \end{aligned} \tag{8.35}$$

Hence, using Theorem 8.2, it follows that

$$\tau_\theta = -\sigma_{b_{3_\theta}}(K_{3_\theta}x_3) - \sigma_{b_{2_\theta}}(K_{2_\theta}x_2) + \sigma_{b_{1_\theta}}(K_{1_\theta}x_1), \tag{8.36}$$

and this implies in closed loop that $x_{\{3,2,1\}}$ is bounded and goes to zero. Similarly, let us consider $y_1 = \dot{y} - V_{y_r}$, $y_2 = \phi$, and $y_3 = \dot{\phi}$, with V_{y_r} being the desired translational velocity in the y axis, then (8.33) yields

$$\begin{aligned} \dot{y}_1 &\approx gy_2 + y_2\epsilon_y V_w, \\ \dot{y}_2 &= y_3, \\ \dot{y}_3 &= \tau_\phi + \epsilon_\phi V_w, \end{aligned} \tag{8.37}$$

and the control law to stabilize the previous subsystem becomes

$$\tau_\phi = -\sigma_{b_{3_\phi}}(K_{3_\phi}y_3) - \sigma_{b_{2_\phi}}(K_{2_\phi}y_2) - \sigma_{b_{1_\phi}}(K_{1_\phi}y_1). \tag{8.38}$$

From (8.36) and (8.38) notice that the angles θ and ϕ are bounded by b_{2_θ} and b_{2_ϕ}, respectively, and we can choose these bounds small enough such that $\sin(\cdot) \approx (\cdot)$ and $\cos(\cdot) \approx 1$. This signifies that the outlined assumptions hold.

[1] This control approach was also implemented to control the horizontal position in Chap. 5.

8.1.2.1 Crosswind

Wind disturbances can be written as a linear combination of N (possibly infinitely many) sinusoidal functions of time. For simulation purposes, we adopt a classical mathematical representation of the disturbance generated by the lateral wind and given by the following model [4]:

$$V_w(t) = V_{w_a} + V_{w_r}(t) + V_{w_g}(t) + V_{w_t}(t),$$

where V_{w_a} is the initial average value of the wind speed, V_{w_r} expresses the ramp component representing the steady increase of wind, V_{w_g} describes the gust component, and finally, V_{w_t} denotes the turbulence component. The ramp and gust components are specified by their amplitude, A_r and A_g, in m/s, their starting time, T_{s_r} and T_{s_g}, in seconds, and their stopping time, T_{e_r} and T_{e_g}, in seconds, respectively. Therefore, the ramp and gust components are expressed by

$$V_{w_r} = \begin{cases} 0 & \forall t < T_{s_r}, \\ A_r \frac{t-T_{s_r}}{T_{e_r}-T_{s_r}} & \forall T_{s_r} \leq t \leq T_{e_r}, \\ A_r & \forall T_{e_r} < t, \end{cases}$$

$$V_{w_g} = \begin{cases} 0 & \forall t < T_{s_g}, \\ A_g - A_g \cos(2\pi \frac{t-T_{s_g}}{T_{e_g}-T_{s_g}}) & \forall T_{s_g} \leq t \leq T_{e_g}, \\ 0 & \forall T_{e_g} < t. \end{cases}$$

In our study we consider the following assumptions:

- The quadrotor is perturbed by two wind gusts, i.e., the initial average is zero and the wind does not increase, thus $V_{w_a} = V_{w_r} = 0$. The first gust begins at $t_{s_g} = 100$ s and finishes at $t_{e_g} = 120$ s with an amplitude of $A_g = 2.5$ m/s whilst the other will have $t_{s_g} = 200$ s and $t_{e_g} = 210$ s with $A_g = 1.5$ m/s.
- To represent the turbulence component, V_{w_t}, a band-limited white noise has been used with $\max(V_{w_t}) < 1$ m/s. In addition, $V_w = 0$ $\forall t > 230$ s.
- Only lateral wind gusts coming from the x axis are considered, and therefore it can be supposed that the lateral wind affects the x, z, and θ dynamics more strongly, i.e., $w_x > w_z > w_\theta > w_y > w_\phi > w_\psi$.

Notice from previous definition that the wind is bounded, in our case $|V_w| < 2A_g + 1$. Fig. 8.1 introduces the perturbation applied in simulation.

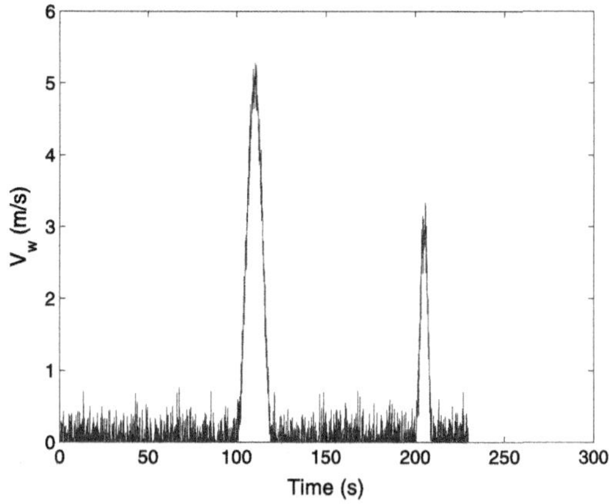

Figure 8.1 Wind gusts applied to the vehicle.

8.1.3 Simulation Results

From the proposed theorems it can be concluded that the control laws (8.29), (8.30), (8.36), and (8.38) stabilize the systems given by (8.27), (8.28), (8.35), and (8.37), respectively. Nevertheless, we will validate our controllers for the nonlinear system (6.19). The simulation procedure is the following:

- Stabilize the quadrotor at a desired altitude with small disturbances. The desired value as $z_d = 10$ m whilst $\psi_d, \theta_d, \phi_d = 0°$ and $V_{y_r}, V_{x_r} = 0$ m/s. The initial positions for the states are zero except for the yaw angle, $\psi(0) = 10°$.
- Apply two lateral wind gusts in the x axis when the vehicle is at hover. Additionally, we will consider $\epsilon_x = 0.99$, $\epsilon_z = 5/6$, $\epsilon_\theta = 4/6$, $\epsilon_y = 3/6$, $\epsilon_\phi = 2/6$, and $\epsilon_\psi = 1/6$.

In Figs. 8.2–8.4 we introduce the simulation performance of the quadrotor vehicle. In Fig. 8.2 the solid line represents the real error value introduced in the controller.

Observe that even if the wind displaces the vehicle several meters, the controller bounds the error in order to consider it a small perturbation and to realize slow convergence to the desired value. Thereby the control algorithm will guarantee the stability of the closed-loop system. Notice that when the gusts are applied, the vehicle is displaced in the altitude and in the horizontal plane. This fact is common because the quadrotor is not equipped with a sensor to measure this perturbation. In Fig. 8.3 where the

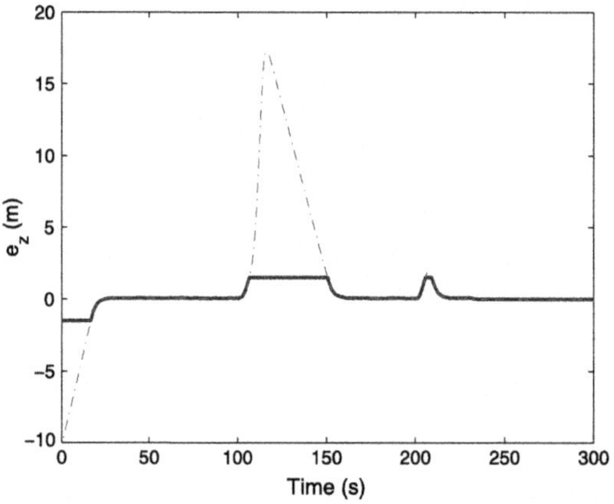

Figure 8.2 Altitude response (dashed–dotted line) and error in z (solid line).

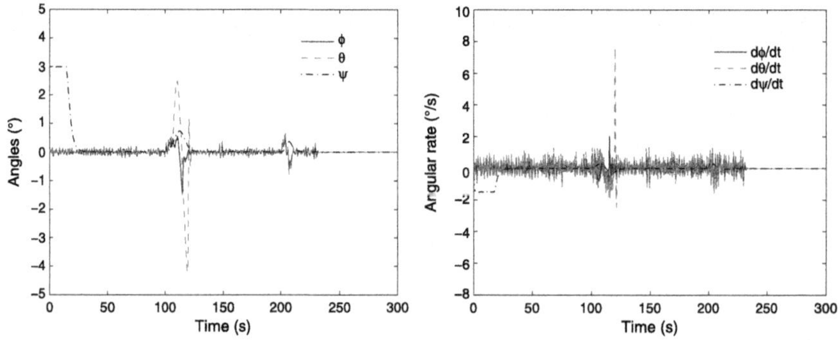

Figure 8.3 Angular positions and angular rates responses.

orientation and angular rate responses are shown, we can observe that the angles in the lateral and longitudinal dynamics are small, this is due to the bounds imposed by the controllers. In other words, the system considers that the vehicle has small errors in the translational velocities, see Fig. 8.4, and therefore the angle to correct this error will also be small. Fig. 8.3 is a consequence of Fig. 8.4, i.e., when the wind gusts appear the vehicle reacts to oppose the angular movement produced by these perturbations. The angular rate is an essential parameter to carefully consider in practical applications.

In Fig. 8.4 the translational velocities are depicted, it can be observed that all velocities computed by the control scheme are bounded. In this way

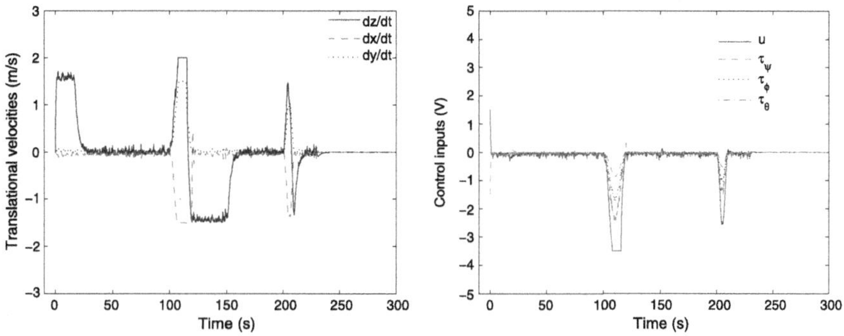

Figure 8.4 Translational velocities and control inputs responses.

we can impose a linear behavior in the system, the previous considerations imply that the computed control inputs will be small and bounded.

8.1.4 Experimental Results

The controls laws were implemented on the same platform described in Sect. 5.2 of Chap. 5. Recall that the platform consists of a quadcopter vehicle equipped with an IGEPv2 board, IMU 3DMGX3-25, an ultrasonic range finder SRF10, and a PS3eye camera used to compute the optical flow and to estimate the translational rates in the horizontal plane.

Several flight tests were carried out with wind gusts. The goal of the experiment was to validate the robustness of the controller. During hover the helicopter was perturbed by wind gusts produced by a commercial fan with a velocity of 4 m/s.

We remind that only measurements of the orientation, angular rate, altitude, and translational velocities in the horizontal plane are available for the vehicle. $\dot{z}$ is estimated using the classical Euler approximation. One advantage of the control scheme is that every state is bounded separately, this fact helps considerably in decoupling the states and allowing each state to be separately tuned.

The practical hierarchy used to tune the subsystems is as follows:

1. Stabilization of the orientation: (a) yaw dynamic and (b) pitch or/and roll movements. The goal is to eliminate the oscillations when adding manual perturbations or changing the desired value. When the orientation is tuned, we evaluate performance with semi-automatic flights, i.e., the orientation is in autonomous mode whilst the altitude and the

translational movements are given by a pilot. Observe that such controllers allow realizing this tuning procedure. Nevertheless, for control schemes that include a combination of all the states in each control algorithm, this procedure is not valid.

Practical applications have shown that even if the orientation is well stabilized there exists a small derivation in the angles that produces small displacements in the horizontal plane.

2. Stabilization of the position or/and velocity in the horizontal plane. This step will depend of the sensor employed. In our case we have only a sensor to estimate the translational velocities. When introducing these measures in the closed-loop system, the vehicle is capable of correcting the deviation in the pitch/roll angles.
3. Stabilization of the altitude.

The simplicity of the algorithm is very useful in practical applications and, in particular, this facilitates the tuning of the control parameters. The control gains and the amplitudes of the saturation functions in the control algorithm were tuned as follows. First of all, the amplitude of $\sigma_{b_{\dot{\phi}}}$ was tuned in such a way that the roll angular velocity $\dot{\phi}$ remained close to zero even when a disturbance was introduced manually. Next, the amplitude of $\sigma_{b_{\phi}}$ was chosen in such a way that the quadrotor roll angle was sufficiently small. In both cases, it was necessary to avoid choosing high amplitudes which normally lead to oscillations. The amplitude of $\sigma_{\dot{y}}$ was chosen such that the effect of a small disturbance in the horizontal velocity $\dot{y}$ was soon compensated. This procedure was also used to tune the x–θ dynamics.

In the experiment the aerial vehicle was exposed to crosswind coming from different directions. Another awkwardness in the experiment was deficiency in the measurement of the x and y position. Nevertheless, without any measure of x and y, the closed-loop system remained stable. In Figs. 8.5–8.7 several graphs showing this fact are introduced. Notice the good behavior of the vehicle in these figures.

Observe the altitude and attitude responses of the vehicle from Figs. 8.5–8.6. Small oscillations in the z-graph (Fig. 8.5) are due to the applied wind gusts. In Fig. 8.7 the translational velocities can be observed.

From the experiments it can be concluded that the obtained results are very promising for several reasons: first, stabilization is achieved without any wind measurement, only the maximum value of the wind strength is used. Wind gusts were applied in different directions, and the closed-loop system remained stable, showing that the control strategy works well in simulations and real-time experiments.

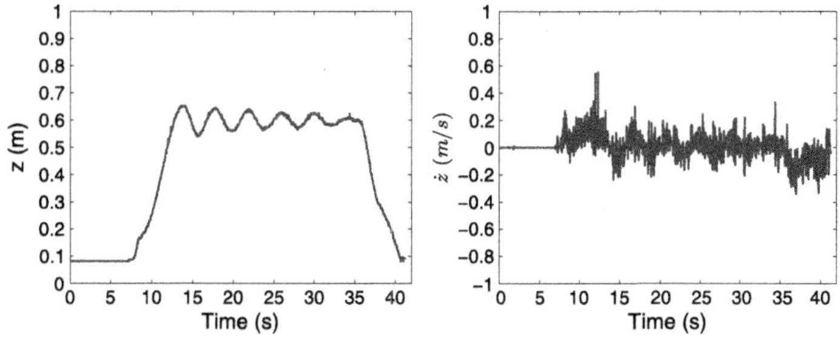

Figure 8.5 Altitude and vertical velocity responses of the quadrotor in the presence of wind.

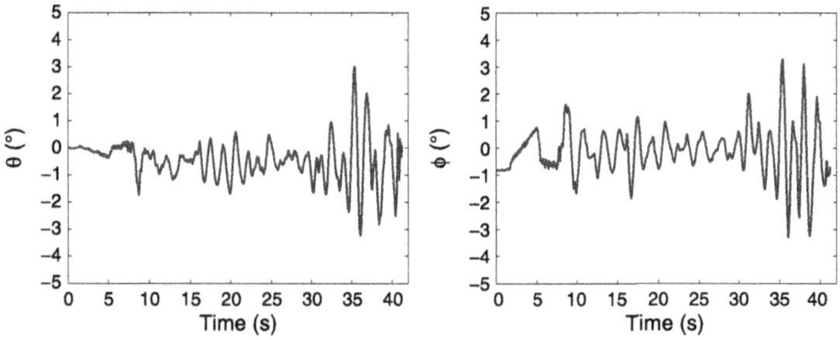

Figure 8.6 Pitch and roll angle responses of the quadcopter with wind.

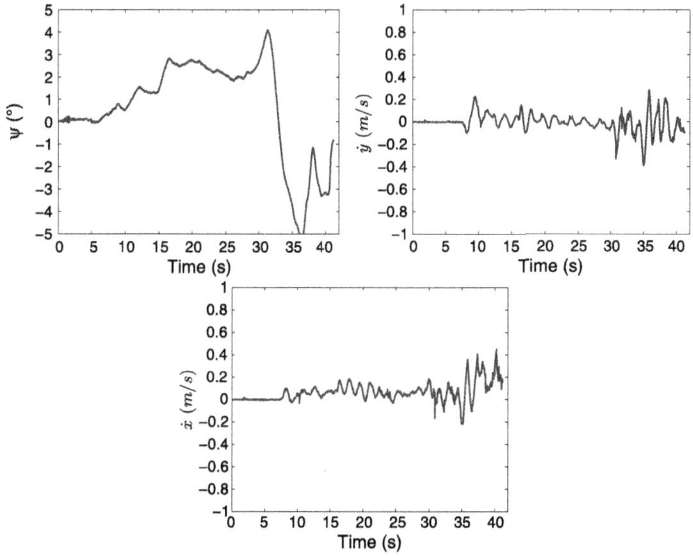

Figure 8.7 Yaw angle and translational velocities responses of the experimental test.

8.2 ROBUST CONTROL BASED ON AN UNCERTAINTY ESTIMATOR

The UDE-based control algorithm, which was originally proposed in [5], is a robust control strategy that is devoted to handling uncertainties and disturbances. Based on the assumption that a signal can be recovered by passing it through a filter of appropriate bandwidth, all uncertainties and disturbances can be treated as a lumped signal to be estimated and compensated in the controller straight away. The UDE strategy has demonstrated good performance in handling uncertainties and disturbances and has been successfully applied to robust input–output linearization [6], linear and nonlinear systems with state delays [7], and uncertain nonlinear systems [8,9].

In this section, the UDE-based control strategy is applied to the attitude and position control of quadrotors. Experimental evidence of this fact is emphasized. Rather than using nonlinear controllers and highly-accurate aerodynamic models based on parameter identification, these experiments demonstrate that a simple linear control strategy with very few tuning parameters such as the UDE-based control is a very interesting choice for controlling quadrotors in real applications, dealing with model uncertainties and persistent disturbances with outstanding performance. This is corroborated in flight tests.

8.2.1 UDE-Based Robust Control Strategy

Consider the following class of nonlinear single-input multiple-output (SIMO) systems:

$$\begin{aligned} \dot{\boldsymbol{x}}(t) &= (\mathbf{A} + \Delta\mathbf{A})\,\boldsymbol{x}(t) + (\mathbf{B} + \Delta\mathbf{B})\,u(t) + \mathbf{f}(\boldsymbol{x}, u, t) + \mathbf{d}(t), \\ \mathbf{y}(t) &= \boldsymbol{x}(t), \end{aligned} \tag{8.39}$$

where $\boldsymbol{x}(t) \in \mathbb{R}^n$ and $u(t) \in \mathbb{R}$ are the state and control variable, respectively, $\mathbf{f}(\boldsymbol{x}, u, t) : \mathbb{R}^n \times \mathbb{R} \times \mathbb{R}^+ \to \mathbb{R}^n$ is a possibly unknown nonlinear function, and $\mathbf{d}(t) : \mathbb{R}^+ \to \mathbb{R}^n$ is the vector of unknown disturbances. The full state is assumed to be measurable. The state and control matrices are split so that $\mathbf{A}$ and $\mathbf{B}$ are known and $\Delta\mathbf{A}$ and $\Delta\mathbf{B}$ are parametric uncertainties. It is assumed that the nonlinear function $\frac{\partial[\mathbf{f}(x,u)+\mathbf{B}u]}{\partial u} \neq 0$ for all $(\boldsymbol{x}, u) \in \mathbb{R}^n \times \mathbb{R}$. Moreover, we assume that the desired dynamics of the closed-loop system are given in terms of the linear model described by

$$\dot{\boldsymbol{x}}_m(t) = \mathbf{A}_m \boldsymbol{x}_m(t) + \mathbf{B}_m r(t), \tag{8.40}$$

where $\mathbf{A}_m \in \mathbb{R}^{n\times n}$, $\mathbf{B}_m \in \mathbb{R}^n$, $\boldsymbol{x}_m \in \mathbb{R}^n$, and $r(t) \in \mathbb{R}$ is a piece-wise continuous and uniformly bounded command to the system.

8.2.1.1 General Case

The control objective is to derive a control law such that the state error between the actual plant state $\boldsymbol{x}$ and the reference model state $\boldsymbol{x}_m$, defined by

$$\mathbf{e} = \boldsymbol{x}_m - \boldsymbol{x}, \tag{8.41}$$

is asymptotically stable. Differentiating (8.41), using (8.39)–(8.40), and adding and subtracting $\mathbf{A}_m\boldsymbol{x}$ results in

$$\dot{\mathbf{e}} = \mathbf{A}_m\mathbf{e} + \mathbf{A}_m\boldsymbol{x} + \mathbf{B}_m r - (\mathbf{A} + \Delta\mathbf{A})\boldsymbol{x} - (\mathbf{B} + \Delta\mathbf{B})u - \mathbf{f} - \mathbf{d}. \tag{8.42}$$

Then the error will be asymptotically stable and satisfy the dynamic equation $\dot{\mathbf{e}} = \mathbf{A}_m\mathbf{e}$ if the following relation is fulfilled

$$\mathbf{A}_m\boldsymbol{x} + \mathbf{B}_m r - (\mathbf{A} + \Delta\mathbf{A})\boldsymbol{x} - (\mathbf{B} + \Delta\mathbf{B})u - \mathbf{f} - \mathbf{d} = 0. \tag{8.43}$$

From (8.43), the control input signal $u(t)$ should satisfy

$$u = \mathbf{B}^{+}[\mathbf{A}_m\boldsymbol{x} + \mathbf{B}_m r - \mathbf{A}\boldsymbol{x}] + u_d, \tag{8.44}$$

where $\mathbf{B}^{+} = (\mathbf{B}^T\mathbf{B})^{-1}\mathbf{B}^T$ is the pseudo-inverse of $\mathbf{B}$, and

$$u_d = \mathbf{B}^{+}[-\Delta\mathbf{A}\boldsymbol{x} - \Delta\mathbf{B}u - \mathbf{f} - \mathbf{d}] \tag{8.45}$$

denotes the unknown terms in (8.44). The unknown terms in the brackets of (8.45) can be solved from the system dynamics described by (8.39), allowing to rewrite (8.45) as

$$u_d = \mathbf{B}^{+}[\mathbf{A}\boldsymbol{x} + \mathbf{B}u - \dot{\boldsymbol{x}}]. \tag{8.46}$$

Intuitively, the latter indicates that the unknown dynamics and disturbances can be estimated from the known dynamics of the systems and control signal. Following the procedures provided in [5], the signal given by (8.46) can be accurately represented in the frequency domain as[2]

$$U_d(s) = G_f(s)\mathbf{B}^{+}[\mathbf{A}\mathbf{X}(s) + \mathbf{B}U(s) - s\mathbf{X}(s)] \tag{8.47}$$

if the filter $G_f(s)$ is a strictly proper low-pass filter with unitary gain and zero phase shift over the spectrum of u_d, and zero gain elsewhere. Taking the Laplace transform of (8.44), using the expression for $U_d(s)$ derived in

[2] The Laplace transformation is introduced to facilitate the manipulation of expressions.

(8.47), and solving for $U(s)$, the UDE-based control law can be derived as [5]

$$U(s) = [\mathbf{I} - G_f\mathbf{B}^+\mathbf{B}]^{-1}\mathbf{B}^+\left[\mathbf{A}_m\mathbf{X} + \mathbf{B}_m R - \mathbf{A}\mathbf{X}(1 - G_f) - sG_f\mathbf{X}\right]. \quad (8.48)$$

Because of the pseudo-inverse, Eq. (8.48) is only an approximation, in general. However, as shown in [10], it is satisfied in some cases. The asymptotic stability of the closed-loop system is established in [11], when the filter $G_f(s)$ is chosen appropriately as a strictly-proper stable filter with unity gain and zero phase shift over the spectrum of the uncertain term $\mathbf{f}(\boldsymbol{x}, u, t) + \mathbf{d}(t)$, and zero gain elsewhere. See [11] for more details.

8.2.1.2 The Case with an SISO System

Assume that the system under consideration is a controllable SISO system. It can be represented in the controllable canonical form, i.e., the matrices involved admit the following partitioning:

$$\mathbf{A} = \begin{bmatrix} 0 & \mathbf{I}_{n-1} \\ & \mathbf{A}_1 \end{bmatrix}, \mathbf{B} = \begin{bmatrix} \mathbf{0}_{n-1} \\ b \end{bmatrix}, \mathbf{A}_m = \begin{bmatrix} 0 & \mathbf{I}_{n-1} \\ & \mathbf{A}_{1m} \end{bmatrix}, \mathbf{B}_m = \begin{bmatrix} \mathbf{0}_{n-1} \\ b_m \end{bmatrix},$$

where $\mathbf{A}_1 = [-a_1, -a_2, \ldots, -a_n]$ and $\mathbf{A}_{1m} = [-a_{m1}, -a_{m2}, \ldots, -a_{mn}]$. Then the transfer functions from the input to the first state variable X_1 (taken as the output Y) for both the known plant model and the reference model are given by $G(s) = \frac{b}{P(s)}$ and $G_m(s) = \frac{b_m}{P_m(s)}$, respectively, with $P(s) = s^n + \sum_{i=1}^{n} a_i s^{i-1}$ and $P_m(s) = s^n + \sum_{i=1}^{n} a_{im} s^{i-1}$. Substituting the above matrices into (8.48), the UDE-based control law for the system (8.39) is derived as:

$$U = \frac{1}{(1 - G_f)b}\left(b_m R + \sum_{i=1}^{n}(a_i - a_{im})\mathbf{X}_i - G_f\sum_{i=1}^{n} a_i\mathbf{X}_i - sG_f\mathbf{X}_n\right). \quad (8.49)$$

Assume that the frequency range of the unknown system dynamics and the external disturbances is limited by ω_f. Then $G_f(s)$ can be approximately chosen as a low-pass filter $G_f(s) = \frac{1}{Ts+1}$ with $T = 1/\omega_f > 0$. In this case $\frac{1}{1-G_f}$ is a PI controller, denoted as $PI(s) = \frac{Ts+1}{Ts}$. Then the control law (8.49) can be rewritten, after some algebraic manipulations, as

$$U(s) = PI(s)\left(\frac{b_m}{b}R(s) - M(s)Y(s)\right) \quad (8.50)$$

with $M(s) = \frac{b_m}{b}G_m^{-1}(s) - PI^{-1}(s)G^{-1}(s)$. This can be shown to be a two-degree-of-freedom control structure. Denoting by $G_p(s)$ the actual transfer function of the plant and considering the nominal case, i.e., $G_p(s) = G(s)$, the relevant closed-loop transfer functions for reference tracking and disturbance are respectively

$$\frac{Y(s)}{R(s)} = \frac{b_m}{b} \frac{PI(s) G_p(s)}{1 + PI(s) M(s) G_p(s)} = G_m(s) \quad \text{and} \tag{8.51}$$

$$\frac{Y(s)}{D(s)} = \frac{G_p(s)}{1 + PI(s) M(s) G_p(s)} = \frac{b}{b_m} G_m(s) \frac{Ts}{Ts+1}. \tag{8.52}$$

As illustrated above, the proposed control structure allows decoupling between reference tracking and disturbance rejection. From (8.52), disturbances can be attenuated arbitrarily fast by reducing T because of the term $Ts/(Ts+1)$. Moreover, the parameters of the control law are "automatically" tuned by choosing the reference model. This makes tuning of the proposed controller very easy. Decreasing the parameter T has a counter-effect. Let us illustrate this point by looking at the transfer functions between the noise at each measurement channel $N_i(s)$ and the control $U(s)$, obtained as

$$\frac{U(s)}{N_i(s)} = PI(s) \frac{1}{b} \left(a_i - a_{mi} - \frac{a_i}{Ts+1} \right), \quad i = 1, \ldots, n-1, \tag{8.53}$$

$$\frac{U(s)}{N_n(s)} = PI(s) \frac{1}{b} \left(a_n - a_{mn} - \frac{a_n + s}{Ts+1} \right). \tag{8.54}$$

One can see that the channel $N_n(s)$ is the most critical in terms of noise. For high frequencies, that is, for $s \to \infty$, $\lim_{s \to \infty} \left| \frac{U(s)}{N_n(s)} \right| = \frac{1}{b} \left| \frac{1}{T} + (a_{mn} - a_n) \right|$. Therefore, the lower the T, the better the disturbance rejection, but the higher the amplification of high frequencies in the measurement noise. The good feature is that this trade-off can be easily met in practice by tuning T.

8.2.1.3 Matching the Quadrotor Model

A fine model of a quadrotor system can be very complicated because of the aerodynamic effects, as explained in Chap. 2. Most of this section is concerned with the attitude and altitude control of quadrotors, which are crucial as discussed in [12]. From the dynamics equations derived using the Newton–Euler approach (2.7)–(2.8), a simplified, yet fairly accurate model of these critical variables is obtained as

$$
\begin{aligned}
\ddot{\phi} &= \frac{I_y - I_z}{I_x}\dot{\theta}\dot{\psi} - \frac{I_r}{I_x}\dot{\theta}\Omega + \frac{l\tau_\phi}{I_x}, \\
\ddot{\theta} &= \frac{I_z - I_x}{I_y}\dot{\psi}\dot{\phi} + \frac{I_r}{I_y}\dot{\phi}\Omega + \frac{l\tau_\theta}{I_y}, \\
\ddot{\psi} &= \frac{I_x - I_y}{I_z}\dot{\theta}\dot{\phi} + \frac{l\tau_\psi}{I_z}, \\
\ddot{z} &= g - \frac{\Gamma}{m}\cos\phi\cos\theta,
\end{aligned}
\tag{8.55}
$$

where ϕ, θ, and ψ are the roll, pitch, and yaw Euler angles, and I_i, $i = x, y, z$ are the moments of inertia, I_r is the inertia of the motor, l means the distance from each motor to the gravity center of the vehicle, $\Omega = \Omega_2 + \Omega_4 - \Omega_1 - \Omega_3$ with Ω_i being the rotor speed of the ith motor, τ_i, $i = \phi, \theta, \psi$ are the input torques along the axes of a body-fixed reference frame, z is the coordinate along the z-axis of the body-fixed frame which points downwards, g is the gravity acceleration, m is the mass of the quadrotor, and $\Gamma = \sum_{i=1}^{4} F_i$ is the total thrust. Since I_x and I_y are almost symmetric and only differ slightly because of construction tolerance, the roll and the pitch axes have very similar dynamics.

The variables in the model (8.55) are referred to as critical, meaning that an unsatisfactory control of those variables can easily lead to a system failure. Furthermore, the displacements in the x–y plane can be controlled by commanding the roll and pitch angles, as explained in [13]. Thus, a reliable and accurate attitude controller is a necessary prerequisite for implementing position tracking controllers. The equations of model (8.55) can be written in the form of (8.39) as

$$
\begin{aligned}
\ddot{\phi} &= f_1(\dot{\theta}, \dot{\psi}) + u_\phi, \\
\ddot{\theta} &= f_2(\dot{\psi}, \dot{\phi}) + u_\theta, \\
\ddot{\psi} &= f_3(\dot{\theta}, \dot{\phi}) + u_\psi, \\
\ddot{z} &= f_4(\theta, \phi) + u_z
\end{aligned}
\tag{8.56}
$$

with the control inputs given by $u_\phi = l\tau_\phi/I_x$, $u_\theta = l\tau_\theta/I_y$, $u_\psi = l\tau_\psi/I_z$, and $u_z = g - \Gamma/m$, and the nonlinear terms being $f_1(\dot{\theta}, \dot{\psi}) = (I_y - I_z)/I_x\dot{\theta}\dot{\psi} - I_r/I_x\dot{\theta}\Omega$, $f_2(\dot{\psi}, \dot{\phi}) = (I_z - I_x)/I_y\dot{\psi}\dot{\phi} + I_r/I_y\dot{\phi}\Omega$, $f_3(\dot{\theta}, \dot{\phi}) = (I_x - I_y)/I_z\dot{\theta}\dot{\phi}$, and $f_4(\theta, \phi) = \Gamma/m(1 - \cos\phi\cos\theta)$.

Then each of the equations in (8.56) represents a single-input dynamic system of a double integrator with an unknown disturbance term.

8.2.1.4 The Case with a 3D Hover System

In this section, the 3D hover system described in Sect. 3.3.6.1 of Chap. 3 is investigated to validate the proposed control law in simulations and experiments. As previously mentioned, for design purposes each axis is considered decoupled and modeled by a double integrator $G(s) = 0.1/s^2$. The output of the system is the measured angle whereas the control action is the voltage applied to the motors. In order to better illustrate the performance of the UDE in comparison with a PID-based controller, simulations of the linearized model are presented. The validation of the proposed control law for model (8.55) will be supported by the experiments.

Controller Design

Denote by $\Theta(s)$ and $\Theta^{\text{ref}}(s)$ the Laplace transform of $\theta(t)$ and θ^{ref}, respectively. The following 2-DOF PID controller having weighted set-point

$$U_{\text{pid}}(s) = \left(\epsilon K_p + \frac{K_i}{s}\right)\Theta^{\text{ref}}(s) - \left(K_p + \frac{K_i}{s} + K_d s\right)\Theta(s)$$

with $\epsilon = 0.6956$, $K_p = 90$, $K_d = 50$, and $K_i = 39.2$; is designed to control the pitch angle, and it is used as a base for comparison. The resulting closed-loop transfer function is

$$\frac{\Theta(s)}{\Theta^{\text{ref}}(s)} = \frac{6.26}{s^2 + 4.37s + 6.26}. \tag{8.57}$$

The roll angle is controlled using the proposed strategy. The same closed-loop model obtained in (8.57) is chosen as the desired reference model, and thus

$$\dot{\boldsymbol{x}}_m(t) = \begin{bmatrix} 0 & 1 \\ -6.26 & -4.37 \end{bmatrix} \boldsymbol{x}_m(t) + \begin{bmatrix} 0 \\ 6.26 \end{bmatrix} \phi^{\text{ref}}(t). \tag{8.58}$$

Plugging the selected parameters into (8.50) results in the following control law:

$$U_{\text{ude}}(s) = \frac{Ts+1}{Ts}\left(\frac{6.25}{0.1}\Phi^{\text{ref}}(s) - M(s)\Phi(s)\right)$$

with $M(s) = \frac{2.22s^2+6.12s+6.26}{Ts+1}$. Furthermore, T is chosen as $T = 0.28$ s to achieve the same robustness index as the 2-DOF PID controller.

Experimental Results

Since the state of the plant is fully accessible (the angles and their first derivatives are measurable), the actual controllers were implemented using

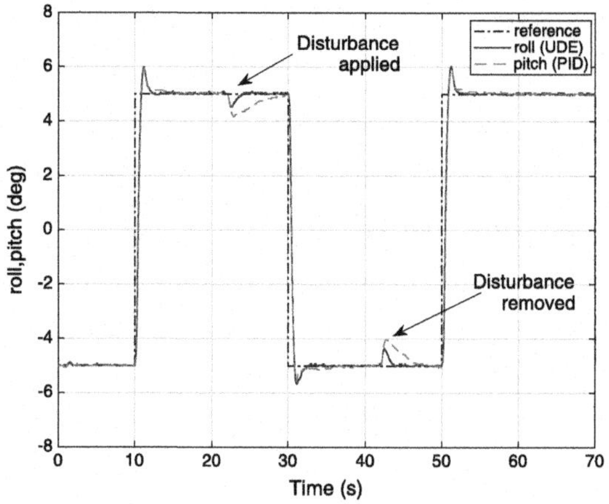

Figure 8.8 Disturbance rejection comparison for similar reference tracking performance: PID (roll), UDE (pitch).

full state-feedback with the same tuning as in the simulations. The control laws is

$$U_{\text{pid}}(s) = \left(62.6 + \frac{39.2}{s}\right)\Theta^{\text{ref}}(s) - \left(90 + \frac{39.2}{s}\right)Y_1(s) - 50\,Y_2(s),$$

$$U_{\text{ude}}(s) = \frac{0.28s+1}{0.28s} \times \left(\frac{6.25}{0.1}\Phi^{\text{ref}}(s) - 6.26Y_1(s) - 4.37Y_2(s) - \frac{s}{0.28s+1}\,Y_2(s)\right),$$

where $Y_1(s) = \{\Theta(s), \Phi(s)\}$ and $Y_2(s) = \{s\Theta(s), s\Phi(s)\}$ depending on the axis under control. An experiment was carried out to demonstrate the performance of disturbance rejection.

Fig. 8.8 shows the output of the system from the first experiment in which a square input signal of ±5 degrees was used as the reference signal. A step load disturbance of 2.5 V, generated by software to offset the control signal, was applied at $t = 23$ s and removed at $t = 43$ s. The UDE results in smaller output perturbations and faster disturbance rejection.

8.2.1.5 Flight Tests

The previous experiments have validated the theoretical results regarding the performance improvement of the UDE strategy with respect to conventional PID controllers in a laboratory platform. The aim of this section is to corroborate such improvement in flight tests. Extensive experiments are reported using two different prototypes.

UPV Prototype

In this section the experimental results from several real flight tests carried out on the quadrotor described in Sect. 3.4.3.1 of Chap. 3 are reported. The UDE strategy is compared with the following PID-like control law that is widely used in real-time implementations [14,15]:

$$u_x(t) = \sigma_{p_x}(k_{p_x}\bar{x}) + \sigma_{d_x}(k_{d_x}\dot{x}) + \sigma_{i_x}(k_{i_x}\int_0^t \bar{x}\,\mathrm{d}t), \tag{8.59}$$

where $\sigma(\cdot)$ defines the saturation function defined in Sect. 8.1, k_{p_x}, k_{d_x}, and k_{i_x} are the proportional, derivative, and integral gains, respectively, and $\bar{x} = x_d - x$ denotes the tracking error with x_d the desired value. The same control law is applied to the roll and pitch axes, i.e., $x = \phi, \theta$, with the parameters given in Table 8.1.

The UDE-based controllers for each DOF were tuned using the double integrator model, as appearing in (8.56), which implies $a_1 = a_2 = 0$. The parameters of the UDE control law for each DOF are given in Table 8.2. Because of the (ideal) symmetry of the quadrotors, both controllers for the roll and pitch axes were tuned to obtain the same closed-loop poles $-1 \pm i$, but a final online adjustment was necessary. The time constant T was selected as small as possible without introducing too much noise in the control signal.

Table 8.1 Parameters of the PID-based controller

DOF	k_{p_x}	k_{d_x}	k_{i_x}	p_x	d_x	i_x
Roll ϕ	3.3	1.2	0.03	100	50	50
Pitch θ	3.3	1.2	0.03	100	50	50
Altitude z	150	1.2	50	120	60	200

Table 8.2 Parameters of the UDE controller

DOF	b	T	a_{m2}	a_{m1}	b_m
Roll ϕ	1.5	0.6	3.2	4	4
Pitch θ	1.5	0.6	3.2	4	4
Yaw ψ	1	5	7.6	16	16
Altitude z	1.7	0.6	3	2.25	2.25

1st Flight Test: Hovering

In this experiment, the UDE-based control law was implemented for one of the roll and pitch axes, while the other axes were controlled using the

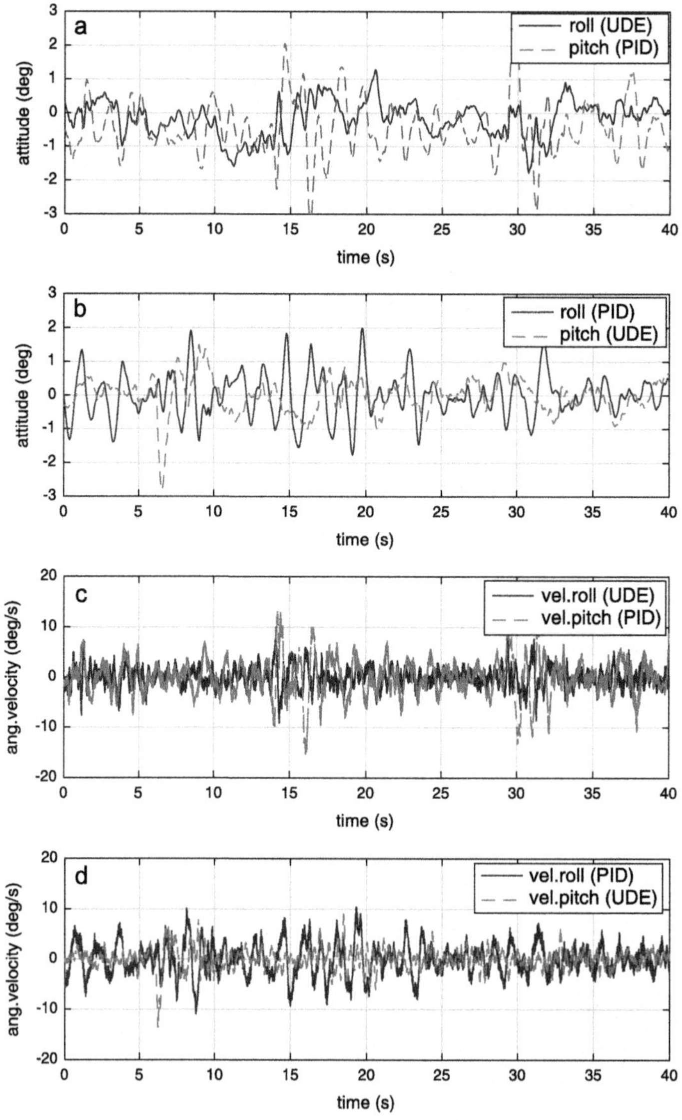

Figure 8.9 1st flight test: UDE for roll and PID for pitch (a, c) and PID for roll and UDE for pitch (b, d) at hover.

PID-based control law in (8.59). Data was recorded while the quadrotor was hovering, i.e., flying freely with zero reference inputs in both axis. The experiment was repeated after swapping the controllers for each axis, in order to provide non-biased results. The results from both experiments are shown in Fig. 8.9.

In both cases the axis controlled with the UDE-based strategy exhibits less deviation with respect to the reference input. This can be observed in both the angular position and angular velocity. Integral Absolute Error (IAE) and Root Mean Squared Error (RMSE) are presented in Table 8.3 for comparison.

Table 8.3 IAE and RMSE of the results in Figs. 8.9 and 8.10

		Hover (Fig. 8.9)		Disturb. (Fig. 8.10)	
		IAE	RMSE	IAE	RMSE
Pitch	UDE	15.5	0.55	21.6	0.82
	PID	29.4	0.93	32.7	1.24
Roll	UDE	17.3	0.56	18.7	0.72
	PID	19.7	0.65	45.0	1.77

2nd Flight Test: Disturbance Rejection

The goal of this experiment is to provide comparative results of the disturbance rejection capability of both control strategies. The experiments were carried out by applying disturbances to the quadrotor while it was hovering, hitting the quadrotor by hand at an intermediate point between both axes. Due to the construction of the prototype, this is easy because such points can be any of the corners of the squared protection frame. Similarly, the experiment was performed twice, with different combinations of controllers. The results are shown in Fig. 8.10.

The UDE-based control strategy results in much faster performance in disturbance rejection than the PID-based controller. This can also be seen from the corresponding IAE and RMSE shown in Table 8.3. Indeed, the UDE-based controller outperforms the widely-used PID control law significantly.

3rd Flight Test: Full Control

The purpose of this experiment is to give an overall view of the behavior of a quadrotor when the 4 DOF in the system (8.56) is fully controlled with

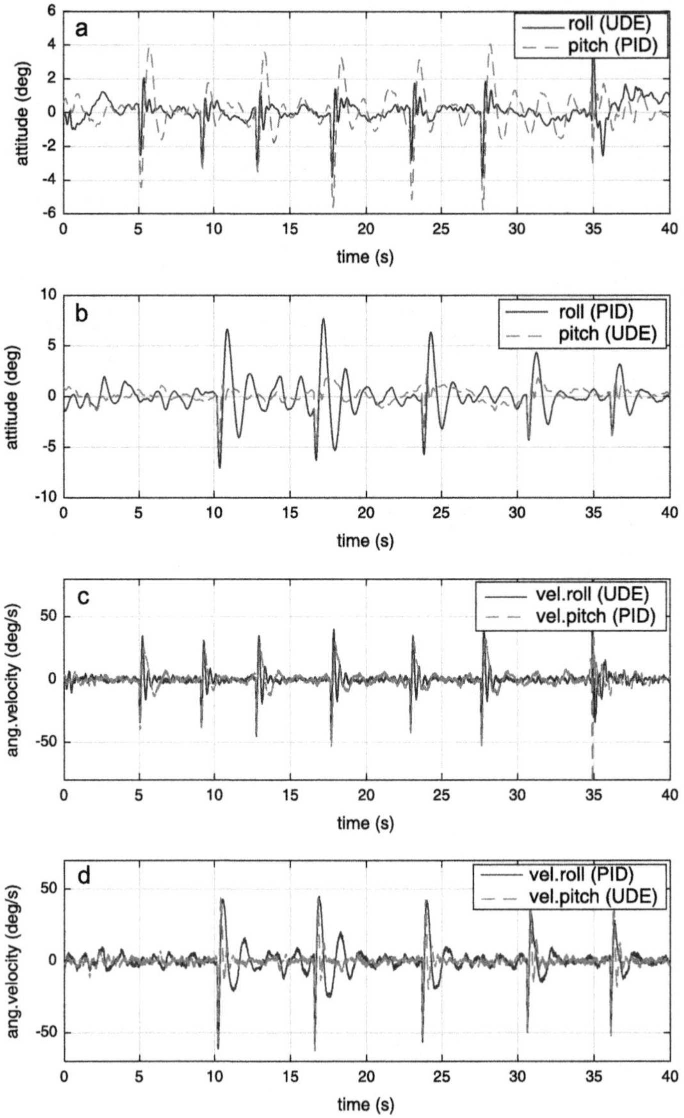

Figure 8.10 2nd flight test: UDE for roll and PID for pitch (a, c) and PID for roll and UDE for pitch (b, d) under disturbances.

the proposed UDE-based control strategy. The roll and pitch references were set at zero and step changes in yaw and altitude were applied. The results are shown in Fig. 8.11. Notice that, even though the quadrotor is a

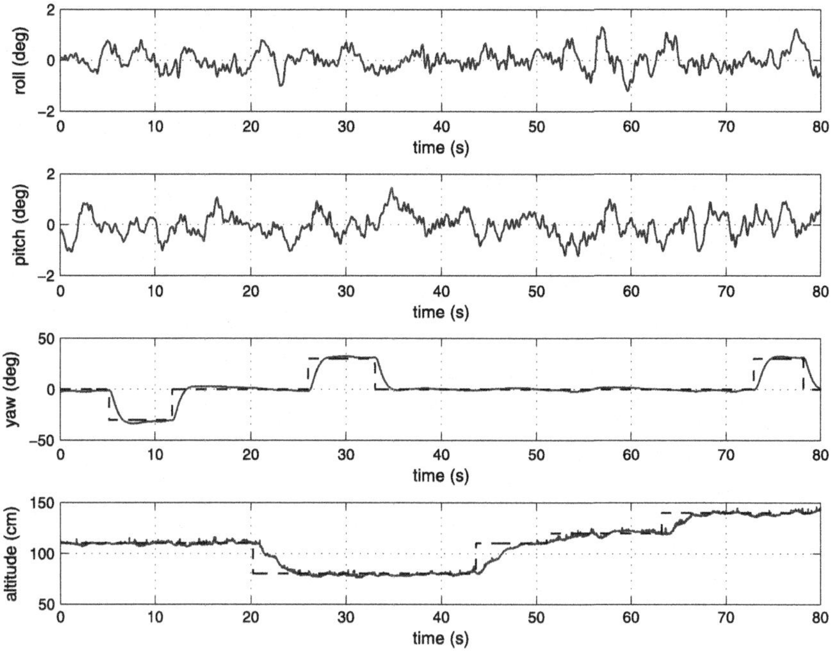

Figure 8.11 3rd flight test: results with the proposed UDE-based control.

coupled nonlinear system, the UDE-based control strategy could deal with it fairly well. No visible coupling effects appeared in the roll and the pitch while the yaw and altitude references were changed. The RMSE of the roll and pitch angles were 0.39 and 0.46 degrees, respectively, while the RMSE of the yaw and altitude tracking the input command were 6.21 degrees and 6.7 cm, respectively. Notice that experiments performed in [16] revealed that the attitude measurements have an accuracy of 0.15 degrees (RMSE) approximately (furthermore, they are highly degraded in real flight tests due to vibrations induced by the motors and the presence of lateral accelerations). Hence, it is remarkable for the proposed UDE control strategy to achieve 0.39 and 0.46 degree RMSE for the roll and pitch angles.

Compiègne AR Drone

Other flight tests were carried out using the prototype presented in Sect. 4.1.5 of Chap. 4. Two experiments are presented next, intended to demonstrate the performance improvement in the presence of wind. The strategy is compared with an equivalent state-feedback controller without

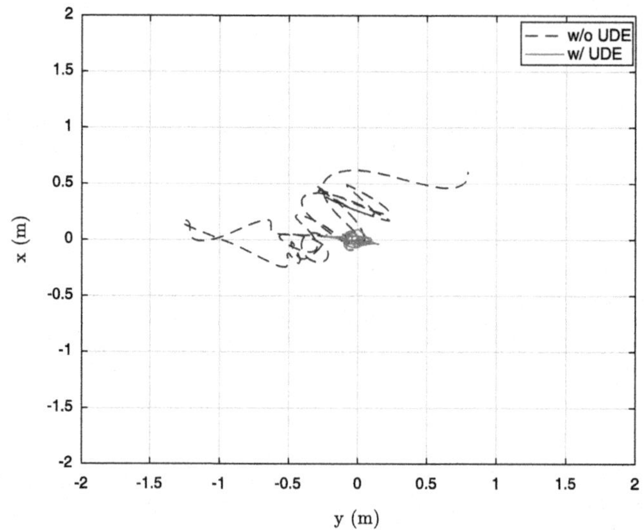

Figure 8.12 4th flight test: *x*–*y* chart comparing a state-feedback controller (dashed line) and the UDE strategy (solid line) during hovering in a windy environment.

uncertainty compensation (each of the experiments described next is realized twice in the same environment, using each of the controllers).

4th Flight Test: Position Hold in a Windy Environment

In this experiment two fans are located facing the point (0, 0) towards the negative y-axis, and initially turned-off. The quadrotor is hovering at (0, 0) and, about 4 s later, the fans are turned on. Comparative results are shown in Fig. 8.12. Notice that the error position is much smaller (solid line) when using the UDE strategy. When using a conventional state-feedback controller (dashed line), the wind induces an oscillatory behavior of the quadrotor, which even becomes unstable.

5th Flight Test: Navigation in a Windy Environment

Now the fans are located in the same way as before but it is already blowing at $t = 0$. The vehicle starts from an initial position $(-2, -2)$, where it is only slightly affected by the windstream, and it is sent to the desired position (0, 0). Figs. 8.13–8.15 introduce the results. Notice that better performance is obtained again when using the UDE-based control scheme. When only a state feedback is used, the aerial vehicle cannot hold the desired position and, ultimately, is destabilized and crashes. From Figs. 8.14–8.15 observe

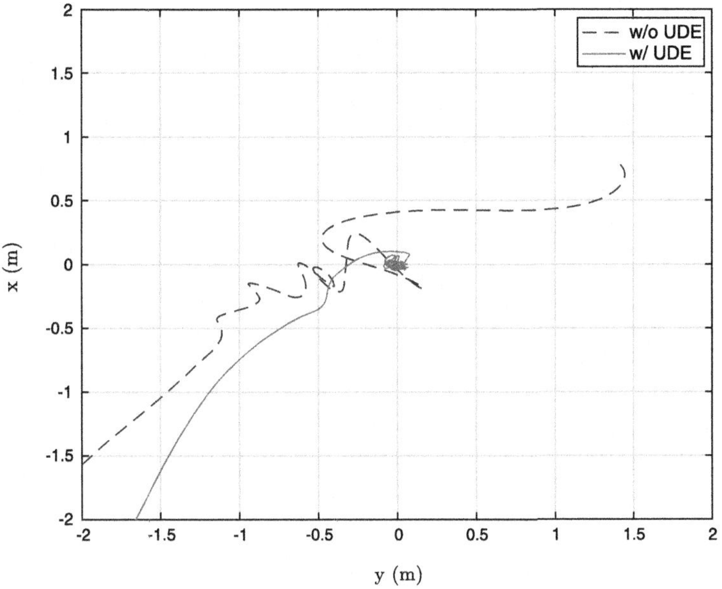

Figure 8.13 5th flight test: *x*–*y* chart comparing a state-feedback controller (dashed line) and the UDE strategy (solid line) during navigation in a windy environment.

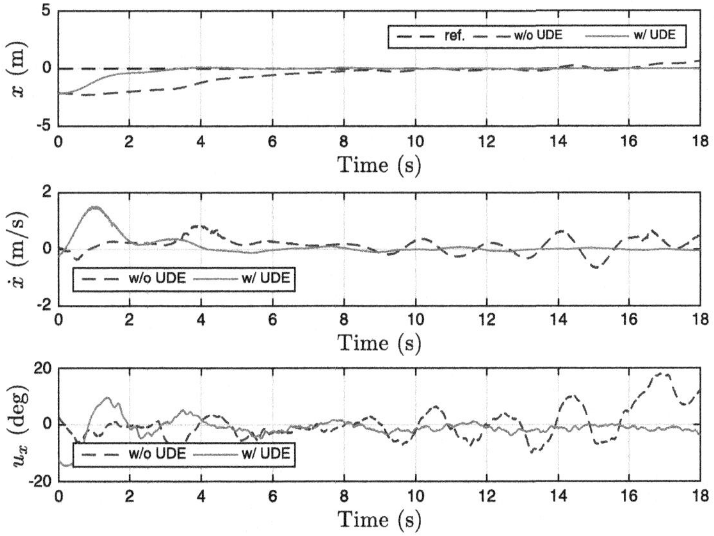

Figure 8.14 5th flight test: evolution of variables in the *x*-axis over time comparing a state-feedback controller (dashed line) and the UDE strategy (solid line) during navigation in a windy environment.

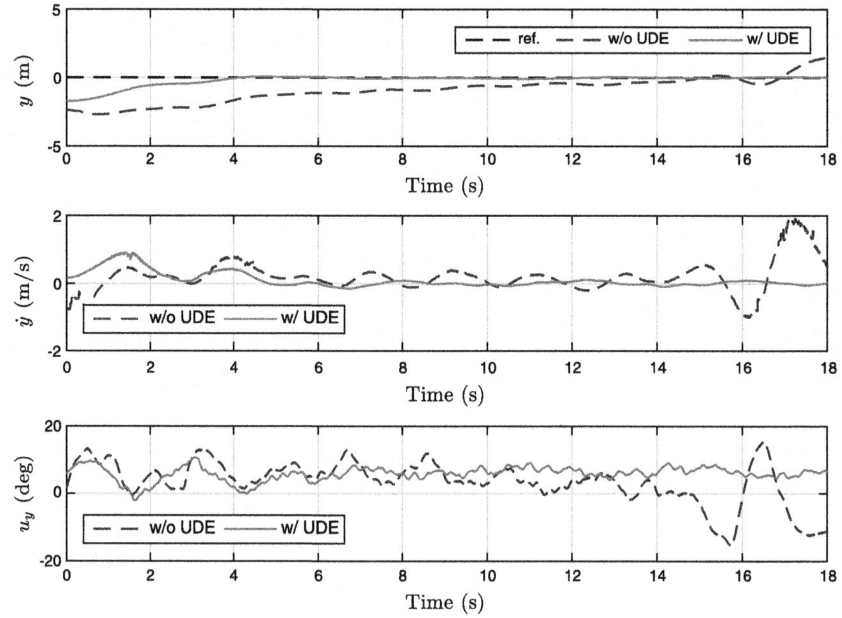

Figure 8.15 5th flight test: evolution of variables in the y-axis over time comparing a state-feedback controller (dashed line) and the UDE strategy (solid line) during navigation in a windy environment.

that the control inputs u_x and u_y are in degrees. This is because a cascade control structure is used, and hence the position controller produces angular references that are tracked by the attitude controller.

6th Flight Test: Rejection of Sinusoidal Input Disturbances

In order to check the performance on input disturbance rejection, an artificial sinusoidal input to the motors is generated by software. The size of this signal is small enough so that the system does not become unstable. As it can be seen at the top of Fig. 8.16, the input disturbance makes the system deviate ± 10 degrees when it is controlled by a simple state-feedback (dashed line). The UDE is able to estimate the disturbance and counteract it very fast so that the deviation is less than ± 1 degree (solid line). The real disturbance is known, as it is generated by software, and it is shown at the bottom of Fig. 8.16 (dashed line) along with the estimation of the UDE (solid line).

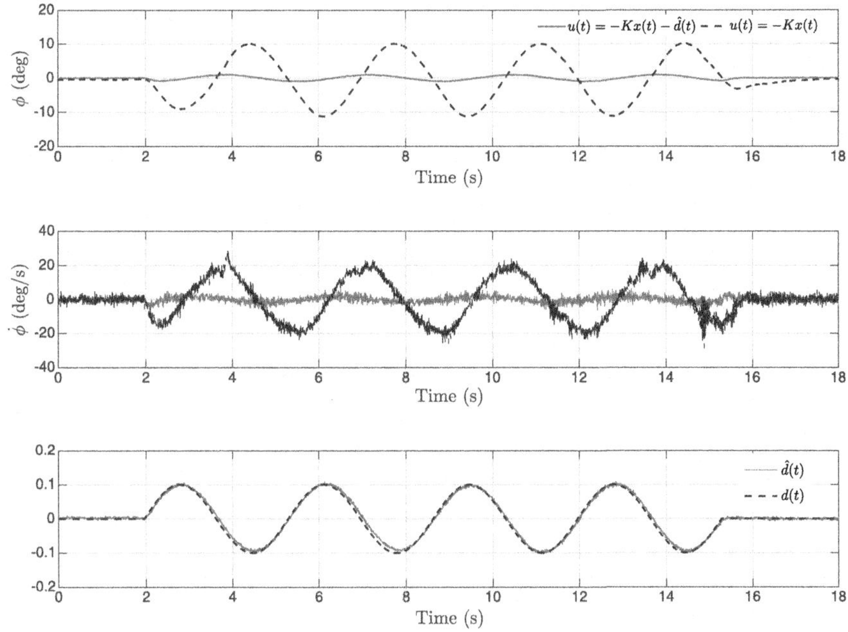

Figure 8.16 6th flight test: comparison of a state-feedback controller (dashed line) and the UDE strategy (solid line) when rejecting a sinusoidal input disturbance.

8.3 DISCUSSION

Two approaches to improve robustness of simple control algorithms were presented. First, a control law based on saturation functions and designed from linear systems was shown to be robust and to impose linear behavior in the nonlinear system. The controllers were analyzed and proved to be robust with respect to unknown perturbations and nonlinear uncertainties in the model. Simulations and flight tests have demonstrated good performance of these algorithms even in the presence of wind. The simplicity of this controller allows for an easy implementation and tuning in practice.

The second methodology was based on an uncertainty estimator. The performance of the UDE-based control strategy has been extensively validated with a Quanser experimental platform and also with a quadrotor prototype in flight tests. The UDE-based strategy applied to the attitude control has demonstrated better performance than the PID-based controller, as evidenced by indices like the integral absolute error, reduced by $36.5 \sim 73.8\%$ when impulse external disturbances had been applied, and the root mean squared error, reduced by $20.3 \sim 56.5\%$ while hovering. Furthermore, outstanding results have been reported using the same technique to navigate in windy environments.

REFERENCES

1. R.A. Freeman, P. Kokotovic, Robust Nonlinear Control Design, Birkhäuser, Boston, London, UK, 1996.
2. R. Sanz, P. Garcia, Q. Zhong, P. Albertos, Robust control of quadrotors based on an uncertainty and disturbance estimator, Journal of Dynamic Systems, Measurement, and Control 138 (2016) 071006–071013.
3. P. Khalil, Nonlinear Systems, Prentice Hall, USA, 1996.
4. J. Slootweg, Modeling wind turbines in power systems dynamics simulation, PhD thesis, Delf University of Technology, 2003.
5. Q.-C. Zhong, D. Rees, Control of uncertain LTI systems based on an uncertainty and disturbance estimator, Journal of Dynamic Systems, Measurement, and Control 126 (2004) 905–910.
6. S. Talole, S. Phadke, Robust input–output linearisation using uncertainty and disturbance estimator, International Journal of Control 82 (2009) 1794–1803.
7. A. Kuperman, Q.-C. Zhong, Robust control of uncertain nonlinear systems with state delays based on an uncertainty and disturbance estimator, International Journal of Robust and Nonlinear Control 21 (2011) 79–92.
8. J.P. Kolhe, M. Shaheed, T. Chandar, S. Talole, Robust control of robot manipulators based on uncertainty and disturbance estimation, International Journal of Robust and Nonlinear Control 23 (2013) 104–122.
9. V. Deshpande, S. Phadke, Control of uncertain nonlinear systems using an uncertainty and disturbance estimator, Journal of Dynamic Systems, Measurement, and Control 134 (2) (2011) 024501, 7 pp.
10. K. Youcef-Toumi, O. Ito, A time delay controller for systems with unknown dynamics, Journal of Dynamic Systems, Measurement, and Control 112 (1990) 133–142.
11. B. Ren, Q.-C. Zhong, J. Chen, Robust control for a class of non-affine nonlinear systems based on the uncertainty and disturbance estimator, IEEE Transactions on Industrial Electronics 62 (2015) 5881–5888.
12. R. Mahony, V. Kumar, P. Corke, Multirotor aerial vehicles: modeling, estimation, and control of quadrotor, IEEE Robotics & Automation Magazine 19 (2012) 20–32.
13. P. Castillo, R. Lozano, A. Dzul, Modelling and Control of Mini-Flying Machines, Springer Science & Business Media, 2006.
14. A. Sanchez, P. Garcia, P. Castillo Garcia, R. Lozano, Simple real-time stabilization of vertical take-off and landing aircraft with bounded signals, Journal of Guidance, Control, and Dynamics 31 (2008) 1166–1176.
15. G. Sanahuja, P. Castillo, A. Sanchez, Stabilization of *n* integrators in cascade with bounded input with experimental application to a VTOL laboratory system, International Journal of Robust and Nonlinear Control 20 (2010) 1129–1139.
16. R. Sanz, L. Rodenas, P. Garcia, P. Castillo, Improving attitude estimation using inertial sensors for quadrotor control systems, in: International Conference on Unmanned Aircraft Systems (ICUAS), Orlando, FL, USA, May 27–30, 2014.

CHAPTER 9

Trajectory Generation, Planning & Tracking*

The autonomy of an UAV can be defined as its capacity to accomplish different tasks with a high level of performance, maneuverability, and with less oversight of human operators [1]. Three fundamental features increase the autonomy of UAVs, these characteristics are reference trajectory generation, planning, and tracking. This chapter describes all these aspects. First of all, we deal with the problem of reference trajectory generation. The generated trajectory is optimal in terms of flying time. Furthermore, it is computed using a translational kinematic model, and satisfies the aerial vehicle's physical restrictions. Then, we develop a strategy for trajectory planning, based on operational research approaches, in which we include the maximum flight time, keeping optimization objective. Finally, we describe a robust control strategy, based on the vehicle full dynamic model, to follow the computed reference trajectory and validate its compatibility with the dynamics of a quadrotor.

The trajectory generation and planning algorithms proposed in this work are based on an optimality notion. They minimize the arrival time to the desired point and take the physical and environmental constraints of the vehicle into account. Physical restrictions can be regarded as limitations on the flight path angle, velocity and their rate of change, respectively, as well as on the rate of change of heading angle, while environmental constraints involve wind effect and obstacles. Moreover, they take into account the feasibility of the trajectory by respecting onboard energy limitations like consumed fuel, flight time, etc. The approach is focused on a quadrotor dedicated to inspection missions, which requires the presence of a camera attached to the vehicle. This camera is supposed to be fixed in a specified position corresponding to the longitudinal axis of the vehicle. Hence three scenarios for simulations have been taken into account. The first deals with an aerial vehicle moving at a constant altitude. The second treats the three-dimensional trajectory generation problem. In both scenarios the aerial

* This chapter was developed in collaboration with Elie Kahale & Yasmina Bestaoui from the IBISC Laboratory, EA 4526, Evry University Val d'Essone, France.

Indoor Navigation Strategies for Aerial Autonomous Systems.
DOI: http://dx.doi.org/10.1016/B978-0-12-805189-4.00012-3

vehicle is supposed to fly in a quiet environment. The obtained results for both schemes incorporate a trajectory with multiple waypoints for a structure inspection task. The third scenario considers the case of a windy environment in which the aerial vehicle flies at a fixed altitude. Furthermore, the flying time is limited. The originality of our work is threefold: first, we adopt a point mass model to generate reference trajectories for a quadcopter; second, we generalize Dubins planning approach to 3D space planning and consider the advantage of the variation of the vehicle's velocity; third, we take into account the wind effect.

The control strategy is based on saturation functions; for more details about this controller see Chap. 8. The method has been successfully applied in the stabilization of the quadrotor at hovering; nevertheless, it has been poorly implemented for tracking purposes. The idea is to explore its effectiveness for trajectory following and its robustness in the case of strongly noisy sensors.

9.1 QUADROTOR MATHEMATICAL DESCRIPTION

The quadcopter model as described in Chap. 2 will be used for the trajectory tracking problem. Nevertheless, for the trajectory generation a point mass model is used, where the aerial vehicle is presented as one point which is its own center of gravity.

9.1.1 Rigid Body Model

From (2.4)–(2.5) the following equations can be written for quasi-stationary maneuvers:

$$m\ddot{\xi} = uR_{\xi}, \tag{9.1a}$$

$$\mathbb{J}\ddot{\eta} = \tau, \tag{9.1b}$$

where $\xi = (x, y, z) \in \mathbb{R}^3$ denotes the position of the vehicle's center of gravity relative to a fixed inertial frame, $\eta = (\psi, \theta, \phi) \in \mathbb{R}^3$ presents the orientation of the vehicle expressed in the Euler angles (ψ for yaw, θ for pitch, and ϕ for roll), m is the mass of the aerial vehicle, and $\mathbb{J}$ represents the constant inertia tensor. With quasi-stationary maneuvers $C(\eta, \dot{\eta}) \approx 0$. In addition,

$$R_{\zeta} = \begin{pmatrix} -\sin\theta \\ \cos\theta\sin\phi \\ \cos\theta - \frac{mg}{u} \end{pmatrix}, \quad \tau = \begin{pmatrix} \tau_{\psi} \\ \tau_{\theta} \\ \tau_{\phi} \end{pmatrix},$$

with g defining the gravity acceleration, and u and τ_i being the control inputs.

9.1.2 Point Mass Model

For the trajectory generation requirements, it will be considered that the quadcopter moves only in the longitudinal axis, therefore the lateral displacement must be close to zero, i.e., $y, \phi \approx 0$. This can be obtained proposing a controller to ensure closed-loop stability of this subsystem. Hence the orientation of the vehicle yields

$$\eta_1 = (\psi, \theta) \in \mathbb{R}^2.$$

Let us define

$$q_i = (\xi(t_i), \eta_1(t_i), V(t_i)) \in \mathbb{R}^6 \tag{9.2}$$

that represents a vector configuration q_i of a quadrotor at a specified time t_i. In the above equation $\xi(t_i)$, $\eta(t_i)$, and $V(t_i)$ define the position, orientation, and translational velocity of the aerial vehicle, respectively.

For trajectory missions, i.e., to reach a specified or final configuration q_f from the actual or initial q_0, the vehicle time profile of x, y, z, θ, ψ, V, $\dot{\theta}$, $\dot{\psi}$, and $\dot{V}$ must be known along all the trajectory. Note that the roll angle is considered as previously stabilized. From this perspective the quadcopter must be described by its Point Mass Model for trajectory generation requirements.

This model, which resulted from Newton's second law of motion, describes the inertial velocity vector $\vec{V}_I$ with respect to an inertial frame and the external forces acting on the vehicle. Moreover, it is mathematically accurate under the assumption of flat earth and symmetric flight; the last supposition means that the sideslip angle is zero [2]. The inertial velocity is given by

$$\vec{V}_I = \vec{V} + \vec{V}_w, \tag{9.3}$$

with $\vec{V}$ representing the relative velocity vector and $\vec{V}_w$ describing the wind velocity vector which is given by $[W_x W_y W_z]^T$. In this model the states are: x for down-range, y for cross-range, z for the altitude, γ for the flight path angle, χ for the heading angle, and V for the translational velocity.

The flight path angle γ is the angle defined between $\vec{V}$ and its projection in the x–y plane. Similarly, the pitch angle contains the sum of the

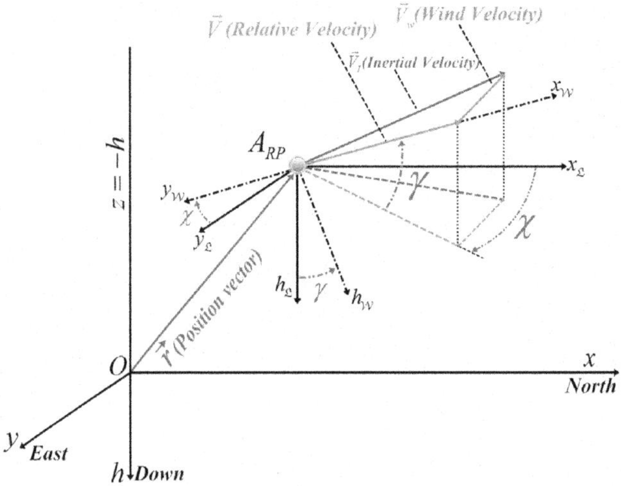

Figure 9.1 Coordinate system for Point Mass Model.

flight path angle and the angle of attack [3], i.e., $\theta = \gamma + \alpha$. For rotorcraft vehicles $\alpha \ll 1$, and it can be neglected. Thus, considering $\alpha \approx 0$, we have $\theta = \gamma$. Furthermore, the heading angle χ can be regarded as the yaw angle ψ because it is measured from the north to the projection of $\vec{V}$ in the local horizontal plane. Then $\chi = \psi$.

Therefore, from the previous discussion and Fig. 9.1, it follows that

$$\dot{x} = V\cos\psi\cos\theta \ + W_x, \tag{9.4a}$$
$$\dot{y} = V\sin\psi\cos\theta + W_y, \tag{9.4b}$$
$$\dot{z} = V\sin\theta + W_z. \tag{9.4c}$$

From Newton's law, we have that $m\dot{\vec{V}} = \sum \vec{F}$, where $\vec{F}$ describes the external forces acting on the vehicle, and $\dot{\vec{V}}$ denotes the accelerations of the vehicle which are given as

$$\dot{\vec{V}} = \begin{bmatrix} a_{long} \\ a_{lat} \\ a_{vert} \end{bmatrix} = \begin{bmatrix} \dot{V} \\ V\dot{\psi}\cos\theta \\ V\dot{\theta} \end{bmatrix},$$

where a_{long}, a_{lat}, and a_{vert} denote longitudinal, lateral, and vertical accelerations, respectively. Notice that θ, ψ, and V define the attitude of the vehicle, consequently and without loss of generality, $\dot{\theta}$, $\dot{\psi}$, and $\dot{V}$ can be considered as virtual control inputs and can be written as

$$\dot{\theta} = u_1, \tag{9.5a}$$
$$\dot{\psi} = u_2, \tag{9.5b}$$
$$\dot{V} = u_3. \tag{9.5c}$$

Real prototypes include some physical constraints that limit their performance. So to be more realistic when validating the algorithms in simulation, some constraints need to be included in the model. These restrictions could take the form of bounds on control and some states variables, and we define these as follows:

$$|u_1| \leq U_{1max}, \qquad |\theta| \leq \theta_{max}, \tag{9.6a}$$
$$|u_2| \leq U_{2max}, \qquad |V| \leq V_{max}, \tag{9.6b}$$
$$|u_3| \leq U_{3max}. \tag{9.6c}$$

For more details see [2–5].

9.2 TIME-OPTIMAL TRAJECTORY GENERATION

The aim of this section is to compute an optimal trajectory under acceleration, heading, and flight path rate limits, together with the associated control demand. Thus the goal is to guide the autonomous aerial vehicle from one configuration (Eq. (9.2)) to another in minimal time. Moreover, the proposed trajectory must take into account wind effects. In general, the wind speed can be modeled as a sum of two components: a nominal deterministic component, available through meteorological forecasts or measured with a Doppler radar, and a stochastic component, representing deviations from the nominal one. Hence, including the deterministic component in trajectory calculation can not only contribute to saving time and energy if the wind comes in the direction of motion, for example, but also to reduce the controller workloads.

This objective can be formulated by introducing the following performance criterion:

$$J = \min \int_{t_0}^{t_f} dt, \tag{9.7}$$

and then the problem is to provide an admissible control $U(t) : [t_0, t_f] \longmapsto \Omega \in \mathbb{R}^3$ which minimizes the performance criterion (9.7), subject to the differential constraints (equations of motion (9.4) and (9.5))

$$\dot{X}(t) = f\left(X(t), U(t), t\right), \tag{9.8}$$

the prescribed initial conditions

$$X(t_0) = \begin{bmatrix} x_0 & y_0 & z_0 & \theta_0 & \psi_0 & V_0 \end{bmatrix}^T, \tag{9.9}$$

and the prescribed final conditions

$$X(t_f) = \begin{bmatrix} x_f & y_f & z_f & \theta_f & \psi_f & V_f \end{bmatrix}^T, \tag{9.10}$$

where $X(t) \in \mathbb{R}^6$ and $U(t) \in \mathbb{R}^3$ are the state and control variables, respectively, which are expressed as

$$X(t) = \begin{bmatrix} x(t) & y(t) & z(t) & \theta(t) & \psi(t) & V(t) \end{bmatrix}^T, \tag{9.11}$$

$$U(t) = \begin{bmatrix} u_1(t) \\ u_2(t) \\ u_3(t) \end{bmatrix} = \begin{bmatrix} \dot{\theta}(t) \\ \dot{\psi}(t) \\ \dot{V}(t) \end{bmatrix}, \tag{9.12}$$

and such that the corresponding state trajectory satisfies the limitations on the control inputs and states expressed in (9.6) in addition to the following restriction:

$$(x - x_c)^2 + (y - y_c)^2 + z^2 \geq r_p^2. \tag{9.13}$$

Note that the previous equation describes Bridge's pillar as a cylinder. A theoretical analysis of this problem, from the point of view of the solution structure, e.g., type of control demand number and duration of active constraints, can be found in [6].

9.2.1 Numerical Solution

The problem described in the previous section is infinite-dimensional, which means its solution is not a finite vector of numbers, but rather a function. However, for a real-life application it is impossible to compute the optimal function, so the use of approximate methods is required. Such techniques attempt to find a finite-dimensional representation of the solution which is accurate at the nodes of the representation, has acceptable error between the nodes, and converges to the true function as the number of nodes tends to infinity [7].

The numerical solution of the optimal control problem can be categorized into two main approaches. The first approach corresponds to the direct method which replaces the continuous time interval with a grid of discrete points, thus approximating it with a finite-dimensional problem, albeit of high dimension (hundreds of discretized variables). This is equivalent to transforming the problem into a sequence of nonlinear constrained optimization problems. The second approach corresponds to the indirect approach which preserves the infinite-dimensional character of the task and uses the theory of optimal control to solve it. Then the first step of this method is to formulate an appropriate *Two Point Boundary Value Problem* (*TPBVP*), and the second step is to solve it numerically.

In this chapter the direct method is preferred. This technique is known as the Direct Collocation Approach [7] and offers the following advantages:

- Simplicity of implementation
- Robustness
- Being not sensitive to the choice of initial conditions
- Ease to include constraints on state variables

Hence we use a Nonlinear Programming solver using MATLAB with respect to the discretized control. The corresponding discretized state variables are determined recursively using a numerical integration scheme (e.g., Euler, Runge–Kutta, etc.) [8].

Therefore the time interval $[t_0, t_f]$ is divided into N nodes as follows:

$$t_0 \leq t_1 \leq t_2 \leq \cdots \leq t_N = t_f \tag{9.14}$$

such that

$$t_k = t_0 + (k-1) \cdot h, \quad h := \frac{t_f - t_0}{N-1}, \quad k = 1, \ldots, N. \tag{9.15}$$

Thus the vector of unknown variables is composed of the control inputs over all nodes and the final time t_f as shown below

$$\zeta = \left[t_f, U_1^T, U_2^T, \ldots, U_N^T\right] \in \mathbb{R}^{N_\xi}, \quad N_\zeta = 3N + 1, \tag{9.16}$$

and the state variables are computed recursively using Euler approximation applied to (9.8), yielding the following:

$$X_{k+1} = X_k + h \cdot f\left(X_k, U_k\right), \quad k = 1, \ldots, N-1, \tag{9.17}$$

with the initial and final conditions given in (9.9) and (9.10).

From the previous discussion the problem of optimal control can be described as follows:

$$
\begin{aligned}
&\text{Minimize} && J = t_f, && \text{(9.18a)}\\
&\text{Subject to} && \dot{X}_k = f\left(X_k, U_k\right), && \text{(9.18b)}\\
& && X(t_1) = X_0, && \text{(9.18c)}\\
& && X(t_N) = X_f, && \text{(9.18d)}\\
& && C_{lw} \le U_k \le C_{up}, && \text{(9.18e)}\\
& && S_{lw} \le S\left(X_k\right) \le S_{up}, && \text{(9.18f)}\\
& && S_1\left(X_k\right) \ge r_p^2, && \text{(9.18g)}
\end{aligned}
$$

where $k = 1, 2, \ldots, N$, C_{lw} and C_{up} are lower and upper bounds on control inputs across all nodes. The kth set of control rate limits, expressed in (9.18e), is given as

$$
\begin{bmatrix} \dot{\theta}_{min} \\ \dot{\psi}_{min} \\ \dot{V}_{min} \end{bmatrix} \le \begin{bmatrix} \dot{\theta}_k \\ \dot{\psi}_k \\ \dot{V}_k \end{bmatrix} \le \begin{bmatrix} \dot{\theta}_{max} \\ \dot{\psi}_{max} \\ \dot{V}_{max} \end{bmatrix}. \tag{9.19}
$$

Moreover, S_{lw} and S_{up} denote the lower and upper constraints enforced on the path. The kth set of path restrictions, presented in Eq. (9.18f), is shown as

$$
\begin{bmatrix} \theta_{min} \\ V_{min} \end{bmatrix} \le \begin{bmatrix} \theta_k \\ V_k \end{bmatrix} \le \begin{bmatrix} \theta_{max} \\ V_{max} \end{bmatrix}. \tag{9.20}
$$

The kth set of path restrictions, presented in (9.18g), is shown as

$$
\left(x_k - x_c\right)^2 + \left(y_k - y_c\right)^2 + z_k^2 \ge r_p^2, \tag{9.21}
$$

with x_c, y_c, and r_p^2 describing the pillar axis and its radius, respectively.

The next section describes a trajectory planning strategy for a quadrotor dedicated for inspection, monitoring, or surveillance missions.

9.3 UAV ROUTING PROBLEM FOR INSPECTION-LIKE MISSIONS

In this section we consider a problem of bridge inspection using a small UAV. Such a problem has been addressed in [9], where the authors have proposed two solutions based on a hybridization between Zermelo's navigation problem and Traveling SalesMan (TSP)/Vehicle Routing Problem (VRP) approaches. The wind was supposed to vary linearly, and the pillar was assumed as points of interest to be visited by the vehicle. In other words, the configuration space was considered to be obstacle-free. In addition, each point was characterized by only its position without including its orientation and velocity into account. Our work is a complement to this paper. In our approach the Venturi effect (wind acceleration) is implemented, and we take the presence of obstacles into consideration. Also each point is defined by its position, orientation, and velocity.

9.3.1 Problem Statement

The UAV routing problem for structure inspection can be formulated as follows:

Let $q = \{q_1, q_2, \ldots, q_n\}$ be a set of points such that $q_2, \ldots, q_n$ are the points situated on the bridge which must be inspected, and q_1 is the departing position representing the ground base.

Suppose that the travel cost matrix C represents the required flight time between every pair q_i, q_j, $i \neq j$.

In addition, assume that T_{Req}, called the inspection vector, is a vector specifying the required time to inspect the point q_i.

Finally, consider that T_{Max} is the maximum time allowed for the vehicle to fly. Then the UAV routing problem for structure inspection is to plan a set of tours in such a way that:

- Each point in $\{q_2, \ldots, q_n\}$ is visited only once.
- All tours start and end at the ground base q_1.
- The maximum time allowed to fly is never violated.
- The total time required to all tours is minimal.

It is important to state here that the travel cost matrix is asymmetric. Then the required time to fly from a point q_i with orientation γ_i, χ_i and velocity V_i to a point q_j with orientation γ_j, χ_j and velocity V_j may not be equal to the required time to fly in the opposite direction.

9.3.2 Capacitated Vehicle Routing Problem

The Capacitated Vehicle Routing Problem is an interesting approach for monitoring-like tasks. In fact, the limited carrying capacity of goods is replaced by an allowed maximum flight time, and the number of vehicles can be regarded as the number of vehicles required in UAVs fleet, **or** the number of flights (for one aerial vehicle) needed to cover the points of interest. To solve such a problem, a wide variety of methods can be found in the literature.

- **Constructive methods:** savings and insertion
- **Improvement methods:** 2-change, 3-change, 2-relocate, 3-relocate, sweep, GENI.

In the following section, we present the **savings method** developed for solving CVRP.

Savings Method

The savings approach is a heuristic algorithm which was first proposed in 1964 by Clarke and Wright [10] to solve a CVRP in which the number of vehicles is free. Its basic idea is very simple: considering a depot D and n demand points, supposing that initially the solution to the VRP consists of using n vehicles, and dispatching one vehicle to each one of the n demand points. Then, obviously, the total tour length of such a solution is $2\sum_{i=1}^{n} C(D,i)$.

Now, if we use a single vehicle to serve two points, e.g., i and j, on a single trip, the total distance traveled is reduced by the following amount

$$\begin{aligned} S(i,j) &= 2C(D,i) + 2C(D,j) - \left[C(D,i) + C(i,j) + C(D,j)\right] \\ &= C(D,i) + C(D,j) - C(i,j). \end{aligned} \tag{9.22}$$

The quantity $S(i,j)$ is known as the ***savings*** resulting from combining points i and j into a single tour. Whenever the value of $S(i,j)$ is larger, combining i and j in a single tour becomes more desirable. However, i and j cannot be combined if the resulting tour violates one or more constraints of the VRP.

Before describing the algorithm we introduce the following definition:

Definition 9.1. A point i is said to be interior to a route if it is not adjacent to the depot D in the order of the traversal of points.

The savings algorithm of Clarke and Wright can now be presented (as expressed in [11]) as follows:

Step 1. Compute the savings $S(i,j) = C(D,i) + C(D,j) - C(i,j)$ for $i,j = 2,\ldots,n$, and $i \neq j$.

Step 2. Rank the savings in descending order of magnitude.

Step 3. For the savings $S(i,j)$ under consideration, include the arc (i,j) in a route if the constraints imposed on the route will not be violated, and if

(i) Neither i nor j have already been assigned to a route. In this case, a new route is initiated including both i and j.

(ii) Exactly one of the two points (i or j) has already been included in an existing route and that point is not interior to that route. In this case, the arc (i, j) is added to the same route.

(iii) Both i and j have already been included in two different existing routes and neither point is interior to its route. In this case, both routes are merged.

Step 4. If the savings list $S(i, j)$ has not been exhausted, return to Step 3 and shift to the next entry in the list; otherwise, stop.

Note that any point that has not been assigned yet to a route during Step 3 must be served by a vehicle route that begins at the depot D, visits the unassigned point, and returns to D.

The Clarke–Wright algorithm can be programmed to run very efficiently and, since it involves very simple manipulations of the data set, it can be used with large-scale problems. Since nodes are added to routes once or twice at a time, an additional advantage of the algorithm is that it is possible to check whether each addition would violate any set of constraints, even when that set is quite complicated. For example, besides the constraints on maximum capacity and maximum distance, other constraints might be included, such as a maximum number of points that any vehicle may visit.

A number of variants of this method were proposed, e.g., [12,13]. The Clarke and Wright algorithm supposes that the cost matrix C is symmetric, i.e., $C(i, j) = C(j, i)$. In addition, it implicitly ignores vehicle fixed costs and fleet size. Vehicle costs f can easily be taken into account by adding this constant to every C_j $(j = 2, \ldots, n)$. Solutions with a fixed number of vehicles can be obtained by repeating Step 3 until the required number of routes have been reached, even if the savings become negative.

Next section deals with the problem of trajectory tracking for a quadrotor.

9.4 TRAJECTORY TRACKING PROBLEM

The algorithm introduced in Sect. 9.2 provides a feasible and flyable reference trajectory. Nevertheless, even if it is obtained from the imposed vehicle restrictions, it is important to examine its applicability and compatibility with the dynamics of the quadrotor (Eq. (9.1)). First, a controller needs to be proposed to stabilize the vehicle dynamics. Then the obtained trajectory can be used as reference signals in the controller to tracking.

Remember that the main objective of the chapter is to propose new solutions to generate trajectories and tracking in minimum time, and not to propose a new control algorithm. Therefore, due to our goals, we prefer a control scheme that is simple to guaranty easy implementing, for this we chose a simple nonlinear algorithm based on saturation functions proposed in Chap. 8.

Observe that this control algorithm was generally used to stabilize the quadrotor at hover, and this has been poorly applied in trajectory tracking with restrictions in angles, angles rate, and translational velocities. These constraints could degrade the closed-loop system performances; however, in simulations we will prove that it is not the case.

Therefore, the controllers for the altitude and yaw angle dynamics from Chap. 8 can be written as

$$u = \sec(\theta)\sec(\phi)\bar{r}, \tag{9.23}$$

$$\bar{r} = -\mathfrak{K}_{z1}\left(\dot{z}-\dot{z}_{ref}\right)-\mathfrak{K}_{z2}\left(z-z_{ref}\right)+mg,$$

$$\tau_{\psi} = -\sigma_{\psi1}\left(\mathfrak{K}_{\psi1}\left(\dot{\psi}-\dot{\psi}_{ref}\right)\right)-\sigma_{\psi2}\left(\mathfrak{K}_{\psi2}\left(\psi-\psi_{ref}\right)\right), \tag{9.24}$$

where $\mathfrak{K}_{z1}, \mathfrak{K}_{z2}, \mathfrak{K}_{\psi1}, \mathfrak{K}_{\psi2}$ are positive constants, $\dot{z}_{ref}$ and z_{ref} define the desired vertical velocity and altitude, respectively, and $\dot{\psi}_{ref}$ and ψ_{ref} denote the reference angular rate and angle, respectively. Notice that these references values come from the trajectory generation algorithm. In addition, θ and ϕ are assumed to be limited so that they can not reach 90 degrees.

On the other hand, for the x and y dynamics the controllers are

$$\begin{aligned}\tau_{\phi} &= -\sigma_{\phi1}\left(\mathfrak{K}_{\phi1}\left(\dot{y}-\dot{y}_{ref}\right)\right)-\sigma_{\phi2}\left(\mathfrak{K}_{\phi2}\left(y-y_{ref}\right)\right)-\sigma_{\phi3}\left(\mathfrak{K}_{\phi3}\dot{\phi}\right)\\ &\quad-\sigma_{\phi4}\left(\mathfrak{K}_{\phi4}\left(\phi-\phi_{ref}\right)\right),\end{aligned} \tag{9.25}$$

$$\begin{aligned}\tau_{\theta} &= \sigma_{\theta1}\left(\mathfrak{K}_{\theta1}\left(\dot{x}-\dot{x}_{ref}\right)\right)+\sigma_{\theta2}\left(\mathfrak{K}_{\theta2}\left(x-x_{ref}\right)\right)-\sigma_{\theta3}\left(\mathfrak{K}_{\theta3}\left(\dot{\theta}-\dot{\theta}_{ref}\right)\right)\\ &\quad-\sigma_{\theta4}\left(\mathfrak{K}_{\theta4}\left(\theta-\theta_{ref}\right)\right),\end{aligned} \tag{9.26}$$

where $\mathfrak{K}_{\phi i}, \mathfrak{K}_{\theta i}$ are positive constants and the reference values coming from the trajectory algorithm. Remember that for the roll dynamics the only constraint is that the roll angle be zero, thus $\phi_{ref} = 0$.

9.5 SIMULATION RESULTS

In order to illustrate the performance of the proposed trajectory generation and planning approaches and validate the tracking strategy, numerical

Table 9.1 Restrictions on state and control variables

Minimum	Variable	Maximum	Units
−30	θ	30	degrees
0.1	V	5	m/s
−5	$\dot{\theta}$	5	degrees/s
−5	$\dot{\psi}$	5	degrees/s
−1.25	$\dot{V}$	1.25	m/s^2

simulations are carried out in this section. Hence different scenarios for suggested algorithms are considered.

The vehicle's restriction are described as lower and upper limitations on the state and control variables which are specified in Table 9.1.

9.5.1 Reference Trajectory Generation and Planning

Three scenarios are examined for this case. In the first, we consider the case in which the quadcopter is flying at a fixed altitude, i.e., only the horizontal x–y plane is studied. Thus both the pitch angle and its derivative are equal to zero. In the second scenario the aerial vehicle is assumed to move in three dimensions, more precisely, to fly around a structure to perform an inspection mission. For both cases, the trajectory is separated into several sectors according to the number of way-configurations to visit. Then the optimization process described in Sect. 9.2.1 is applied for every segment separately. While in the third scenario we consider the case of trajectory planning for a bridge inspection mission in a windy environment.

Trajectories at Fixed Altitude (2D)

In this case the equations of motion presented in (9.4) and (9.5) are reduced to the following:

$$\dot{x} = V\cos\psi, \qquad \dot{\psi} = u_1,$$
$$\dot{y} = V\sin\psi, \qquad \dot{V} = u_2.$$

The grid of discrete points for each sector is fixed to 84 nodes (N). The initial configuration q_0, the final q_f, and the way-configurations q_i in this case are presented in Table 9.2. The interior point algorithm included in *fmincon solver* in MATLAB is selected to solve our nonlinear optimization problem. This choice is motivated by the fact that this approach can handle

Table 9.2 Configurations to be visited during the traveled trajectory

	x [m]	y [m]	ψ [degrees]	V [m/s]
q_0	0	0	0	0.1
q_1	10	5	90	2
q_2	0	15	90	2
q_3	10	25	90	2
q_4	0	30	90	2
q_f	10	35	0	0.1

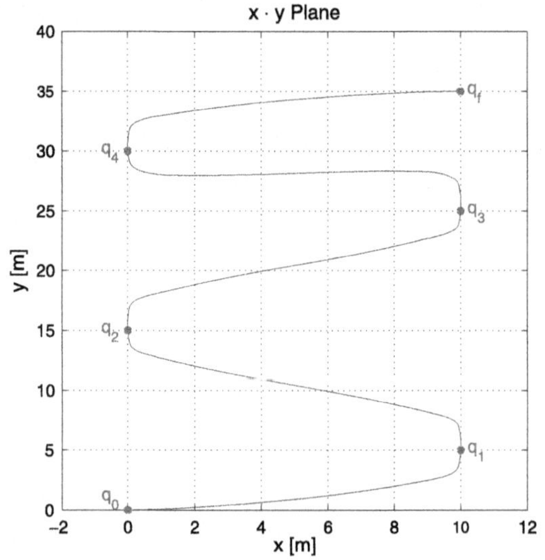

Figure 9.2 Flying at a constant altitude trajectory.

large-scale problems. Moreover, it satisfies the imposed constraints (bounds) at all iterations. The position, orientation, velocity, and control inputs of the aerial vehicle are illustrated in Figs. 9.2 and 9.3.

Notice the final configuration is reached in $t_f = 323.2$ s.

Trajectories in 3D Space

Each segment is discretized over 70 nodes, and the interior point algorithm which is included in *fmincon* solver in MATLAB is used to generate the minimal time trajectory. The initial configuration q_0, the final q_f, and the ten way-configurations q_i, $i = 1, \ldots, 10$ are presented in Table 9.3. The

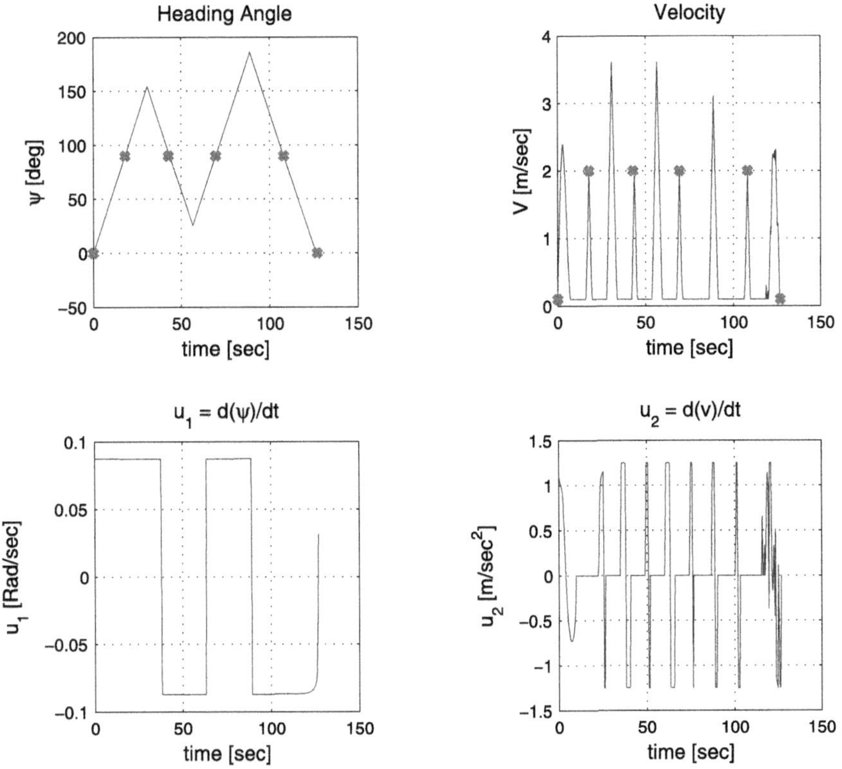

Figure 9.3 Time profile of yaw angle ψ, vehicle's velocity, and their rates of change $\dot{\psi}$ and $\dot{V}$, respectively.

Table 9.3 Configurations to be visited during the traveled trajectory

	x [m]	y [m]	z [m]	θ [degrees]	ψ [degrees]	V [m/s]
q_0	0	0	0	0	0	0.1
q_1	10	10	5	5	90	1
q_2	0	20	5	0	180	1
q_3	−10	10	5	0	270	1
q_4	0	0	5	0	360	1
q_5	10	10	5	0	450	1
q_6	0	20	10	0	540	1
q_7	−10	10	10	0	630	1
q_8	0	0	10	0	720	1
q_9	10	10	10	0	810	1
q_{10}	20	10	10	0	900	1
q_{11}	−10	10	5	0	990	1
q_f	−5	0	0	0	1080	0.1

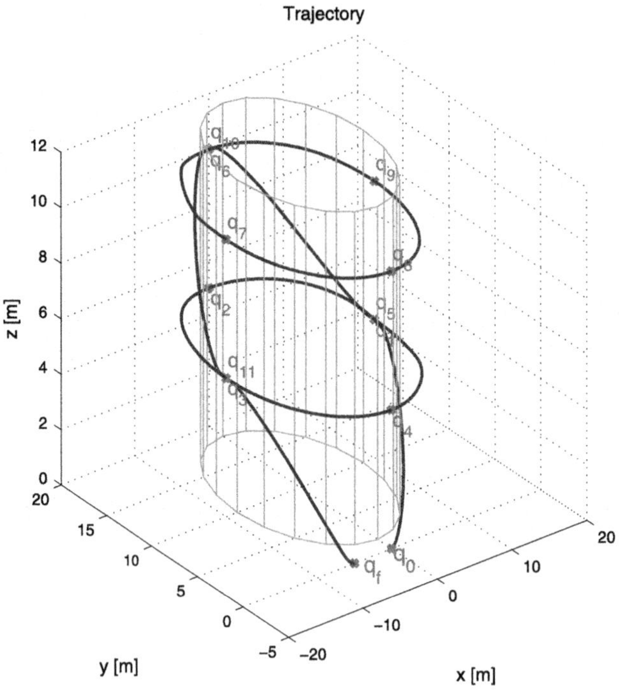

Figure 9.4 3D view of structure inspection scenario.

vehicle's position, orientation, velocity, and control inputs are illustrated in Figs. 9.4, 9.5, and 9.6.

Notice that the required time to complete the trajectory is $t_f = 224.4$ s.

We also remark that the choice of the number of nodes plays an important role in determining the size of the optimization problem. The number of nodes N is chosen such that the resulting trajectory is smooth and calculation time is still reasonable.

Observe that the previous results can be also obtained using Sequential-Quadratic-Programming (SQP) method included in *fmincon* solver in MATLAB which is more advantageous in terms of computation time and algorithm convergence.

Trajectory Planning in a Windy Environment

In this scenario we introduce wind in the trajectory generation process. In fact, the vicinity of bridges is an ideal environment to the appearance of Venturi effect due to its structural design. Furthermore, the passages between pillars can be responsible for increased wind speed, see Fig. 9.7.

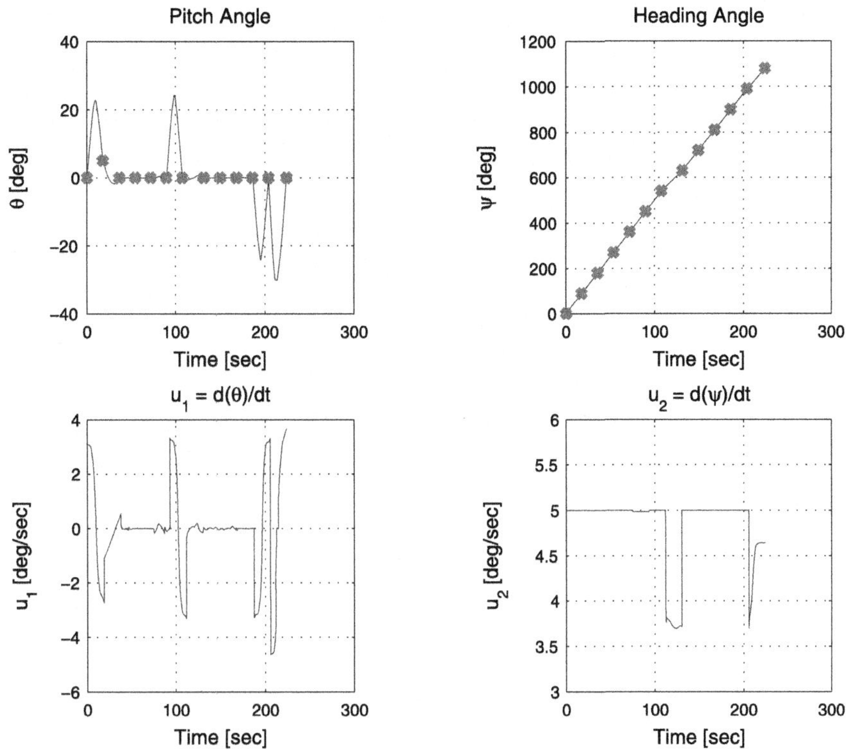

Figure 9.5 Time profile of pitch θ, yaw ψ, and their rate of change $\dot{\theta}$ and $\dot{\psi}$, respectively.

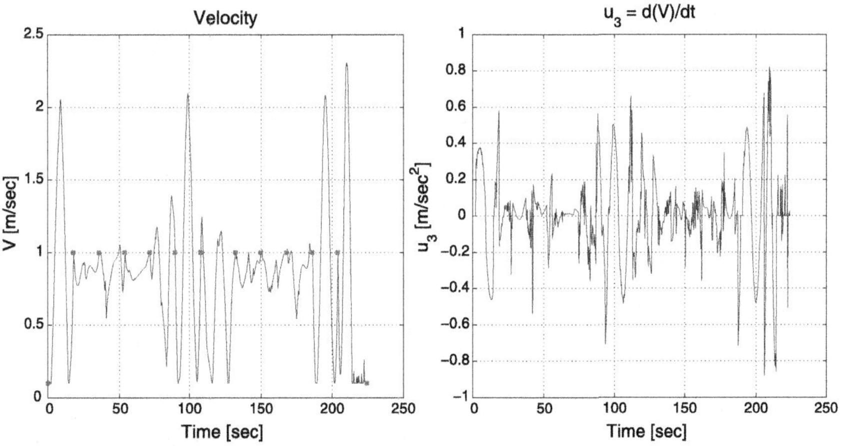

Figure 9.6 Time profile of vehicle's velocity V and acceleration $\dot{V}$.

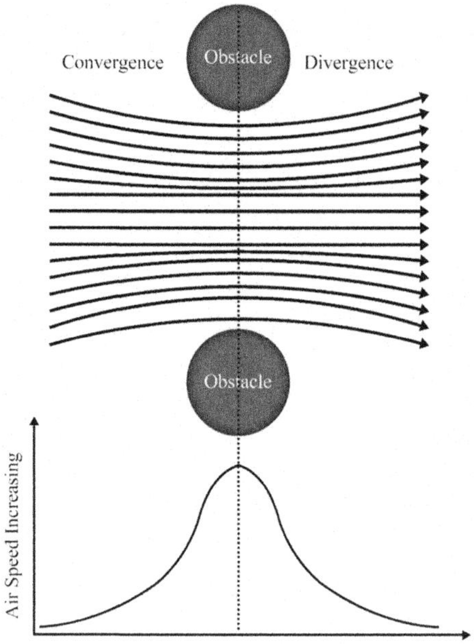

Figure 9.7 Venturi effect.

Table 9.4 Base station and way-points to be visited during the traveled trajectory

	x [m]	y [m]	χ [degrees]	V [m/s]
q_1	0	0	0	0.1
q_2	10	2	90	0.1
q_3	14	6	180	0.1
q_4	10	10	−90	0.1
q_5	6	6	0	0.1
q_6	10	18	90	0.1
q_7	14	22	180	0.1
q_8	10	26	−90	0.1
q_9	6	22	0	0.1

From Fig. 9.7 observe that Venturi effect consists of:

- An increment in wind speed magnitude
- Divergence and convergence of air flow around the obstacle (pillar)

The vehicle is assumed to start from a base station (q_1) and to visit eight points situated on two pillars. These points are provided in Table 9.4, the vehicle's constraints in Table 9.1, and the entire environment is described

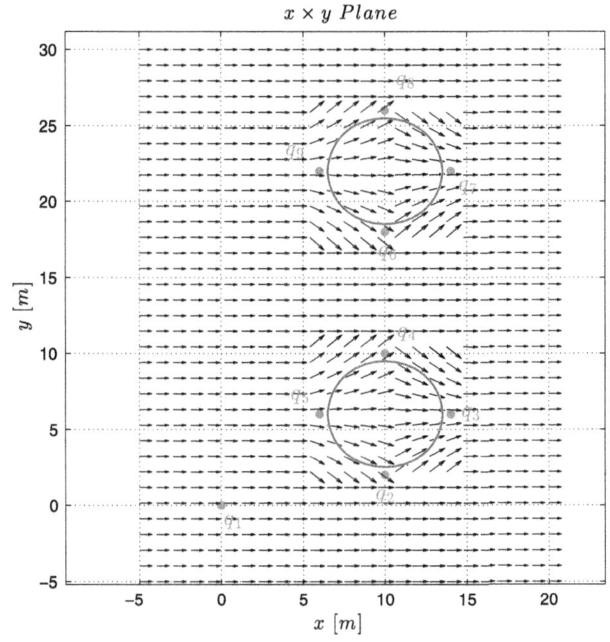

Figure 9.8 Simulation's environment.

in Fig. 9.8. The wind is composed of two parts: the first is constant, covering the whole environment with a magnitude of 0.2 m/s, and the second characterizes Venturi effect with a maximum magnitude of 0.7 m/s. At first glance it seems that the wind is very weak with a magnitude of 0.27 m/s (about 5.4% of V_{max}). But we remark that the desired velocity at each point is equal to 0.1 m/s, which makes gives the wind speed an important factor with respect to vehicle's velocity (2.7 times bigger).

The required time for data collection at each point is estimated to be $T_{Req_i} = 5$ units of time, while the vehicle is capable of flying during $T_{Max} = 68$ unit of time.

The cost matrix C is defined to be the required minimal time to connect all possible pairs (q_i, q_j), $i \neq j$ which gives a 9×9 matrix with 72 elements to be determined (excluding the diagonal terms). For this the trajectory generation approach presented in Sect. 9.2 is employed for each case of C. Therefore, as in the previous simulation scenario, each trajectory is discretized over 70 nodes, and the interior point algorithm included in *fmincon* nonlinear programming solver in MATLAB is used to find the optimal trajectory.

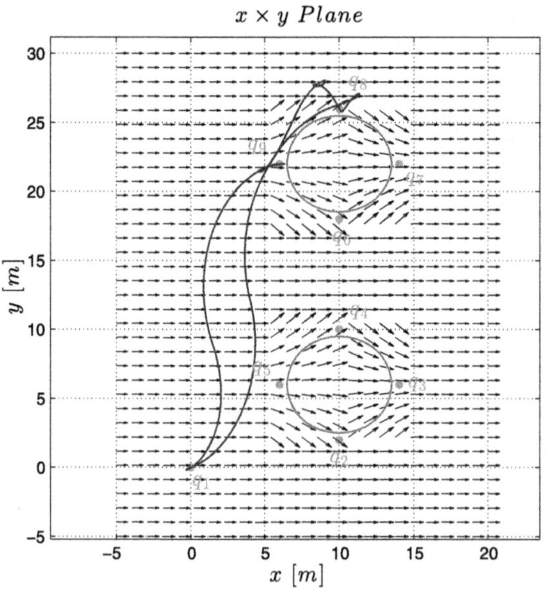

Figure 9.9 Flight 1, $q_1 \rightarrow q_9 \rightarrow q_8 \rightarrow q_1$.

The resulted flight plan consists of three flights shown in Figs. 9.9–9.11.

The obtained solution includes three flight routes:

The first route, illustrated in Fig. 9.9, is the cycle $q_1 \rightarrow q_9 \rightarrow q_8 \rightarrow q_1$ with:

- Required time from q_1 to $q_9 = 10.96$ units of time
- Required time from q_9 to $q_8 = 19.08$ units of time
- Required time from q_8 to $q_1 = 13.57$ units of time
- Total flight time on the route $T = 53.62$ units of time

The second route, presented in Fig. 9.10, is the cycle $q_1 \rightarrow q_2 \rightarrow q_6 \rightarrow q_3 \rightarrow q_1$ with:

- Required time from q_1 to q_2 = 6.42 units of time
- Required time from q_2 to q_6 = 10.43 units of time
- Required time from q_6 to q_3 = 8.72 units of time
- Required time from q_3 to q_1 = 8.91 units of time
- Total flight time on the route T = 49.49 units of time

The third route, described in Fig. 9.11, is the cycle $q_1 \rightarrow q_5 \rightarrow q_7 \rightarrow q_4 \rightarrow q_1$ with:

- Required time from q_1 to q_5 = 6.86 units of time
- Required time from q_5 to q_7 = 12.97 units of time

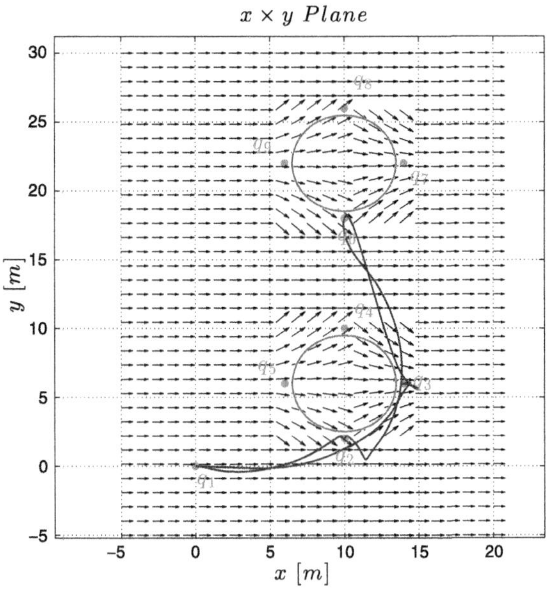

Figure 9.10 Flight 2, $q_1 \rightarrow q_2 \rightarrow q_6 \rightarrow q_3 \rightarrow q_1$.

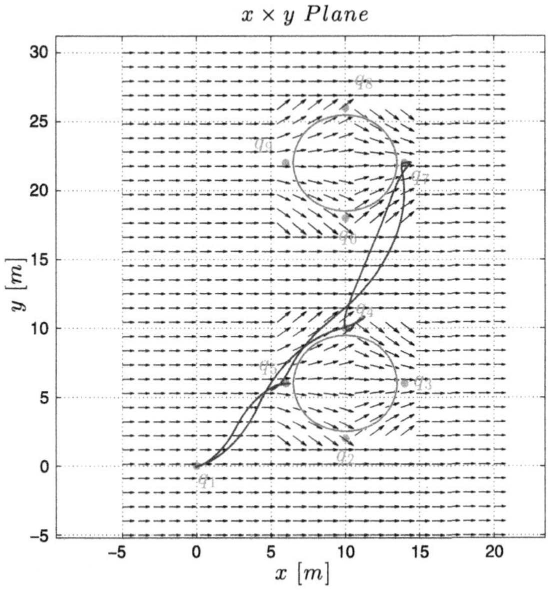

Figure 9.11 Flight 3, $q_1 \rightarrow q_5 \rightarrow q_7 \rightarrow q_4 \rightarrow q_1$.

- Required time from q_7 to q_4 = 7.55 units of time
- Required time from q_4 to q_1 = 10.00 units of time
- Total flight time on the route $T = 52.39$ units of time

Notice that the total time for each route respects the vehicle's capacity.

9.5.2 Trajectory Tracking

In this part we deal with the tracking problem. Our aim is to illustrate two things: first, the applicability and the compatibility of the trajectory computed in Sect. 9.5.1; second, the performance of the control strategy and its robustness facing sensor's uncertainties. For this we proposed two scenarios. Both are based on the results obtained in Sect. 9.2 and taken as references for the full dynamic quadrotor model presented in (9.1). In the first we assume that all existing sensors in the aerial vehicle are ideal, while in the second we consider that some sensors provide noisy data. The control parameters and the limitations of saturation functions are identical for the two cases, and their values were chosen to ensure a stable well-damped response, especially for x, y, and ψ variables. Those values are provided in Table 9.5.

Ideal Sensor Data

The performance of the designed autopilot is illustrated in Figs. 9.12–9.17. In these figures the solid line represents the system response and the dashed line describes the desired value or trajectory. The time profile of the yaw angle and its derivative are shown in Fig. 9.12. The time profile of the x displacement and its absolute error are presented in Fig. 9.13, while the y displacement and its absolute error are presented in Fig. 9.14. In Figs. 9.15 and 9.16 the time profiles of pitch and roll angles are depicted, whereas the control inputs are described in Fig. 9.17.

Table 9.5 Controller's gains

$\mathfrak{K}_{\phi 1} = \mathfrak{K}_{\theta 1} = 7$	$\lvert\sigma_{\phi 1}\rvert = \lvert\sigma_{\theta 1}\rvert = 1$
$\mathfrak{K}_{\phi 2} = \mathfrak{K}_{\theta 2} = 2.5$	$\lvert\sigma_{\phi 2}\rvert = \lvert\sigma_{\theta 2}\rvert = 1.7$
$\mathfrak{K}_{\phi 3} = \mathfrak{K}_{\theta 3} = 2$	$\lvert\sigma_{\phi 3}\rvert = \lvert\sigma_{\theta 3}\rvert = 1.2$
$\mathfrak{K}_{\phi 4} = \mathfrak{K}_{\theta 4} = 7.5$	$\lvert\sigma_{\phi 4}\rvert = \lvert\sigma_{\theta 4}\rvert = 2.5$
$\mathfrak{K}_{\psi 1} = 2.5$	$\lvert\sigma_{\psi 1}\rvert = 1$
$\mathfrak{K}_{\psi 2} = 1.5$	$\lvert\sigma_{\psi 1}\rvert = 1$

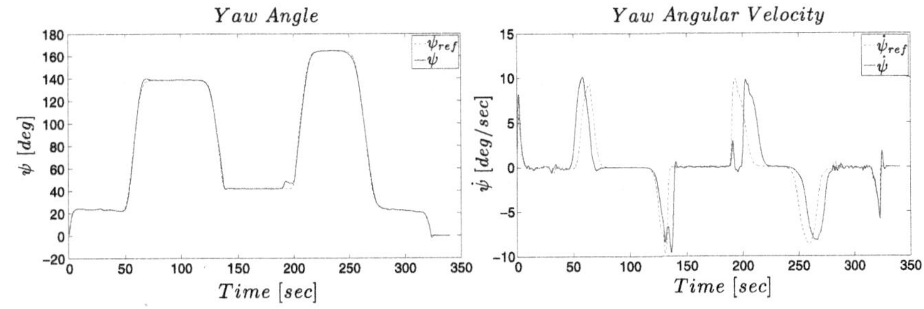

Figure 9.12 Yaw angle and rate of the quadrotor.

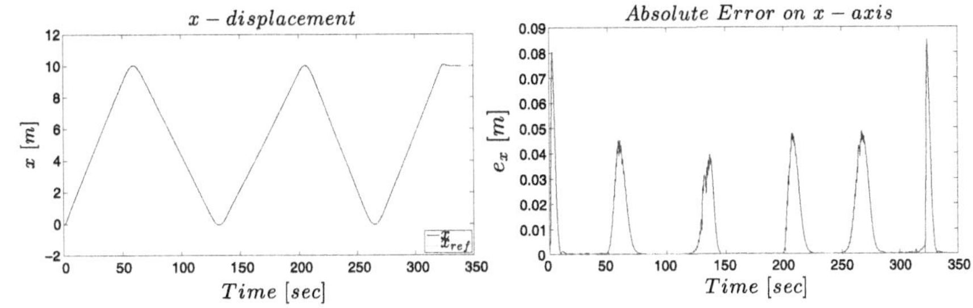

Figure 9.13 x-displacement and its absolute error on the x-axis of the quadrotor.

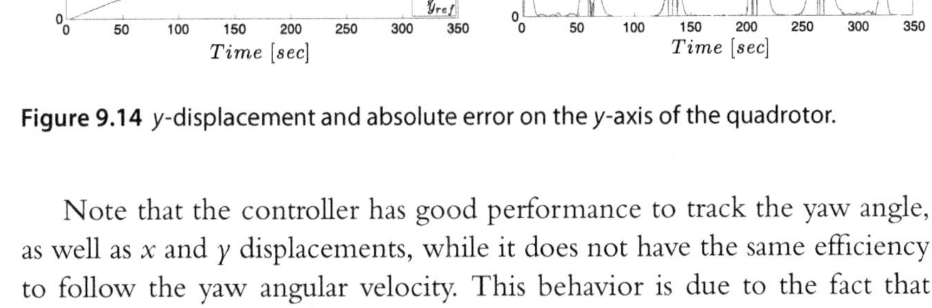

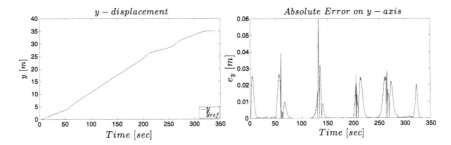

Figure 9.14 y-displacement and absolute error on the y-axis of the quadrotor.

Note that the controller has good performance to track the yaw angle, as well as x and y displacements, while it does not have the same efficiency to follow the yaw angular velocity. This behavior is due to the fact that the gain assigned to the yaw angle, i.e., $\mathfrak{K}_{\psi 2}$, is more important than the one dedicated to its derivative, i.e., $\mathfrak{K}_{\psi 1}$. On the other hand, observe that the desired pitch and roll angles are set to zero, but the real θ and ϕ differ for some periods of time, which is due to movements on x and y axis and acceleration/deceleration effects.

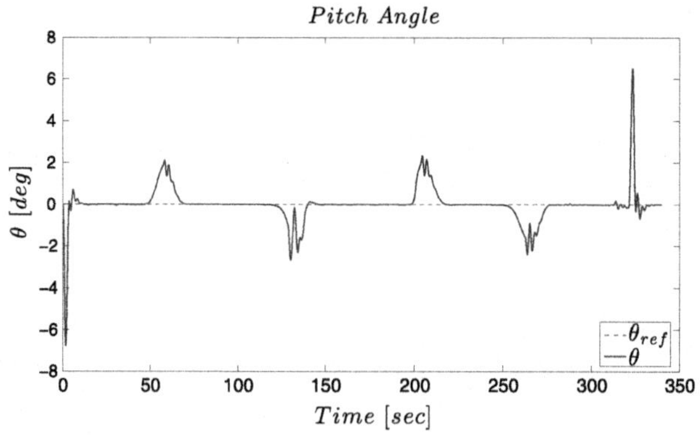

Figure 9.15 Time profile of pitch (θ) angle.

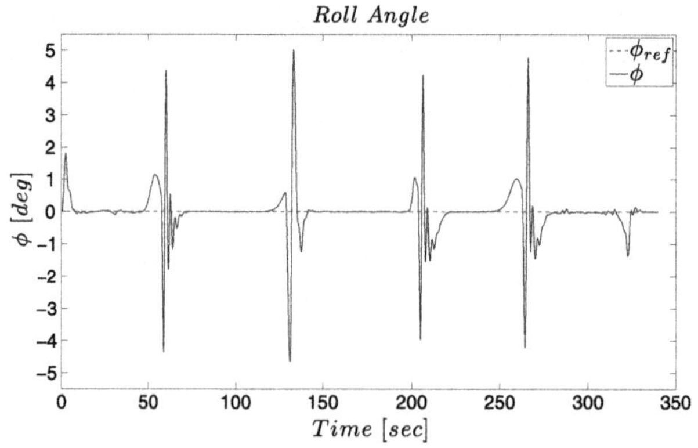

Figure 9.16 Time profile of roll (ϕ) angle.

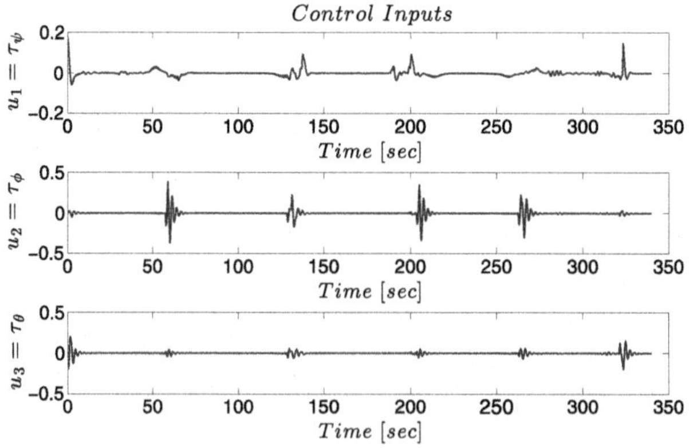

Figure 9.17 Control inputs τ_ψ, τ_ϕ, and τ_θ.

9.5.2.1 Noisy Sensor Data

In this scenario we consider the sensors measuring the vehicle position and the yaw angular velocity as providing noisy data. The objective is to show the robustness of the control strategy facing uncertainties. These disturbances are modeled as a white noise. Fig. 9.18 presents the noise added to position data (x and y) and to yaw angular velocity ($\dot{\psi}$), respectively.

Notice that the precision of these sensors is around ± 1 m for the position measurement and around ± 10 degrees/s for $\dot{\psi}$ measurement, which makes the data provided by those sensors strongly uncertain.

The performance of the control strategy is presented in the Figs. 9.19–9.24. In these figures, as in the previous scenario, the solid line describes the system response, and the dashed line represents the reference trajectory. The time profile of the yaw angle and its derivative are shown in Fig. 9.19. The movement on the x-axis and its absolute error are illustrated in Fig. 9.20, while the movement on the y-axis and its absolute error are presented in Fig. 9.21. In Figs. 9.22 and 9.23 the time profiles of

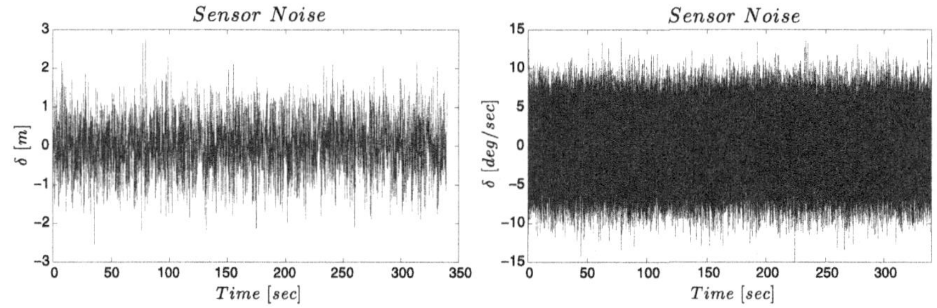

Figure 9.18 Position and yaw angular velocity sensors' noises.

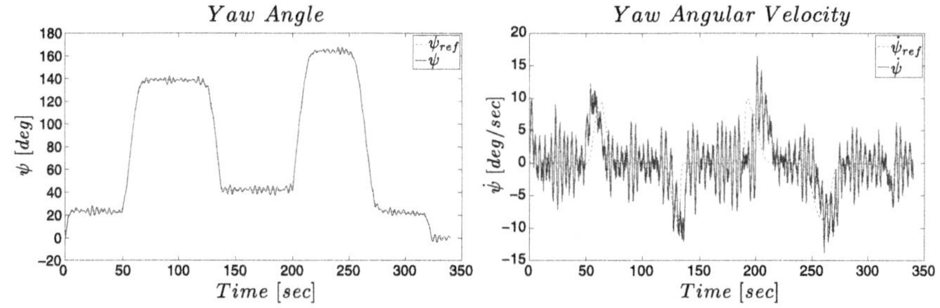

Figure 9.19 Yaw angle of the quadrotor.

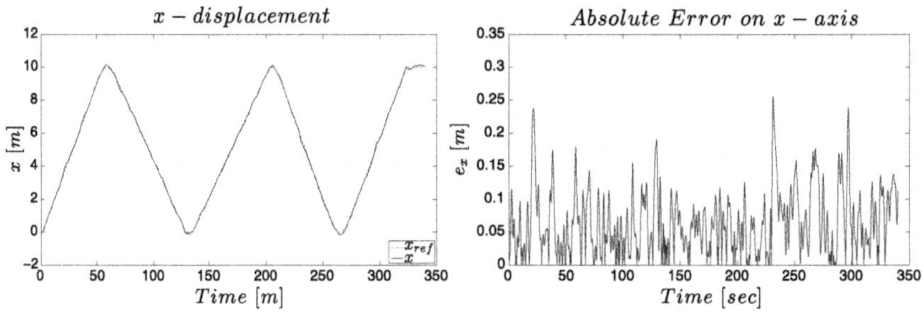

Figure 9.20 *x*-displacement and its absolute error in the *x*-axis of the quadrotor.

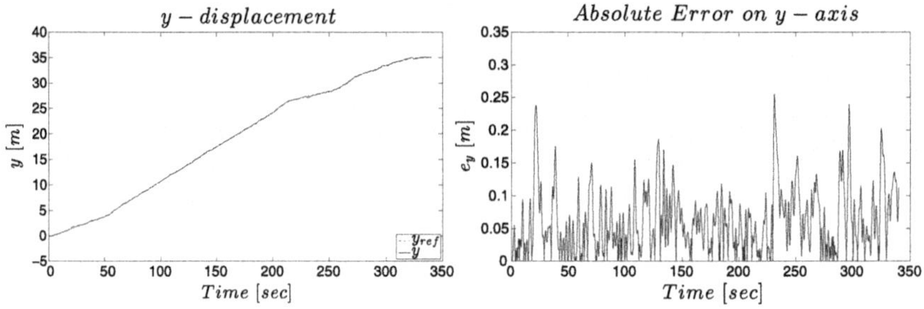

Figure 9.21 *y*-displacement and its absolute error in the *y*-axis of the quadrotor.

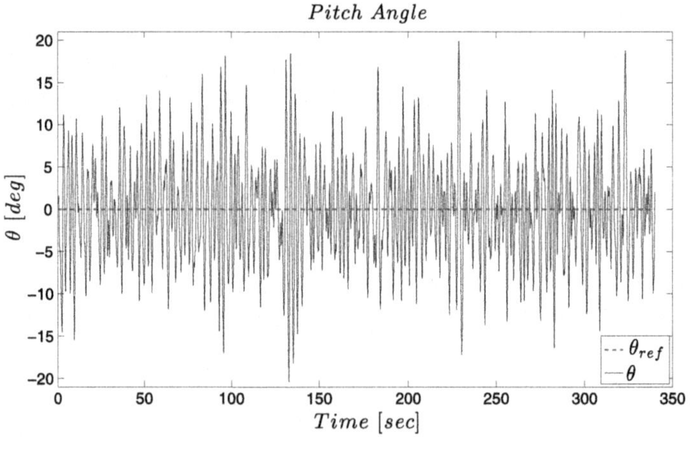

Figure 9.22 Time profile of pitch (θ) angle.

pitch and roll angles are depicted. Finally, the control inputs are described in Fig. 9.24.

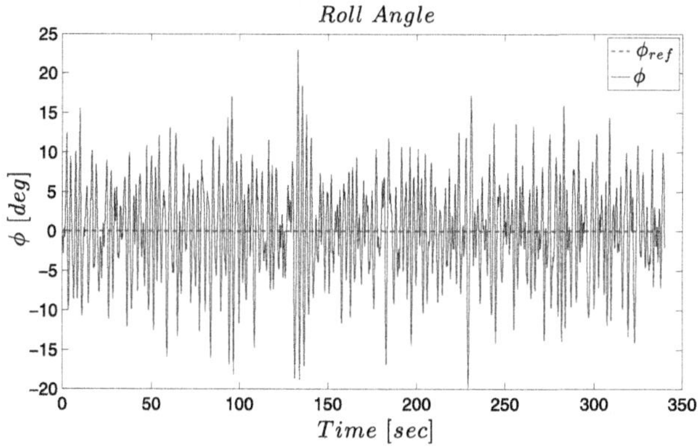

Figure 9.23 Time profile of roll (ϕ) angle.

These figures reveal the system's reaction facing unknown uncertainties. Observe that in spite of the high rate of noise magnitude, the quadrotor fulfills its mission with an absolute error of about 0.12 m on the x-axis and of about 0.1 m on the y-axis, while the impact of the noise is more important on $\dot{\psi}$. This attitude, as in the previous scenario, is due to the fact that the gain assigned to the yaw angle, i.e., $\mathfrak{K}_{\psi 2}$, is more important than the one dedicated to its derivative, i.e., $\mathfrak{K}_{\psi 1}$. However, the effect of noisy data is clear on the time profile of the pitch (θ) and roll (ϕ) angles, as well as control inputs τ_ϕ and τ_θ.

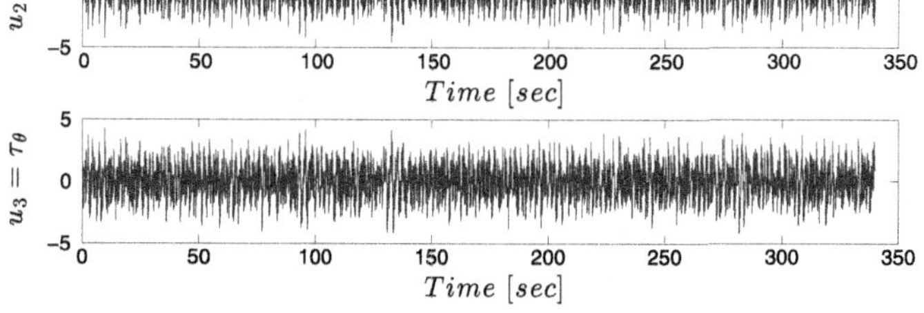

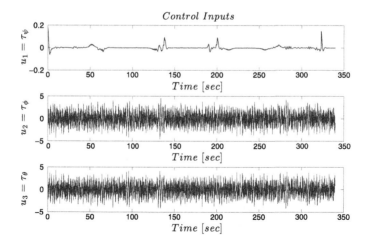

Figure 9.24 Control inputs τ_ψ, τ_ϕ, and τ_θ.

9.6 DISCUSSION

This chapter focused on optimized reference trajectory generation, planning, and robust tracking for an autonomous quadrotor flying vehicle. Some notions about point mass model were introduced. The model based on Newton's second law was deduced to deal with navigation and guidance control systems.

Moreover, an algorithm based on an optimality notion for reference trajectory generation was proposed. The objective was to optimize the traveling time. Hence the minimum-time flight problem was formulated as a problem of the calculus of variations. Then this formulation was converted to a nonlinear constrained optimization problem, and it was solved through a direct collocation approach.

In addition, a trajectory planning strategy was developed. The objective was to include maximum flight time while keeping the optimization objective. Thus the proposed strategy was based on operational research approaches, more specifically a capacitated vehicle routing problem.

Finally, in order to validate the proposed methods, some simulation were carried out using a simple nonlinear controller based on saturation functions. The presented figures have shown that the reference trajectory (computed using the translational kinematics model) was suitable for the designed controller (based on the rigid body model). In addition, they proved that the designed autopilot is robust and has good performance even in the case of noisy data.

REFERENCES

1. A. Barrientos, P. Gutierrez, J. Colorado, Advanced UAV trajectory generation: planning and guidance, in: T.M. Lam (Ed.), Aerial Vehicles, InTech, 2009.
2. D.G. Hull, Fundamentals of Airplane Flight Mechanics, Springer, Berlin, 2007.
3. B.L. Stevens, F.L. Lewis, Aircraft Control and Simulation, Wiley, Canada, 2003.
4. F. Imado, Y. Heike, T. Kinoshita, Research on a new aircraft point-mass model, Journal of Aircraft 48 (2011) 1121–1130.
5. E. Kahale, P. Castillo Garcia, Y. Bestaoui, Autonomous path tracking of a kinematic airship in presence of unknown gust, Journal of Intelligent & Robotic Systems 69 (2013) 431–446.
6. Y. Bestaoui, E. Kahale, Time optimal trajectories of a lighter than air robot with second order constraints and a piecewise constant velocity wind, Journal of Aerospace Computing, Information, and Communication 10 (2013) 155–171.
7. S. Subchan, R. Zbikowski, Computational Optimal Control Tools and Practice, John Wiley & Sons Inc., 2009.

8. C. Büskens, H. Maurer, SQP-methods for solving optimal control problems with control and state constraints: adjoint variables, sensitivity analysis and real-time control, Journal of Computational and Applied Mathematics 120 (2000) 85–108.
9. J. Guerrero, Y. Bestaoui, UAV path planning for structure inspection in windy environments, Journal of Intelligent & Robotic Systems 69 (2013) 297–311.
10. G. Clark, J. Wright, Scheduling of vehicles from a central depot to a number of delivery points, Operations Research 12 (1964) 568–581.
11. R. Larson, A. Odoni, Urban Operations Research, Prentice Hall, Englewood Cliffs, NJ, USA, 1981.
12. T. Gaskell, Bases for vehicle fleet scheduling, Journal of the Operational Research Society 18 (1967) 281–295.
13. H. Paessens, The savings algorithm for the vehicle routing problem, European Journal of Operational Research 34 (1988) 336–344.

CHAPTER 10

Obstacle Avoidance*

Some missions of aerial vehicles could need the autonomous obstacle avoidance to guarantee its self-health. This subject is relatively new and only a few researchers have explored it experimentally. The challenge here is that the vehicle itself detects or knows the position (and dimension) of the obstacle. Several teams have solved this problem considering that the vehicle knows its position in advance. This does not signify that the problem is simple and easy to solve. The most common works found in the literature are based on generating a trajectory in order to avoid the obstacle. And then, the problem is reduced to a path tracking.

The Artificial Potential Field (or APF) method is the common tool to solve this problem. In this approach the obstacle is an undesired point that produces a repulsive potential field to reject an object approaching to it. Even if this is a solution that can be more intuitive, it could generate minimal points where the repulsive force ("originated" by the object) is equal to the attractive force given by the controller to converge to the desired values. This effect can produce oscillations or situations where the vehicle pauses and stays there.

In this chapter we will use these ideas to propose a nonlinear controller to avoid obstacles. The algorithm is based on the backstepping technique with some components coming from the APF technique that are used to generate a new trajectory when the vehicle is approaching the obstacle.

10.1 ARTIFICIAL POTENTIAL FIELD METHOD

Potential fields were introduced in [1]. This method uses the gradient of potential fields to compute virtual forces on the quadrotor when it is close to the obstacle. These forces are included in the position control input of the vehicle. There are two kinds of artificial potential fields, the attractive and the repulsive field. The Potential Field Method (PFM) is based on a simple idea: the vehicle moves in a field of forces. The desired position will be an attractive pole for the vehicle, and the obstacles will be surrounded

* This chapter was developed in collaboration with P. Flores and F. Castaños from Automatic control department at CINVESTAV IPN, Mexico.

Indoor Navigation Strategies for Aerial Autonomous Systems.
DOI: http://dx.doi.org/10.1016/B978-0-12-805189-4.00013-5

by a repulsive field that will push the vehicle away form the obstacle. These potential fields are added to get the complete Artificial Potential Field.

For simplicity let us assume that the obstacle avoidance takes place on the x–y plane. Let $q = \begin{bmatrix} x & y \end{bmatrix}^T$ be the position in the workspace of the quadrotor center of mass and its desired point be $q_d = \begin{bmatrix} x_d & y_d \end{bmatrix}^T$ and suppose that the nth obstacle position is $q_{n_{obs}} = \begin{bmatrix} x_{n_{obs}} & y_{n_{obs}} \end{bmatrix}^T$. Therefore the Artificial Potential Field (U_{tot}) is given by

$$U_{tot} = U_{att} + U_{rep}, \tag{10.1}$$

were U_{att} and U_{rep} define the attractive and repulsive potential field, respectively. The most common definition of the attractive field is given by

$$U_{att}(q) = \frac{1}{2} k \rho^j (q, q_d) \qquad \forall k, j \in \mathbb{R}^+, \tag{10.2}$$

where k is a scaling constant, the distance between the quadrotor's position q and the desired position q_d is $\rho = ||q_d - q||$, and $j \in \mathbb{R}^+$. Note that if $j = 1$ then the Artificial Potential Field is a cone, and the attractive resultant force in the gradient increases indefinitely, see Figs. 10.1 and 10.2. The attraction

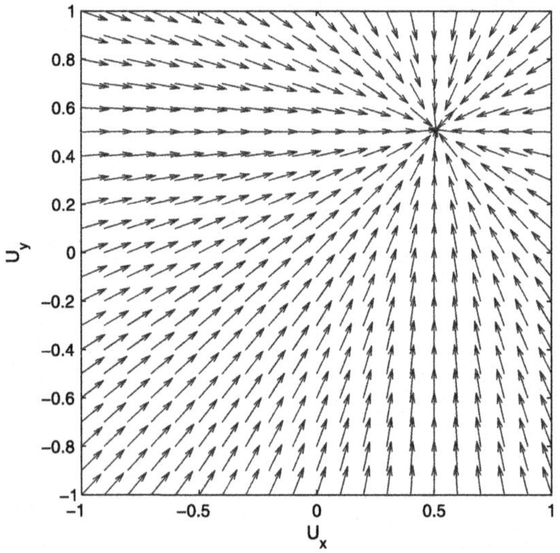

Figure 10.1 Direction of the Attraction Potential Field with $j = 1$.

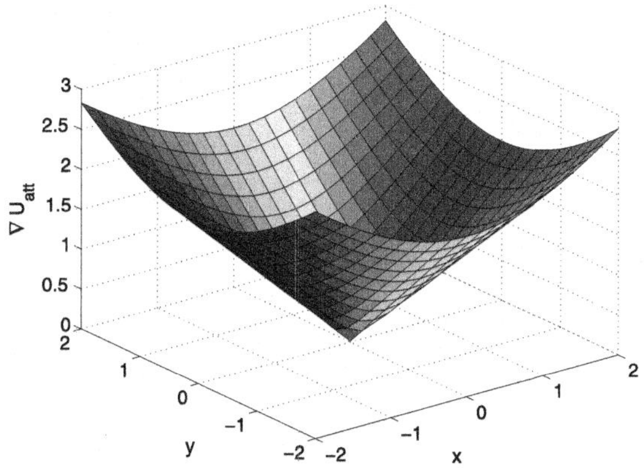

Figure 10.2 Magnitude of the Attraction Potential Field with $j = 1$.

force is given by the negative gradient of the Attraction Potential Field:

$$\begin{aligned} F_{att} &= -\nabla U_{att}(q) = -\frac{1}{2}k\begin{bmatrix} \frac{\partial\rho(q,q_d)}{\partial x} & \frac{\partial\rho(q,q_d)}{\partial y} \end{bmatrix} \\ &= -\frac{1}{2}k\frac{(q-q_d)}{||q-q_d||}. \end{aligned} \tag{10.3}$$

Observe that the force is constant except at q_d where U_{att} is singular.

When $j = 2$ the potential field is quadratic and its shape is parabolic; see [2] and Figs. 10.3 and 10.4. Its attractive force converges linearly if the robot is achieving the desired position, i.e., when the vehicle is near the goal position, the vehicle will move slowly. This is very important because it reduces overshooting at q_d, and because of that, it has become one of the most popular methodologies for the obstacle avoidance task due stability characteristics [3]. The resulting attractive force is given by

$$F_{att} = -\nabla U_{att}(q) = -\frac{1}{2}k(q - q_d). \tag{10.4}$$

The total potential field U_{tot} must be a positive, continuous, and differentiable function, and it must contain a minimum at $q = q_d$; and for the repulsive force, the equation field should be a nonnegative, continuous and differentiable function, and its value must increase while the vehicle approaches the obstacle, and becomes smaller further away from the obstacle. As for the attractive potential field, there are many options for the repulsive

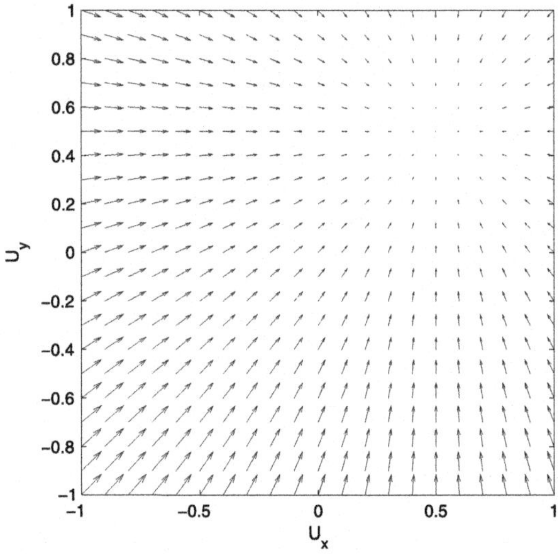

Figure 10.3 Direction of the Attraction Potential Field with $j = 2$.

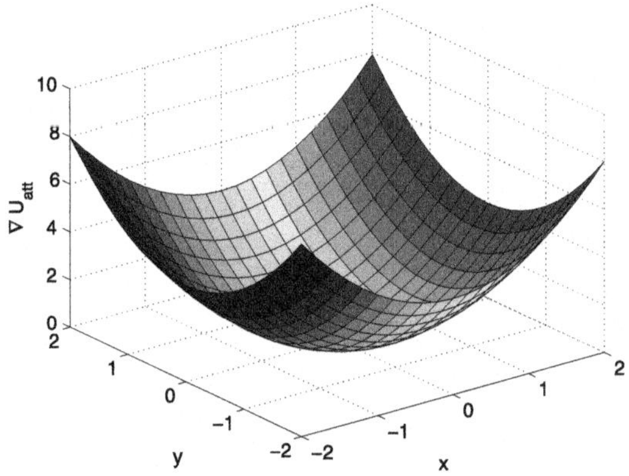

Figure 10.4 Direction of the Attraction Potential Field with $j = 2$.

potential field. Let U_{rep} be given by

$$U_{rep} = \begin{cases} \frac{1}{2}\beta\left(\frac{1}{\rho(q, q_{obs})} - \frac{1}{\rho_o}\right)^2 & \text{if} \quad \rho(q, q_{obs}) \leq \rho_o, \\ 0 & \text{if} \quad \rho(q, q_{obs}) > \rho_o, \end{cases} \qquad \forall \beta, \rho_o \in \mathbb{R}^+, \quad (10.5)$$

where β denotes a scaling factor, the distance between the obstacle and the quadrotor is given by $\rho(q, q_{obs})$, and ρ_o defines a constant related to the efficient distance in which the rejection field can act, see Fig. 10.5.

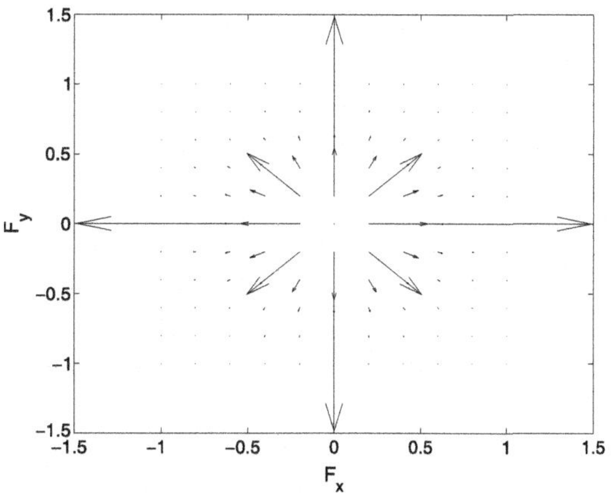

Figure 10.5 Repulsive Field direction and magnitude.

Another expression for the repulsive potential field can be found in [2]. This expression is used when the desired position is nearby or in the active zone of the repulsive zone. This function is known as *Goal Non-reachable with Obstacles Nearby* (GNRON) and reads

$$U_{rep} = \begin{cases} \frac{1}{2}\left(\frac{1}{\rho(q,q_{obs})} - \frac{1}{\rho_o}\right)^2 \rho^{\beta}(q,q_d) & \text{if} \quad \rho(q,q_{obs}) \leq \rho_o, \\ 0 & \text{if} \quad \rho(q,q_d) > \rho_o. \end{cases} \tag{10.6}$$

Thus the resultant force that acts directly to the vehicle is

$$F_{tot} = -\nabla U_{tot} = -\nabla U_{att} - \nabla U_{rep}. \tag{10.7}$$

This method does not work quite right, there are some disadvantages when it is applied in the APF method. One of these disadvantages is the possibility to get trapped in a local minimum. This happens especially in environments that are cluttered with obstacles. There are four common conditions when local minima occur [4]:

- When an obstacle is located between the robot and the goal, and the centers of the robot, obstacle, and goal are co-linear;
- When the desired position is within the active region of an obstacle, the repulsive force of the obstacle will push the robot away from the goal. In [2] the authors propose a solution for this problem;
- When the robot encounters a non-convex obstacle (U-shaped);
- When the robot passes through a narrow passage between two obstacles, this can lead to oscillations.

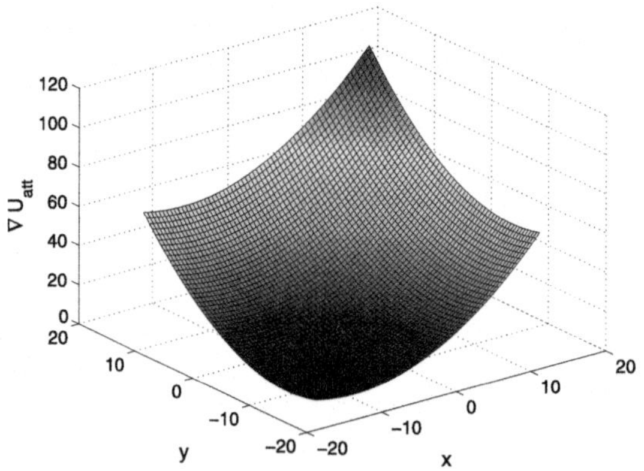

Figure 10.6 Quadratic attractive potential field.

There are several methods based on the APFs that deal with the local minima problem. For example, [4] developed a method based on the Virtual Force Field (VFF), called Enhanced Virtual Force Field, that uses improved functions to detour forces to overcome the minima problem. The main drawback of these techniques is, however, that while avoiding local minima, they have many heuristic parameters that need to be adjusted for each problem [5].

As an example of combining the attractive and repulsive potential fields consider the desired position at $\begin{bmatrix} x_d & y_d \end{bmatrix}^T = \begin{bmatrix} -10 & -10 \end{bmatrix}^T$ with the obstacle at $\begin{bmatrix} x_{obs} & y_{obs} \end{bmatrix}^T = \begin{bmatrix} 5 & 5 \end{bmatrix}^T$. In Figs. 10.6–10.8 the attractive, repulsive, and sum of fields are shown, respectively, and that might be enough to generate a vehicle's trajectory at any initial position in the field.

In Fig. 10.9 the effect of local minima can be seen, where the quadrotor might be hovering.

10.2 OBSTACLE AVOIDANCE ALGORITHM

The previous methodology is used to autonomously navigate an aerial vehicle with obstacles in the trajectory. For simulations and experimental purposes we have considered a quadcopter vehicle, and the control algorithm used was based on the backstepping technique.

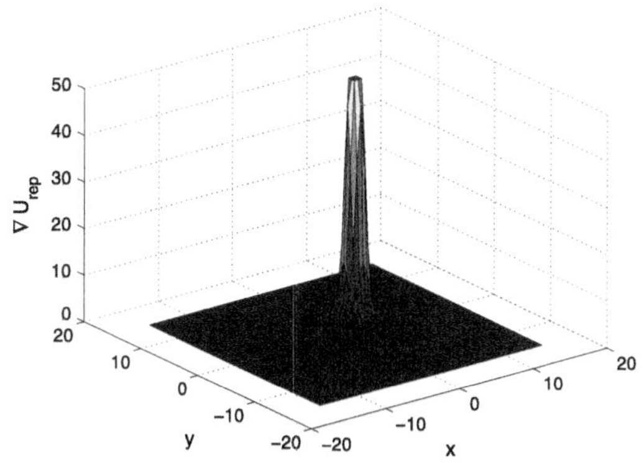

Figure 10.7 Repulsive field.

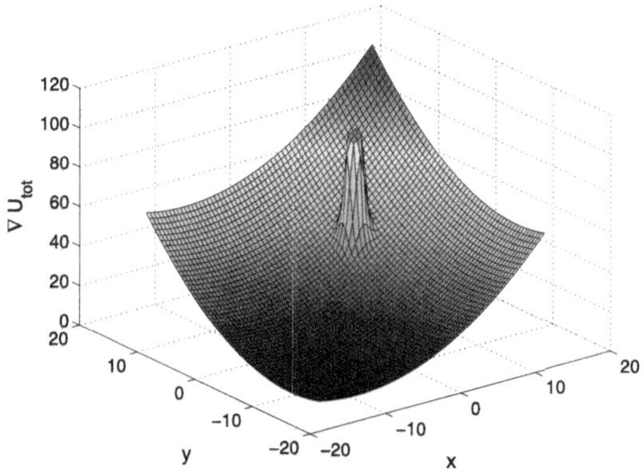

Figure 10.8 Combined field.

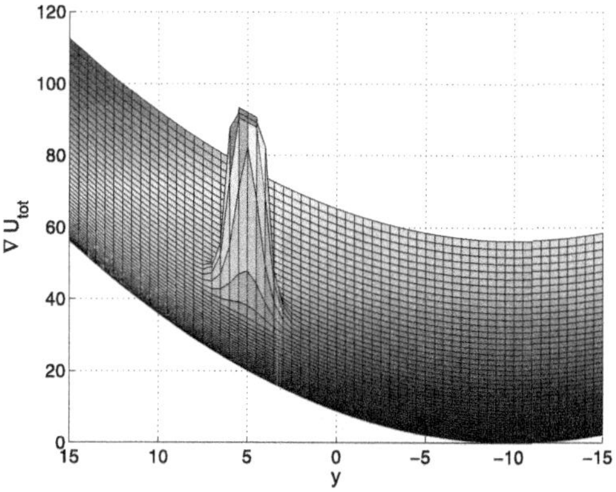

Figure 10.9 Minima effect.

From Chap. 2 and Eqs. (2.17)–(2.18) the nonlinear dynamic equations for the quadrotor can be written as:

$$
\begin{aligned}
\ddot{x} &= -(\sin\theta)\frac{1}{m}u_z, & \ddot{\theta} &= \dot{\phi}\dot{\psi}\left(\frac{I_z - I_x}{I_y}\right) - \frac{I_r}{I_y}\dot{\phi}\Omega + \frac{l}{I_y}u_\theta,\\
\ddot{y} &= (\cos\theta\sin\phi)\frac{1}{m}u_z, & \ddot{\phi} &= \dot{\theta}\dot{\psi}\left(\frac{I_y - I_z}{I_x}\right) - \frac{I_r}{I_x}\dot{\theta}\Omega + \frac{l}{I_x}u_\phi,\\
\ddot{z} &= (\cos\theta\cos\phi)\frac{1}{m}u_z - g, & \ddot{\psi} &= \dot{\theta}\dot{\phi}\left(\frac{I_x - I_y}{I_z}\right) + \frac{l}{I_z}u_\psi.
\end{aligned}
$$

The position in the Euclidean space is $\xi = \begin{bmatrix} x & y & z \end{bmatrix}^T$, the attitude using the Euler angles is $\eta = \begin{bmatrix} \phi & \theta & \psi \end{bmatrix}^T$, the parameters inherent to the quadrotor are the mass (m) and the length between the actuators and the center of mass (l). The inertia tensors are represented as I_x, I_y, and I_z, Ω means the body rate, I_r signifies the inertia of the motor and the control inputs as u_i.

For control purposes the quadrotor dynamic is rewritten in the state spaces as

$$
\begin{aligned}
&\begin{bmatrix} \dot{x}_1 & \dot{x}_2 & \dot{y}_1 & \dot{y}_2 & \dot{z}_1 & \dot{z}_2 \end{bmatrix}^T\\
&\quad= \begin{bmatrix} x_2 & \frac{U_x u_z}{m} & y_2 & \frac{U_y u_z}{m} & z_2 & \frac{\cos\theta_1 \cos\phi_1 u_z}{m} - g \end{bmatrix}^T, \qquad (10.8)\\
&\begin{bmatrix} \dot{\phi}_1 & \dot{\phi}_2 & \dot{\theta}_1 & \dot{\theta}_2 & \dot{\psi}_1 & \dot{\psi}_2 \end{bmatrix}^T\\
&\quad= \begin{bmatrix} \phi_2 & f_\phi + g_\phi u_\phi & \theta_2 & f_\theta + g_\theta u_\theta & \psi_2 & f_\psi + g_\psi u_\psi \end{bmatrix}^T, \qquad (10.9)
\end{aligned}
$$

with $f_\phi = \theta_2\psi_2\left(\frac{I_y - I_z}{I_x}\right) - \frac{I_r}{I_x}\theta_2\Omega$, $f_\theta = \phi_2\psi_2\left(\frac{I_z - I_x}{I_y}\right) - \frac{I_r}{I_y}\phi_2\Omega$, $f_\psi = \theta_2\phi_2\left(\frac{I_x - I_y}{I_z}\right)$, $g_\phi = \frac{l}{I_x}$, $g_\theta = \frac{l}{I_y}$, and $g_\psi = \frac{l}{I_z}$, where $f_i(\cdot)$, $g_i(\cdot)$ are nonlinearities related to the Coriolis forces and the moments generated by the motors, $U_x = -\sin\theta_1$ and $U_y = \cos\theta_1 \sin\phi_1$ will be considered as virtual control inputs used for position control.

The control goal is to make the UAV tracks a desired geometric path denoted by $\xi_d = \begin{bmatrix} x_d & y_d & z_d \end{bmatrix}^T$. Notice that from (10.8) and (10.9) it is easy to see that the attitude dynamic could be considered independent from the position subsystem; nevertheless, this dynamics (ξ) is related to the attitude. For simplicity in simulations let us assume that $\psi_d = 0$, then the control problem for the quadrotor could be divided into two levels: the lower level consists of the attitude control and the upper level of the position control as can be seen in Fig. 10.10.

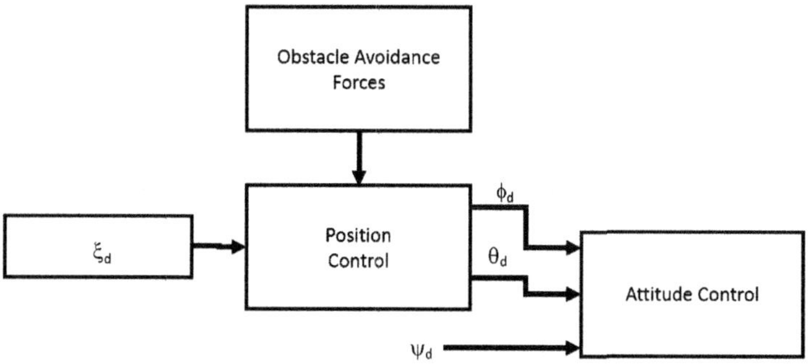

Figure 10.10 Control diagram.

For control design the quadcopter dynamics have been separated into many subsystems consisting of two integrators in a cascade. Each subsystem is stabilized using a backstepping control assuring the convergence to the desired values. We consider that for the attitude dynamics the APFs have any influence directly; nevertheless, they are affected indirectly when the vehicle moves. The following controllers stabilize the orientation of the quadcopter:

$$\begin{aligned} u_\phi &= \frac{I_x}{l}\left[-\alpha_1(\bar{e}_2+\alpha_1\bar{e}_1)-\theta_2\psi_2\left(\frac{I_y-I_z}{I_x}\right)+\frac{I_r}{I_x}\theta_2\Omega+\bar{e}_1-\alpha_2\bar{e}_2\right], \\ u_\theta &= \frac{I_y}{l}\left[-\alpha_3(\bar{e}_4+\alpha_3\bar{e}_3)-\phi_2\psi_2\left(\frac{I_z-I_x}{I_y}\right)+\frac{I_r}{I_y}\phi_2\Omega+\bar{e}_3-\alpha_4\bar{e}_4\right], \\ u_\psi &= \frac{I_z}{l}\left[-\alpha_5(\bar{e}_6+\alpha_5\bar{e}_5)-\theta_2\phi_2\left(\frac{I_x-I_y}{I_z}\right)+\bar{e}_5-\alpha_6\bar{e}_6\right]. \end{aligned} \tag{10.10}$$

Stability analysis of the previous algorithms can be found in [6], the tracking errors are $\bar{e}_i$, and α_i are positive constants.

For the altitude we have considered that it will be constant along the trajectory. Thus the control law is

$$u_z = \frac{m}{\cos\theta_1\cos\phi_1}\left[g-\alpha_7(\bar{e}_8+\alpha_7\bar{e}_7)+\bar{e}_7-\alpha_8\bar{e}_8\right]. \tag{10.11}$$

Once the attitude of the quadcopter is stabilized, the position will be controlled by the virtual inputs U_x and U_y, they are given by

$$\begin{aligned} U_x &= \frac{m}{u_z}\left[-\alpha_9(\bar{e}_{10}+\alpha_9\bar{e}_9)+\bar{e}_9-\alpha_{10}\bar{e}_{10}\right], \\ U_y &= \frac{m}{u_z}\left[-\alpha_{11}(\bar{e}_{12}+\alpha_{11}\bar{e}_{11})+\bar{e}_{11}-\alpha_{12}\bar{e}_{12}\right]. \end{aligned} \tag{10.12}$$

We remark that $\bar{e}_i$ is the error, where $i \in \{1, 3, 5, 7, 9, 11\}$ means the position error for ϕ, θ, ψ, z, x, and y, respectively, while $i \in \{2, 4, 6, 8, 10, 12\}$ denotes the angular or translational velocity error for $\dot{\phi}$, $\dot{\theta}$, $\dot{\psi}$, $\dot{z}$, $\dot{x}$, and $\dot{y}$, respectively.

The previous controller stabilizes a quadcopter vehicle and could allow path tracking. Nevertheless, if an obstacle is on the trajectory, even if the vehicle knows its position, it will crash into it. Now to realize obstacle avoidance, the APF technique will be included in the controller. Notice that we only consider obstacle avoidance in the x–y plane.

From (10.3) observe that the force is a kind of proportional control where the tracking error has been normalized. Similarly, in (10.4) it has the form of a proportional tracking error. Hence the APF algorithm contains an attractive and repulsive potential forces as a linear algorithm but it cannot realize path tracking and is not robust with respect to unknown uncertainties. The proposed backstepping control can be also seen as a tracking control, and it contains inherent attractive force; nevertheless, it does not contain the repulsive component to avoid the obstacle.

To improve such a control, it is possible to include into the backstepping controller the repulsive force in the x and y axis. Then it follows that

$$\begin{aligned} U_x &= \frac{m}{u_z}\left[-\alpha_9(\bar{e}_{10} + \alpha_9\bar{e}_9) + e_9 - \alpha_{10}\bar{e}_{10}\right] + F_{rep_x}, \\ U_y &= \frac{m}{u_z}\left[-\alpha_{11}(\bar{e}_{12} + \alpha_{11}\bar{e}_{11}) + e_{11} - \alpha_{12}\bar{e}_{12}\right] + F_{rep_y}, \end{aligned} \tag{10.13}$$

where F_{rep_x} and F_{rep_y} are the components of the force in the x and y axis, respectively. Notice that the rejection potential field is radial, meaning that the forces begin at the center and move outward, but the movement describes straight lines.

In this approach the existence of local minima could be a problem to deal with. Local minima could occur when the control input is equal to zero, this means

$$\begin{aligned} F_{rep_x} &= \frac{m}{u_z}\left[-\alpha_9(\bar{e}_{10} + \alpha_9\bar{e}_9) + e_9 - \alpha_{10}\bar{e}_{10}\right], \\ F_{rep_y} &= \frac{m}{u_z}\left[-\alpha_{11}(\bar{e}_{12} + \alpha_{11}\bar{e}_{11}) + e_{11} - \alpha_{12}\bar{e}_{12}\right]. \end{aligned} \tag{10.14}$$

This phenomenon happens when the quadrotor follows a trajectory that passes through the center of the obstacle, as shown in Fig. 10.11 where we consider a desired trajectory parametrized in the x–y plane; the z_d the trajectory is given by a step filtered to avoid abrupt changes, the desired

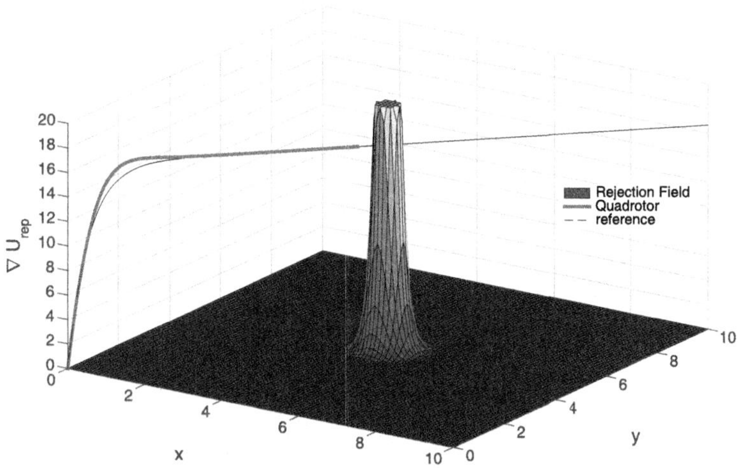

Figure 10.11 Minima effect when the trajectory passes through the center of the obstacle.

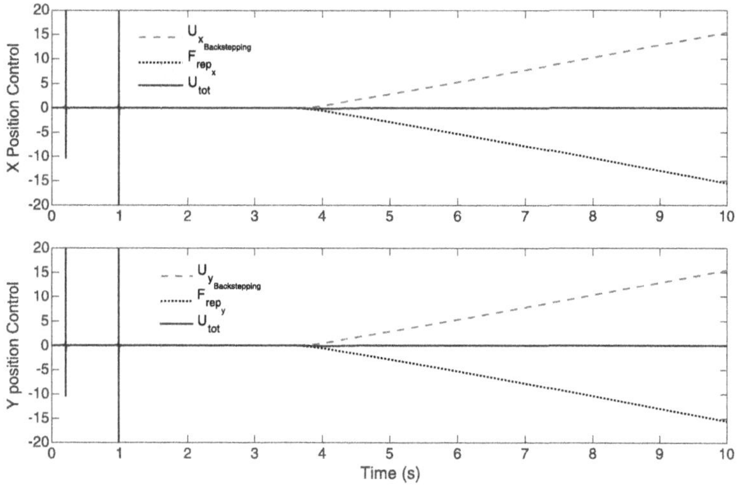

Figure 10.12 Backstepping control and repulsive forces for obstacle avoidance in *x* and *y* in the local minima trap.

values are given by $\begin{bmatrix} x_d & y_d \end{bmatrix}^T = \begin{bmatrix} t & t \end{bmatrix}^T$, $t \in (0, 10)$ and the position of the obstacle is at $\begin{bmatrix} x_{obs} & y_{obs} \end{bmatrix}^T = \begin{bmatrix} 5 & 5 \end{bmatrix}^T$, $\beta = 10$, $\rho_o = 2$.

The control signal response for x and y, when using the proposed controllers, can be analyzed in Fig. 10.12 where the behavior of the repulsive force can be seen although $\rho_o = 2$. Notice that the repulsive force is not strong enough, and the controller is able to neglect the repulsive force, but at $t = 3.58$ the repulsive force increases, and with it the backstepping control; U_x which is the sum of both signals is zero. The same phenomenon happens with U_y, and for these reasons the quadrotor keeps hovering. To avoid the local minima the backstepping controller might be bigger than the repulsive forces; nevertheless, this is not possible in this configuration due to the way the repulsive field has been defined.

Simulations were also carried out when the quadcopter follows a trajectory passing close to the center of the obstacle. All the parameters are equal to those of the previous simulations, but the obstacle position is now at $\begin{bmatrix} x_{obs} & y_{obs} \end{bmatrix}^T = \begin{bmatrix} 5 & 5.1 \end{bmatrix}^T$.

In Fig. 10.13 the behavior of the quadrotor can be seen. Observe that as the trajectory is not at the center of the obstacle the sum of the backstepping controller with the repulsive forces is not null, see also Fig. 10.14.

One approach to deal with the local minima problem is to generate a virtual object where the quadrotor keeps hovering. Nevertheless, the virtual object should be generated when the quadrotor has fallen into the trap caused by the minima, see Fig. 10.15. This object will help to move the vehicle away from the minima position, though it cannot always guarantee avoidance of the local minima. On the other hand, in real-time implementations it has shown good behavior and allowed avoiding the obstacle.

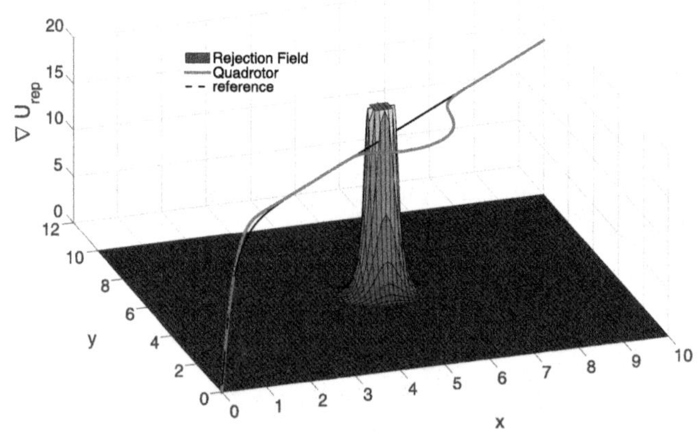

Figure 10.13 Obstacle avoidance when the obstacle is near the trajectory.

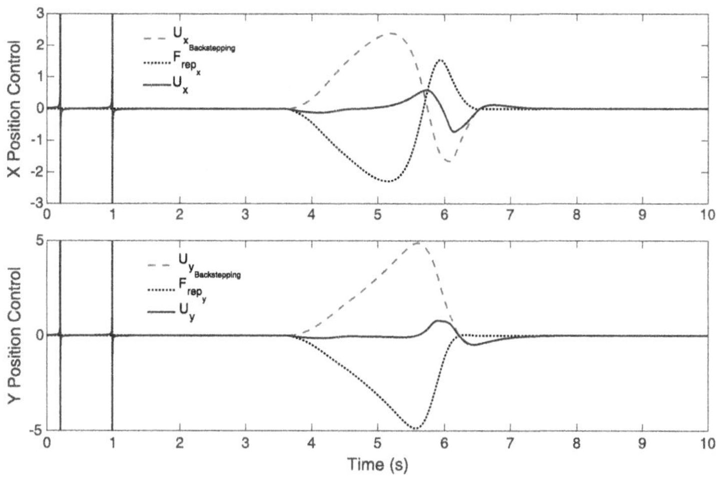

Figure 10.14 Control, Repulsive and Total Control when the obstacle is not on the trajectory.

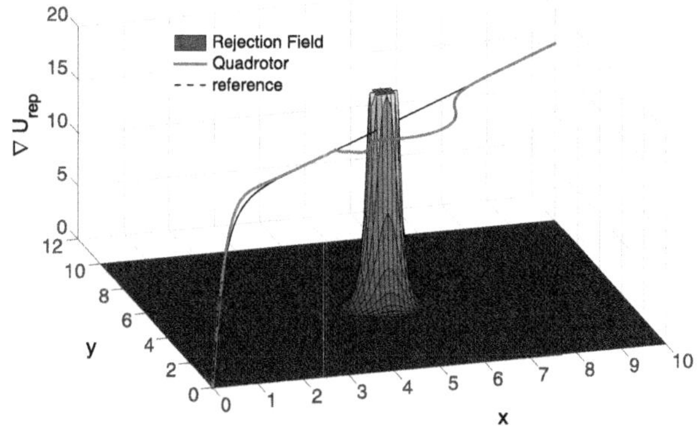

Figure 10.15 Obstacle avoidance with virtual object and obstacle on the desired trajectory.

10.2.1 Implementation

The ideas exposed in the previous section were implemented in a quadrotor prototype providing the next results in position control for obstacle avoidance. In Fig. 10.16 the trap for the minima problem is shown in 3D; notice that, when comparing the experimental results with the simulation results, the quadrotor did not keep hovering in the flight test, it had an oscillatory behavior in the x–y plane.

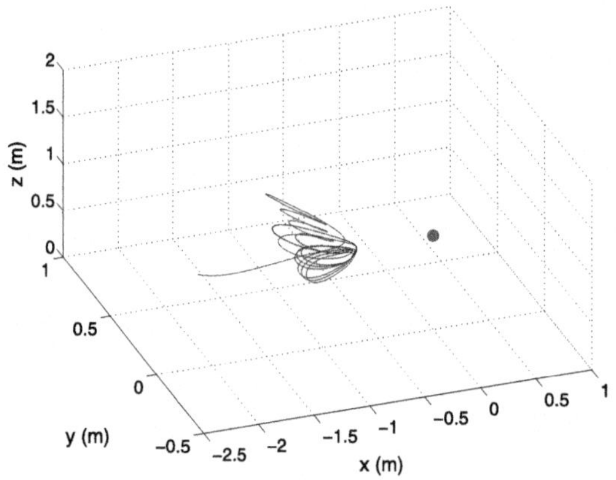

Figure 10.16 Minima effect implemented on a quadrotor UAV in $\mathbb{R}^3$.

When a local minima is reached, the quadrotor flies in an oscillatory way instead of hovering. This happens because (10.14) does not hold, in addition, the oscillations appear because the control algorithm is not precise enough to track the trajectory and the repulsive force is not strong enough to ward off the vehicle from the obstacle, this effect can be observed in Fig. 10.16.

A flight test was realized again, and a virtual obstacle was added to avoid this local minima problem. In Fig. 10.17 the results are illustrated; notice in this graph that, when implementing the APF Method with the virtual obstacle, the obstacle avoidance is realized with good results.

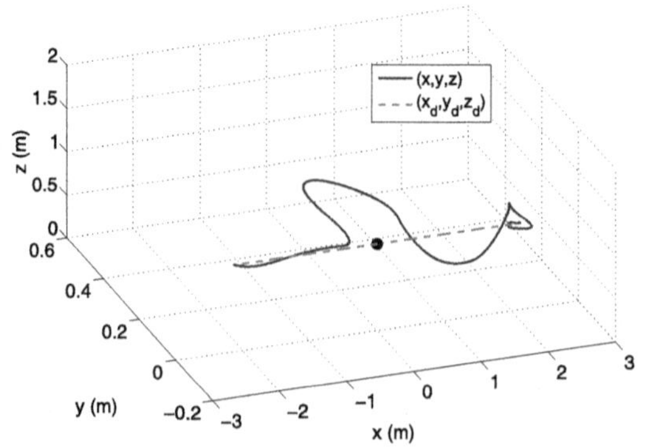

Figure 10.17 Obstacle avoidance in $\mathbb{R}^3$ with the virtual object method.

10.3 LIMIT-CYCLE OBSTACLE AVOIDANCE

The Artificial Potential Field method is a reactive method in which the robot needs to known little or not the model of the robot's surroundings [7]. The APF method adopts the attractive and repulsive surfaces to guide the robot; however, there are some disadvantages: the APF could produce a local minima problem and it does not bring any information to avoid it. There are many efforts to solve this problem such as the virtual object approach that has been previously introduced; nevertheless, it does not guarantee the local minima avoidance because even with the virtual obstacle the vehicle could fall into a local minimum generated by the repulsion forces of the virtual and real obstacle.

Observe from Figs. 10.1 or 10.3 that the potential field is radial, hence the minima problem is inherent to this method, another approach to deal with the minima trap is to change the way of the potential field and to use tangential potential fields around the object. The latter solution is possible when using the limit cycle methodology.

Using limit cycles in motion planning was proposed in [7] where the authors applied it in the context of robot soccer to retrieve the ball while avoiding other robots. This method can be used for obstacle avoidance which could be considered as a subproblem of the motion planning [5].

Limit cycles can arise in 2nd-order nonlinear systems, and they are closed orbits that are stable or unstable. Physically stable limit cycles represent self-sustained oscillations. There are three kinds of limit cycles:

- **Stable limit cycles**, i.e., all trajectories in the vicinity of the limit cycle converging to it as $t \to \infty$;
- **Unstable limit cycles**, i.e., all trajectories in the vicinity of the limit cycle diverging from it as $t \to \infty$;
- **Semi-stable limit cycles**, i.e., some of the trajectories in the vicinity converging to the limit cycle while others diverging from it as $t \to \infty$.

Here t is the time.

For the analysis of the limit cycle, consider the following system:

$$\begin{aligned} \dot{\eta}_1 &= \eta_2 + \eta_1(r^2 - \eta_1^2 - \eta_2^2), \\ \dot{\eta}_2 &= -\eta_1 + \eta_2(r^2 - \eta_1^2 - \eta_2^2), \end{aligned} \tag{10.15}$$

where the η_1 and η_2 are the states of the system, and r is the radius of the circle in which the limit cycle converges. Let $V(\eta_1, \eta_2)$ be a candidate Lyapunov function and $\dot{V}(\eta_1, \eta_2)$ its derivative given by

$$V(\eta_1, \eta_2) = \eta_1^2 + \eta_2^2, \tag{10.16a}$$
$$\dot{V}(\eta_1, \eta_2) = 2\eta_1\dot{\eta}_1 + 2\eta_2\dot{\eta}_2. \tag{10.16b}$$

Introducing (10.15) into (10.16b), it follows that

$$\begin{aligned}
\dot{V}(\eta_1, \eta_2) &= 2\eta_1(\eta_2 + \eta_1(r^2 - \eta_1^2 - \eta_2^2)) + 2\eta_2(-\eta_1 + \eta_2(r^2 - \eta_1^2 - \eta_2^2)) \\
&= 2\eta_1\eta_2 + 2\eta_1^2(r^2 - \eta_1^2 - \eta_2^2) - 2\eta_2\eta_1 + 2\eta_2(r^2 - \eta_1^2 - \eta_2^2) \\
&= 2(\eta_1^2 + \eta_2^2)(r^2 - (\eta_1^2 + \eta_2^2)) \\
&= 2(V(\eta_1, \eta_2))(r^2 - V(\eta_1, \eta_2)).
\end{aligned}$$

Notice that $\dot{V}(\eta_1, \eta_2)$ is negative if $V(\eta_1, \eta_2) > 1$. Eq. (10.15) represents a clockwise limit cycle, see Fig. 10.18.

Note in (10.15) that the term $r^2 - \eta_1^2 - \eta_2^2$ is the circle equation, hence r is the circle radius; as an analogy, its function can be compared to the ρ_o parameter in the radial APF method, that is, the distance from the obstacle in which the rejection potential field is active.

For a counterclockwise limit cycle the analysis is similar, see Fig. 10.19. Let us consider the next system

$$\begin{aligned}
\dot{\eta}_1 &= -\eta_2 + \eta_1(r^2 - \eta_1^2 - \eta_2^2), \\
\dot{\eta}_2 &= \eta_1 + \eta_2(r^2 - \eta_1^2 - \eta_2^2).
\end{aligned} \tag{10.17}$$

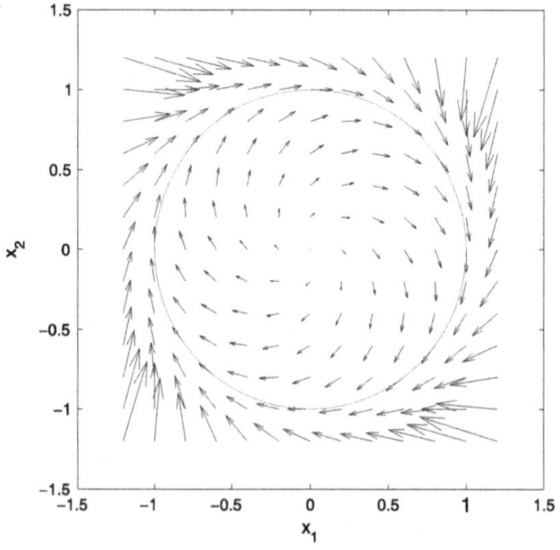

Figure 10.18 Clockwise circle cycle.

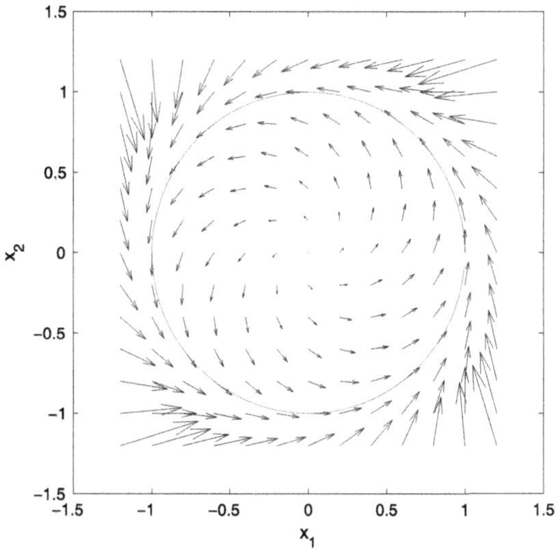

Figure 10.19 Counterclockwise circle limit cycle.

Taking the same Lyapunov function as in (10.16a) with its derivative (10.16b), we obtain

$$\begin{aligned}\dot V(\eta_1,\eta_2) &= 2\eta_1(-\eta_2+\eta_1(r^2-\eta_1^2-\eta_2^2))+2\eta_2(\eta_1+\eta_2(r^2-\eta_1^2-\eta_2^2))\\ &= -2\eta_1\eta_2+2\eta_1^2(r^2-\eta_1^2-\eta_2^2)+2\eta_2\eta_1+2\eta_2(r^2-\eta_1^2-\eta_2^2)\\ &= 2(\eta_1^2+\eta_2^2)(r^2-(\eta_1^2+\eta_2^2))\\ &= 2(V(\eta_1,\eta_2))(r^2-V(\eta_1,\eta_2)).\end{aligned}$$

As in the previous case the derivative is negative if $V(\eta_1,\eta_2)>1$. In both cases this condition holds for stable limit cycles. For obstacle avoidance only stable limit cycles are considered.

10.3.1 Horizontal Obstacle Avoidance

In a case where obstacle avoidance is needed in the x–y plane, the equations to generate the limit cycle are

$$\begin{aligned}\dot x &= y\gamma + x(r^2-x^2-y^2),\\ \dot y &= -x\gamma + y(r^2-x^2-y^2),\end{aligned}\tag{10.18}$$

where x and y are the position control of the quadrotor in the x–y plane. Notice that if $\gamma=1$ the limit cycle is the same as in (10.15) and has a clockwise direction, and if $\gamma = -1$ then it is a counterclockwise limit cycle. We propose a combination of the APF with the limit cycle approach.

Remember that q_{obs} denotes the distance between the quadrotor and the obstacle, and $q_{x_{obs}}$ and $q_{y_{obs}}$ are the components in their respective axes. The control input will be the same as developed in previous sections but with some modifications in the repulsive force. This will have a limit cycle and

$$F_{rep_x} = \begin{cases} q_{y_{obs}} + q_{x_{obs}}(r^2 - q_{x_{obs}}^2 - q_{y_{obs}}^2) & \text{if} \quad ||q - q_{obs}|| < r, \\ 0 & \text{if} \quad ||q - q_{obs}|| > r, \end{cases} \tag{10.19}$$

$$F_{rep_y} = \begin{cases} q_{x_{obs}} + q_{y_{obs}}(r^2 - q_{x_{obs}}^2 - q_{y_{obs}}^2) & \text{if} \quad ||q - q_{obs}|| < r, \\ 0 & \text{if} \quad ||q - q_{obs}|| > r. \end{cases} \tag{10.20}$$

Note that when the rejection field is active the force has the limit cycle shape but it contains the distance between the obstacle and the quadcopter. Hence the control input will be given by (10.13).

The proposed approach was validated in simulation and the performance of the closed-loop system can be observed in Fig. 10.20. Notice that the horizontal flight performance is improved and obstacle avoidance is realized pretty well.

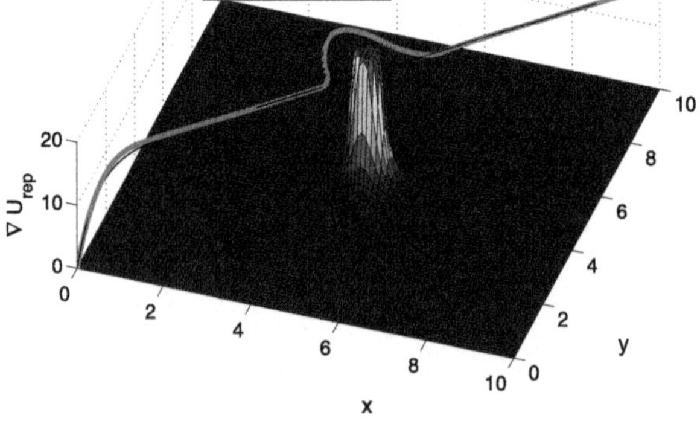

Figure 10.20 Obstacle avoidance with circular limit cycle method.

The references are the same as in the previous simulations, i.e., $\begin{bmatrix} x_d & y_d \end{bmatrix}^T = \begin{bmatrix} t & t \end{bmatrix}$ where $t \in (0, 10)$ and z_d is a filtered step to avoid abrupt changes in the reference. The obstacle is at $\begin{bmatrix} x_{obs} & y_{obs} \end{bmatrix}^T = \begin{bmatrix} 5 & 5 \end{bmatrix}^T$.

10.4 DISCUSSION

Autonomous navigation of aerial vehicles is becoming closer day by day. Advances in sensors, electronics, or embedded systems help researches validate their theoretical results in flight tests. For autonomous navigation it is necessary that the aerial vehicle can perceive its environment. For navigation in unstructured spaces it is essential that the vehicle can react quickly to avoid obstacles present in its trajectory. The APF method is used the most to solve this problem; nevertheless, it could produce a local minimum and the vehicle can stay there. In this chapter, we presented an approach to deal with possible problems and one solution to improve it. The algorithms were validated in simulations and in flight tests, giving pretty good results which motivate us to continue in this way. The limit-cycle technique has been used to avoid the minimal point, showing good results in simulations. Future work will include practical validation of this approach.

REFERENCES

1. O. Khativ, Real-time obstacle avoidance for manipulators and mobile robots, in: IEEE International Conference on Robotics and Automation, Proceedings, vol. 2, St. Louis, Missouri, USA, 1985, pp. 500–505.
2. S.S. Ge, Y.J. Cui, New potential functions for mobile robot path planning, IEEE Transactions on Robotics and Automation 16 (5) (2000) 615–620.
3. J.-C. Latombe, Robot Motion Planning, Springer Science + Business Media, LLC, 1991.
4. L. Zeng, G.M. Bone, Mobile robot navigation for moving obstacles with unpredictable direction changes, including humans, Advanced Robotics 26 (12) (2012) 1841–1862.
5. A. Albers, Obstacle avoidance using limit cycles, Master's thesis, TU Delft, The Netherlands, 2013.
6. P. Flores Palrmeros, P. Castillo, F. Castanos, Backstepping control for a quadrotor vehicle, in: Proceedings of the RED UAS 2015 Workshop on Research, Education and Development of Unmanned Aerial Systems, Nov. 23–25, 2015, Cancún, México.
7. D.-H. Kim, J.-H. Kim, A real-time limit-cycle navigation method for fast mobile robots and its applications to robot soccer, Robotics and Autonomous Systems 42 (1) (2003) 17–30.

CHAPTER 11

Haptic Teleoperation*

Due to security constrains and high degree of difficulty to obtain solutions in hazardous scenarios, fully autonomous navigation continues to be an unsolved problem, where the lack of an overall solution for position estimation and the continuously changing conditions are some of the main challenges to be overcome. Henceforth, the notion of UAVs remotely operated by a human pilot remains a good alternative in several situations. An experienced human user may be able to safely operate a UAV if enough information is provided; however, it is always desirable to ease the task for the user and add robustness against human mistakes, simplifying the system and making it operable even to unexperienced users.

In this chapter we are interested in the case when no reliable position estimation is available, and the human pilot is completely in charge of the mission, manually controlling the drone with the help of an autopilot in attitude stabilization mode. Despite the lack of localization feedback, the pilot must be able to perform a wide range of missions. The key is to assist him with important information about the state of the system, the environmental conditions, and possible risks. Visual feedback from a wireless video camera streaming in real time is a powerful tool often employed for these purposes. However, the abuse of visual information from videos and real-time plots of the system's state may overwhelm an unexperienced pilot, confusing and even distracting him/her from the main goal. One interesting alternative to complement the feedback to the human user is using force feedback by means of a haptic interface [1]. Haptics is a tactile feedback technology which recreates the sense of touch by applying forces, vibrations, or motions to the user. Such technology provides the pilot with important information through the sense of forces, considerably improving flight experience and pilot's awareness. For example, the haptic device can emulate external forces affecting the UAV such as wind gusts or other external perturbations. Another interesting application consists in preventing the user from potential dangers such as obstacles [2], [3] or forbidding him to maneuver in unsafe zones where the drone can damage itself or go out

* This chapter was developed in collaboration with D. Mercado and R. Lozano from Heudiasyc lab, UMR 7253 – Université de Technologie de Compègne, France.

Indoor Navigation Strategies for Aerial Autonomous Systems.
DOI: http://dx.doi.org/10.1016/B978-0-12-805189-4.00014-7

of the communication range, and overall where the UAV may pose a risk for other humans.

We are particularly interested in investigating the use of haptic devices for safe UAV teleoperation by preventing the user from crashing the quadrotor and assisting him/her in obstacle avoidance. The most common techniques for haptic feedback to avoid collisions in UAVs are force feedback (for example, artificial force field) and stiffness feedback using a virtual spring. A comparative analysis of these techniques can be found in [4] and [5]. An adjustable feedback control strategy which accounts for the stiffness and damping effects in the haptic interface is proposed in [6].

11.1 EXPERIMENTAL SETUP

Fixing an experimental prototype is a time-consuming and expensive task because of the high cost and fragility of the sensors and other electronic components embedded on an UAV. In addition, evaluating new algorithms results is a risky mission where the experimental platform is continuously in danger of falling and crashing due to bugs in the implementation or to a bad parameter tuning. Henceforth, an inexpensive navigation test-bed based on the *Robot Operative System* (*ROS*) was built to reduce the implementation time and economic cost. It is conformed by an inexpensive commercial quadrotor type *AR.Drone*, wirelessly connected to a ground computer where all the extra computations are executed using the middleware *ROS*, as illustrated in Fig. 11.1. Such a test-bed follows and extends the ideas presented in [7], and offers an excellent alternative for fast and safe validation of new proposed control strategies, state observers, and computer vision algorithms.

11.1.1 Ground Station

A ground computer is responsible for recovering the drone's information, monitoring in real-time the system states, online parameter tuning, computing the control laws, state observers and computer vision algorithms, as well as giving high level commands and switching between the different operation modes, among others. The ground station is based on *ROS* which is a set of open-source libraries and tools that help to build robot applications, from drivers to the last developed algorithms, with powerful developer tools. It serves as a middleware to assist the different programs, so-called nodes, to interact and communicate among themselves and the

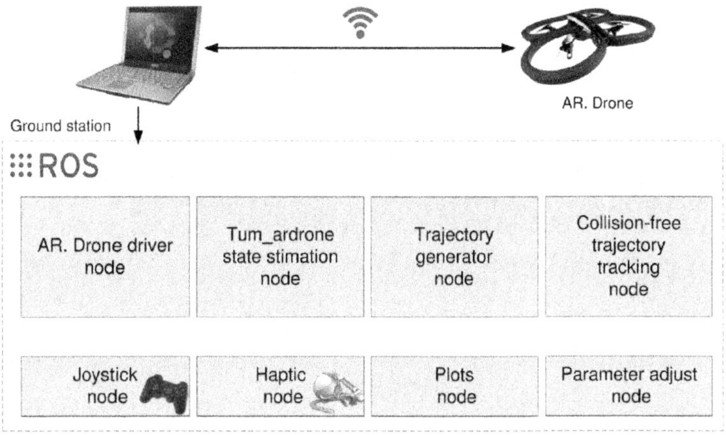

Figure 11.1 Navigation testbed on ROS. *(Courtesy of Parrot images.)*

operative system, by means of messages and services. As can be appreciated in Fig. 11.1, the main nodes utilized in this work are:

- ***AR.Drone* driver node.** This node is in charge of communicating with the *AR.Drone* via WiFi. Here information from the quadrotor embedded sensors, as well as video streams from the two attached cameras, is recovered at a rate of 200 Hz. Also this node controls the drone by sending the desired references to the internal autopilot, along with take-off and landing commands.
- **Tum_ardrone state estimation node [7].** It applies the PTAM algorithm to estimate the quadrotor pose with respect to a visual scene.
- **Haptic node.** This node allows assisted teleoperation with force feedback using a haptic device.
- **Plots node.** The *rqt_plot* node provided with *ROS* facilitates the monitoring task, vital for most mobile robot applications. In it, it is possible to easily plot in real time any available variable.
- **Parameter tuning node.** Online parameter tuning is possible thanks to the *rqt_reconfigure* node included with *ROS*.

These nodes represent the basic operation of the test-bed. However, other nodes can be programmed and easily added to the system to expand or even substitute some of the described nodes, depending on the application and desired behavior of the system.

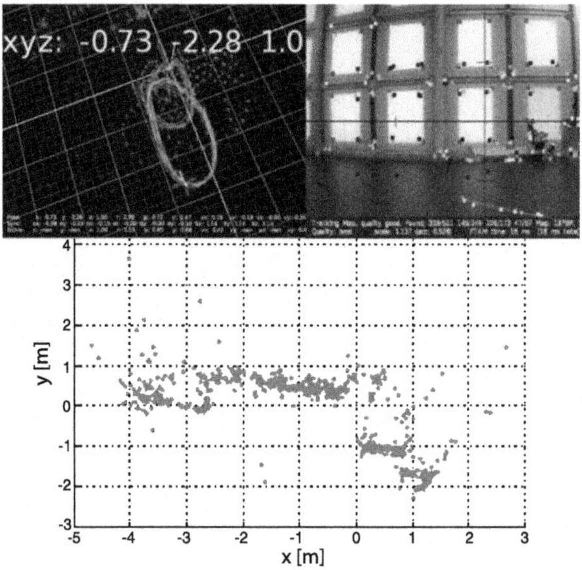

Figure 11.2 UAV localization w.r.t. the sparse depth map (top left). Characteristic features on the image (top right). Horizontal projection of the sparse depth map obtained by PTAM (bottom).

11.1.2 Monocular Vision Localization

The aerial vehicle position is obtained using computer vision and inertial data fused with an EKF algorithm. The vision algorithm, based on Parallel Tracking and Mapping (PTAM), estimates camera pose in an unstructured scene [7–9]. The algorithm executes in parallel the vision information for the tracking and mapping. It also constructs a sparse depth map (see Fig. 11.2) which is used in this work to estimate distance to frontal objects.

Even if the PTAM algorithm is a good solution for pose estimation, it was conceived for mostly static and small scenes, and an absolute scale for the map is not provided. This could be considered as a drawback for MAV (Micro Aerial Vehicle) applications. Nevertheless, in [7] and [10] the authors proposed a nice solution fusing data measurements coming from an IMU, a camera, and ultrasound sensors, and using a scale estimator and an EKF. One advantage of this solution is that the vision approach can be obtained as open-source for *ROS* (Robot Operating System).

For this work the control algorithm code in [7] was replaced with a new one in order to easily implement and validate different control strategies and to help tuning the required gains. Finally, the localization algorithm was modified to recover the point-cloud of the depth map generated by the PTAM algorithm, and send it to another node to estimate the distance to potential collisions, as explained in Sect. 11.2. This way the operator can select online between the different programmed trajectories, control laws, and operation modes, as well as modify in real time any parameter for tuning.

11.1.3 Prototype

The AR.Drone 2.0 is a well-known commercial quadrotor (price $\approx$ 300 USD) which can be safely used close to people and is robust to crashes. It measures 53 $\times$ 52 cm and weighs 0.42 kg. It is equipped with three-axis gyroscopes and accelerometers, an ultrasound altimeter, an air pressure sensor, and a magnetic compass. It also provides video streams from two cameras, the first is looking downwards with a resolution of 320 $\times$ 240 pixels at a rate of 60 fps and is used to estimate the horizontal velocities with an optic flow algorithm while the second camera is looking forward with a resolution of 1080 $\times$ 720 at 30 fps and is used by the monocular vision algorithm. Fig. 11.3 introduces the main technical characteristics of the AR.Drone 2.0.

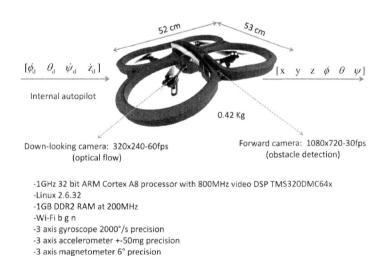

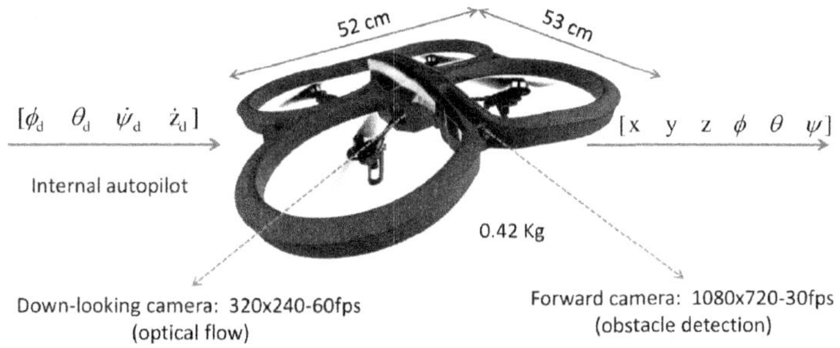

-1GHz 32 bit ARM Cortex A8 processor with 800MHz video DSP TMS320DMC64x
-Linux 2.6.32
-1GB DDR2 RAM at 200MHz
-Wi-Fi b g n
-3 axis gyroscope 2000°/s precision
-3 axis accelerometer +-50mg precision
-3 axis magnetometer 6° precision
-Pressure sensor +/- 10 Pa precision
-Ultrasound sensors for ground altitude measurement

Figure 11.3 AR.Drone 2.0 technical specifications. *(Courtesy of Parrot images.)*

However, neither the software nor the hardware from the AR.Drone can be modified easily. It includes an internal onboard autopilot to control roll, pitch, altitude velocity, and yaw rotational speed (ϕ, θ, $\dot{z}$, and $\dot{\psi}$), according to external references. These references are considered as control inputs and are computed and sent at a frequency of 100 Hz. All sensor measurements are sent to a ground station at a frequency of 200 Hz, where the vision localization and the state estimation algorithms are calculated in real-time on ROS.

11.1.4 Haptic Interface

The low-cost Novint Falcon haptic device is used to remotely control the UAV while providing the pilot with force feedback. It is composed of a three-degree-of-freedom robot arm in delta configuration, with a touch workspace of about 10 × 10 × 10 cm and 400 dpi of resolution (see Fig. 11.4). This haptic device is capable of exerting forces up to 8.9 N in the three axes. The Novint Falcon is connected to the ground station via USB and is controlled with the help of ROS. The position of the final effector and information from the bottoms are recovered and used to control the drone's behavior, while information from the computer vision algorithm is used to determine the exerted forces, improving the flying experience and awareness of the pilot. More details about the control of the UAV from the arm position and the generation of the forces applied by the haptic device are given in Sect. 11.3.

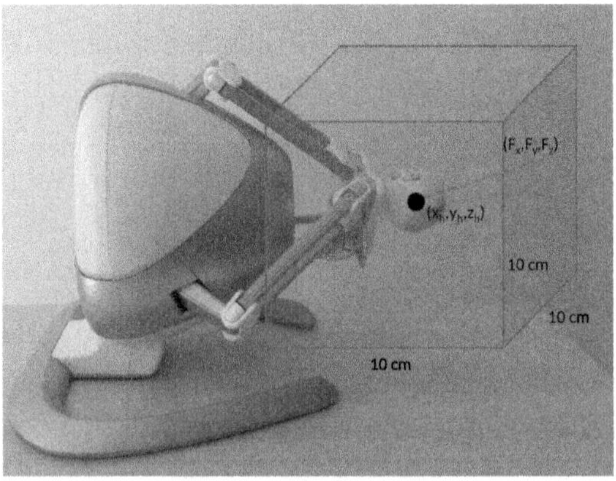

Figure 11.4 Novint Falcon, delta configuration haptic device.

11.2 COLLISION AVOIDANCE

The continuous change in the operation conditions produced by the environmental factors and the interactions with other unknown agents pose a mayor challenge in the pursuit of autonomous and semi-autonomous navigation of mobile robots, with the added difficulty of the payload limitation for UAVs. This drives in the need of real-time perception of the unknown environment and an appropriate reaction, both being a major concern in the design of autonomous UAVs. In the present work, we are interested in the development of an effective strategy for detection and avoidance of collisions in UAV navigation. Particularly, we are interested in the use of the already available information from the embedded sensors in the inexpensive commercial quadrotor AR.Drone 2.0.

We intend to employ computer vision algorithms using information from the frontal camera to detect feature points on the image and use them to estimate the distance to possible obstacles. This is accomplished taking advantage of the sparse depth map generated by the PTAM localization algorithm. Since only a sparse map is available, some possible obstacles can be ignored if not enough visible characteristic points are present. In order to ensure collision avoidance and safe operation of the system, only the horizontal projection of the point-cloud is considered, as depicted in Fig. 11.2. This means that the height of the obstacles is ignored, and obstacle evasion is only possible in the horizontal plane. This approach, although conservative, is still useful for several missions where obstacles such as walls or columns are present. However, it still results in a noisy depth-map, and special attention should be taken for obstacles presenting low-texture surfaces (flat one-color walls).

As stated, distance to possible obstacles is estimated from the horizontal projection of the point-cloud. It is composed of a set P of n characteristic points on the image $p_i(x_i, y_i, z_i)$, $i = 1, \dots, n$, computed by the PTAM algorithm. We define the estimated distance to frontal obstacles d_y as the average depth, w.r.t. the position of the quadrotor along y, of the η points inside certain lateral range ε from the lateral position of the quadrotor x, namely

$$\begin{aligned} d_y &= y - \tfrac{1}{\eta_y} \sum_{i \in \Omega_y} y_i, \\ \Omega_y &= \{p_i(x_i, y_i, z_i) \in P \mid x_i \in [x - \varepsilon, x + \varepsilon]\}. \end{aligned} \tag{11.1}$$

We can define the estimated distance to lateral obstacles d_x in the same way.

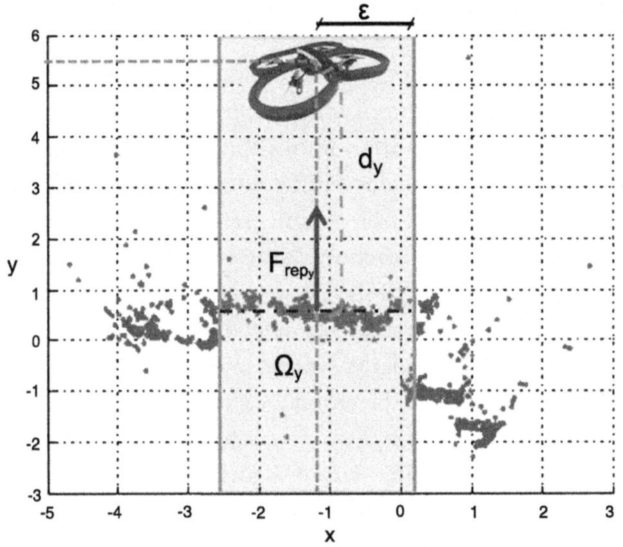

Figure 11.5 Repulsive force scheme. *(Courtesy of Parrot images.)*

In order to avoid collisions, a potential field is applied such that if distance d_i $(i = x, y)$ falls bellow certain safe distance d_s, then a repulsive force F_{rep_i} will be exerted as follows (see Fig. 11.5):

$$F_{rep_i} = \begin{cases} 0 & \text{if } d_i > d_s, \\ -k_{rep_i}\left(\frac{1}{d_i} - \frac{1}{d_s}\right)\left(\frac{1}{d_i^2}\right) & \text{if } d_i \leq d_s. \end{cases} \tag{11.2}$$

Note that lateral obstacle detection becomes much more challenging since only a frontal camera is used for this purpose.

11.3 HAPTIC TELEOPERATION

In this work we successfully applied the Novint Falcon haptic device to assist in teleoperation and to prevent the user from crashing the quadrotor against an obstacle. To do so, the position of the final effector of the haptic device (x_h, y_h, z_h) provides a linear relation between the desired roll and pitch angles (ϕ_d, θ_d) and the desired altitude velocity $(\dot{z}_d)$. It is given as

$$\begin{aligned} \phi_d &= k_{hy}(y_h - o_y), \\ \theta_d &= k_{hx}(x_h - o_x), \\ \dot{z}_h &= k_{hz}(z_h - o_z), \end{aligned} \tag{11.3}$$

where k_{hx}, k_{hy}, k_{hz}, o_x, o_y, and o_z are suitable gains and offsets. This strategy is useful for quadrotors with an internal orientation and altitude controllers. Observe from (11.3) that to keep the quadrotor hovering in a desired position, the user should keep the haptic device at the center of its workspace. In order to assist the user in this task and simplify the manual control of the UAV, a proportional derivative controller is applied to regulate the haptic's final effector position at the origin if no force from the operator is exerted. Similarly, a repulsive force is applied to the haptic device once the quadrotor approaches an obstacle. Then the feedback forces applied to the haptic device are defined as

$$\begin{aligned}
F_x &= \begin{cases} -k_{phx}x - k_{dhx}\dot{x} & \text{if } d_x > d_s, \\ -k_{hrep_x}\left(\frac{1}{d_x} - \frac{1}{d_s}\right)\left(\frac{1}{d_x^2}\right) & \text{if } d_x \le d_s, \end{cases} \\
F_y &= \begin{cases} -k_{phy}y - k_{dhy}\dot{y} & \text{if } d_y > d_s, \\ -k_{hrep_y}\left(\frac{1}{d_y} - \frac{1}{d_s}\right)\left(\frac{1}{d_y^2}\right) & \text{if } d_y \le d_s, \end{cases} \\
F_z &= -k_{phz}z - k_{dhz}\dot{z},
\end{aligned} \tag{11.4}$$

with the control gains k_{phx}, k_{phy}, k_{phz}, k_{dhx}, k_{dhy}, k_{dhz}, $k_{hrep_i} \in \mathbb{R}^+$.

11.4 REAL-TIME EXPERIMENTS

Extensive experiments were executed to validate the proposed algorithms. The parameters used were tuned by trial-and-error and are presented in Table 11.1.

Some experiments can be found on video at

https://www.youtube.com/watch?v=fr0dTSm6Go8

For the teleoperation scenario a human pilot controls the position of the quadrotor through a haptic device. The practical goal here is to use the haptic device to feed back information from the vehicle to the pilot and prevent him from colliding, via opposite forces in the haptic device

Table 11.1 Parameters

$\varepsilon[m]$	k_{rep}	$d_s[m]$	$k_{hrep_{x,y}}$	
0.5	4	5	2	
k_{hx}	k_{hy}	k_{hz}	$k_{phx,y,z}$	$k_{dhx,y,z}$
12.7	14.3	18.2	100	500

when the quadrotor goes out of a safety zone. This is useful, for example, in wall-inspection missions where the operator's visibility is limited.

Collision-free haptic teleoperation is studied through Figs. 11.6–11.9. In these flight tests the user flew the UAV in semi-autonomous mode using a haptic device. Here only the orientation is in autonomous mode. The user attempts to deliberately crash the vehicle against a frontal wall a few times. The first tries were realized slowly, but the last was done at high speed, see Fig. 11.6.

The position responses in the x and y states of the UAV through the experiments are depicted in Fig. 11.7. Observe in this figure (for y axis) at times 6, 9, and 13.5 s, how the reactive collision avoidance algorithm prevented the user from driving the quadrotor to a dangerous area too close to the wall. Furthermore, at the time of 16 s the pilot deliberately tried to crash the UAV in a fast maneuver toward the wall, and even though it got to touch the wall it never crashed thanks to the good performance of the proposed algorithm even in this extreme case. For the x position, it remains quasi-constant since no lateral obstacles were present.

Finally, observe the feedback force applied to the haptic device and the control inputs send to the UAV, respectively shown in Figs. 11.8 and 11.9. When approaching the wall, the haptic device exerts a repulsive force

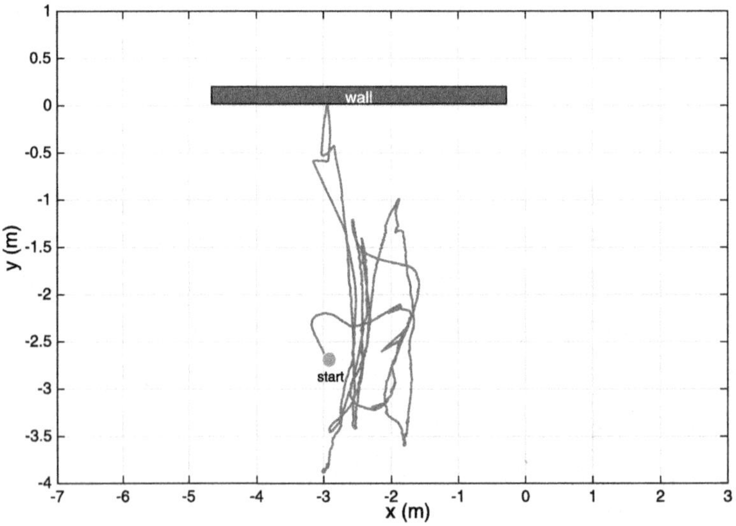

Figure 11.6 x–y performance when the user tries to crash the quadcopter into the wall.

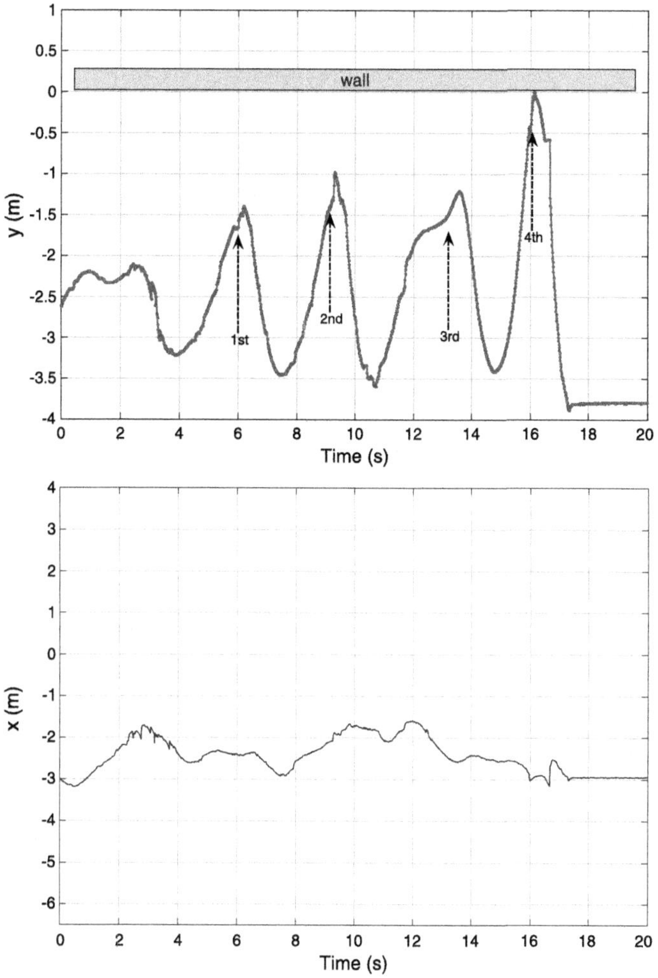

Figure 11.7 *y* and *x* responses (top and bottom, respectively).

alerting the human operator of the danger and even forbidding him from crashing the quadrotor, see the 16th second in Figs. 11.8 and 11.9.

It is important to point out that the obtained results are quite satisfactory, taking into account that only a monocular camera is used to locate the quadrotor and to detect collisions, instead of using an expensive motion capture system and/or extra range sensors as did other teams. Henceforth, the obtained results can be easily reproduced in outdoor flight tests.

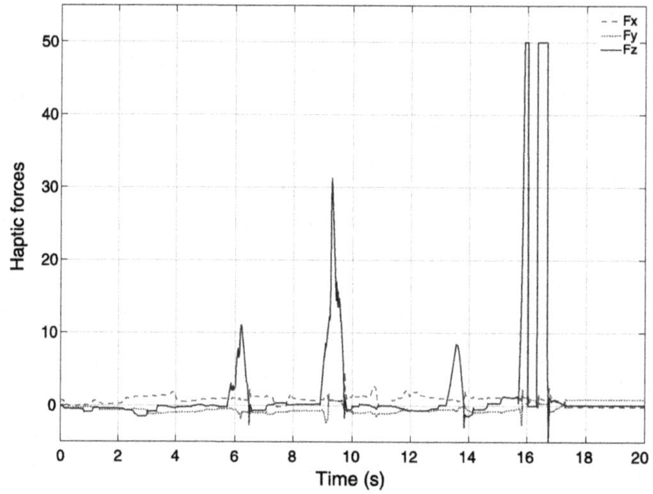

Figure 11.8 Haptic feedback forces.

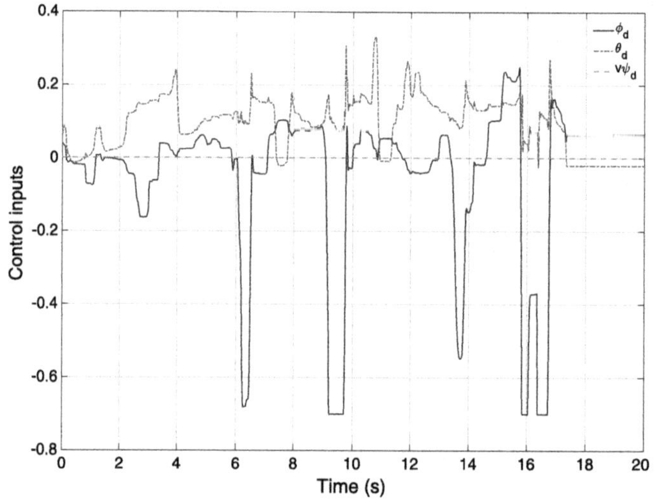

Figure 11.9 UAV control inputs.

11.5 DISCUSSION

In this chapter we have studied the use of a haptic device for collision-free UAV teleoperation, by using the force feedback to drive the human user off from potential obstacles. The use of remote operation of UAVs by a human pilot remains a good alternative to the not always feasible completely autonomous navigation. The use of force feedback through a haptic device to assist the human operator allows improving the flight experience for the pilot, as well as increasing the safety of the mission. In this case we have proposed using reactive collision avoidance by means of potential fields applied to the haptic device. This way the human user is able to feel forces when he is trying to drive the drone into a dangerous area.

Force feedback is also applied to ease the piloting task by automatically directing the final effector of the haptic device to its center, which also implies the attitude stabilization of the UAV.

We also presented an obstacle detection strategy based on computer vision and a frontal camera. Such methodology allows us to detect possible collisions from only visual information. However, as only a sparse depth-map projected to the horizontal plane is used, the height of the obstacles remains unknown. Still this is useful for safe operation in several scenarios, for example, to avoid crashes against walls or columns in indoor missions where pilot's visibility is limited.

REFERENCES

1. P. Stegnano, M. Basile, H. Bthoff, A. Franchi, A semi-autonomous UAV platform for indoor remote operation with visual and haptic feedback, in: International Conference on Robotics and Automation (ICRA), Hong Kong, China, IEEE, 2014.
2. H. Rifa, M. Hua, T. Hamel, P. Morin, Haptic-based bilateral teleoperation of under-actuated unmanned aerial vehicles, in: Proceedings of the 18th World Congress of the International Federation of Automatic Control (IFAC), Milano, Italy, IEEE, 2011.
3. S. Alaimo, L. Pollini, J. Bresciani, H. Bthoff, Evaluation of direct haptic aiding in an obstacle avoidance task for tele-operated systems, in: Proceedings of the 18th World Congress of the International Federation of Automatic Control (IFAC), Milano, Italy, IEEE, 2011.
4. T. Lam, M. Mulder, M. van Paasen, Haptic feedback for UAV tele-operation – force offset and spring load modification, in: International Conference on Systems, Man, and Cybernetics, Taipei, Taiwan, IEEE, 2006.
5. A. Brandt, M. Colton, Haptic collision avoidance for a remotely operated quadrotor UAV in indoor environments, in: International Conference on Systems, Man, and Cybernetics, Istanbul, Turkey, IEEE, 2010.
6. S. Fu, H. Saeidi, E. Sand, B. Sadrfaidpour, J. Rodriguez, Y. Wang, J. Wagner, A haptic interface with adjustable feedback for unmanned aerial vehicles (UAVs) – model, control and test, in: American Control Conference (ACC), Boston, MA, USA, IEEE, 2016.
7. J. Engel, J. Sturm, D. Cremers, Camera-based navigation of a low-cost quadrocopter, in: Intl. Conf. on Intelligent Robot Systems (IROS), Vilamora, Algarve, Portugal, IEEE, 2012, pp. 2815–2821.
8. G. Klein, D. Murray, Parallel tracking and mapping for small AR workspaces, in: Intl. Symposium on Mixed and Augmented Reality (ISMAR), Nara, Japan, IEEE, 2007, pp. 225–234.

9. S. Weiss, D. Scaramuzza, R. Siegwart, Monocular-SLAM-based navigation for autonomous micro helicopters in GPS-denied environments, Journal of Field Robotics 28 (2011) 854–874.
10. J. Engel, J. Sturm, D. Cremers, Scale-aware navigation of a low-cost quadrocopter with a monocular camera, Robotics and Autonomous Systems 62 (2014) 1646–1656.

INDEX

C

D

E

F

M

N

O

P

Q

T

W

Y

Printed in the United States
By Bookmasters